Geometric Analysis of Nonlinear Partial Differential Equations

Geometric Analysis of Nonlinear Partial Differential Equations

Editors

Valentin Lychagin
Joseph Krasilshchik

MDPI • Basel • Beijing • Wuhan • Barcelona • Belgrade • Manchester • Tokyo • Cluj • Tianjin

Editors
Valentin Lychagin
The Arctic University of Norway
Norway

Joseph Krasilshchik
V.A. Trapeznikov Institute of Control Sciences of Russian Academy of Sciences
Russia

Editorial Office
MDPI
St. Alban-Anlage 66
4052 Basel, Switzerland

This is a reprint of articles from the Special Issue published online in the open access journal *Symmetry* (ISSN 2073-8994) (available at: https://www.mdpi.com/journal/symmetry/special_issues/Geometric_Analysis_Nonlinear_PDEs).

For citation purposes, cite each article independently as indicated on the article page online and as indicated below:

LastName, A.A.; LastName, B.B.; LastName, C.C. Article Title. *Journal Name* **Year**, *Volume Number*, Page Range.

ISBN 978-3-0365-1046-0 (Hbk)
ISBN 978-3-0365-1047-7 (PDF)

© 2021 by the authors. Articles in this book are Open Access and distributed under the Creative Commons Attribution (CC BY) license, which allows users to download, copy and build upon published articles, as long as the author and publisher are properly credited, which ensures maximum dissemination and a wider impact of our publications.

The book as a whole is distributed by MDPI under the terms and conditions of the Creative Commons license CC BY-NC-ND.

Contents

About the Editors vii

Preface to "Geometric Analysis of Nonlinear Partial Differential Equations" ix

Călin-Ioan Gheorghiu
Accurate Spectral Collocation Solutions to 2nd-Order Sturm–Liouville Problems
Reprinted from: *Symmetry* **2021**, *13*, 385, doi:10.3390/sym13030385 1

Chuu-Lian Terng and Zhiwei Wu
Lagrangian Curve Flows on Symplectic Spaces
Reprinted from: *Symmetry* **2021**, *13*, 298, doi:10.3390/sym13020298 21

Alexei Kushner and Valentin Lychagin
Generalized Navier–Stokes Equations and Dynamics of Plane Molecular Media [†]
Reprinted from: *Symmetry* **2021**, *13*, 288, doi:10.3390/sym13020288 59

Alexey Samokhin
On Monotonic Pattern in Periodic Boundary Solutions of Cylindrical and Spherical Korteweg–De Vries–Burgers Equations
Reprinted from: *Symmetry* **2021**, *13*, 220, doi:10.3390/sym13020220 69

Daniel J. Arrigo and Joseph A. Van de Grift
Contact Symmetries of a Model in Optimal Investment Theory
Reprinted from: *Symmetry* **2021**, *13*, 217, doi:10.3390/sym13020217 79

Anna Duyunova, Valentin Lychagin and Sergey Tychkov
Quotients of Euler Equations on Space Curves
Reprinted from: *Symmetry* **2021**, *13*, 186, doi:10.3390/sym13020186 87

Jan L. Cieśliński and Zbigniew Hasiewicz
Iterated Darboux Transformation for Isothermic Surfaces in Terms of Clifford Numbers
Reprinted from: *Symmetry* **2021**, *13*, 148, doi:10.3390/sym13010148 97

Valentin Lychagin and Mikhail Roop
Singularities in Euler Flows: Multivalued Solutions, Shockwaves, and Phase Transitions
Reprinted from: *Symmetry* **2021**, *13*, 54, doi:10.3390/sym13010054 109

Jørn Olav Jensen and Boris Kruglikov
Differential Invariants of Linear Symplectic Actions
Reprinted from: *Symmetry* **2020**, *12*, 2023, doi:10.3390/sym12122023 121

Fredrik Andreassen and Boris Kruglikov
Joint Invariants of Linear Symplectic Actions
Reprinted from: *Symmetry* **2020**, *12*, 2020, doi:10.3390/sym12122020 147

Iosif Krasil'shchik
Nonlocal Conservation Laws of PDEs Possessing Differential Coverings [†]
Reprinted from: *Symmetry* **2020**, *12*, 1760, doi:10.3390/sym12111760 163

Stephen Anco and Bao Wang
Geometrical Formulation for Adjoint-Symmetries of Partial Differential Equations
Reprinted from: *Symmetry* **2020**, *12*, 1547, doi:10.3390/sym12091547 175

About the Editors

Valentin Lychagin is a chief researcher at the V.A. Trapeznikov Institute of Control Sciences of the Russian Academy of Sciences and professor emeritus at the Arctic University, Norway. Their main publications are devoted to differential equations, their symmetries and invariants.

Joseph Krasilshchik is a chief researcher at the V.A. Trapeznikov Institute of Control Sciences of the Russian Academy of Sciences and a full professor at the Independent University of Moscow. They were born in Moscow in 1948, and graduated from the Faculty of Mechanics and Mathematics of Moscow State University in 1971. Their main publications are related to geometry and the homological algebra of nonlinear differential equations.

Preface to "Geometric Analysis of Nonlinear Partial Differential Equations"

The origin of the geometrical theory of the differential equation can be traced back to the scientific heritage of Sophus Lie, the great Norwegian mathematician of the 19th century. Forgotten for almost 100 years, it was revived in the work of modern scientists, of whom Lev Ovsyannikov and Alexandre Vinogradov should be mentioned first of all. The theory received new attention and consideration during the 1960s during the "solitonic boom", when the theory of integrable systems with an infinite number of degrees of freedom was created and coined.

Essentially, the geometrical approach to differential equations consists in understanding them as submanifolds (smooth or with singularities) in jet spaces. Such an interpretation allows one to apply powerful methods of modern differential geometry and homological algebra and to look at numerous classical problems from a completely new and unorthodox viewpoint. In particular, it provided a rigorous basis for extremely important concepts such as symmetries, conservation laws, equivalence and differential invariants. Moreover, inside this conceptual scheme, one obtains efficient methods to compute the necessary invariants (both geometrical and algebraic) of differential equations.

The current collection contains twelve papers published in the Special Issue Analysis of Nonlinear Partial Differential Equations of the Symmetry journal and may serve as an illustration of some modern applications of the geometrical methods of partial differential equations. It comprises miscellaneous topics of the local and nonlocal geometry of differential equations and applications of the corresponding methods in hydrodynamics, symplectic geometry, optimal investment theory, etc.

Valentin Lychagin, Joseph Krasilshchik
Editors

Article

Accurate Spectral Collocation Solutions to 2nd-Order Sturm–Liouville Problems

Călin-Ioan Gheorghiu

Tiberiu Popoviciu Institute of Numerical Analysis, Romanian Academy, P.O. Box 68-1, 400110 Cluj-Napoca, Romania; ghcalin@ictp.acad.ro; Tel.: +40-264-582-207

Abstract: This work is about the use of some classical spectral collocation methods as well as with the new software system Chebfun in order to compute the eigenpairs of some high order Sturm–Liouville eigenproblems. The analysis is divided into two distinct directions. For problems with clamped boundary conditions, we use the preconditioning of the spectral collocation differentiation matrices and for hinged end boundary conditions the equation is transformed into a second order system and then the conventional ChC is applied. A challenging set of "hard" benchmark problems, for which usual numerical methods (FD, FE, shooting, etc.) encounter difficulties or even fail, are analyzed in order to evaluate the qualities and drawbacks of spectral methods. In order to separate "good" and "bad" (spurious) eigenvalues, we estimate the drift of the set of eigenvalues of interest with respect to the order of approximation N. This drift gives us a very precise indication of the accuracy with which the eigenvalues are computed, i.e., an automatic estimation and error control of the eigenvalue error. Two MATLAB codes models for spectral collocation (ChC and SiC) and another for Chebfun are provided. They outperform the old codes used so far and can be easily modified to solve other problems.

Keywords: Sturm–Liouville; clamped; hinged boundary condition; spectral collocation; Chebfun; chebop; eigenpairs; preconditioning; drift; error control

MSC: 34B09; 34B40; 34L16; 65L15; 65L20; 65L60; 65L70

Citation: Gheorghiu, C.-I. Accurate Spectral Collocation Solutions to 2nd-Order Sturm–Liouville Problems. *Symmetry* **2021**, *13*, 385. https://doi.org/10.3390/sym13030385

Academic Editors: Calogero Vetro and Mariano Torrisi

Received: 25 January 2021
Accepted: 22 February 2021
Published: 27 February 2021

Publisher's Note: MDPI stays neutral with regard to jurisdictional claims in published maps and institutional affiliations.

Copyright: © 2021 by the author. Licensee MDPI, Basel, Switzerland. This article is an open access article distributed under the terms and conditions of the Creative Commons Attribution (CC BY) license (https://creativecommons.org/licenses/by/4.0/).

1. Introduction

Due to the spectacular evolution of advanced programming environments, a special curiosity arose in the numerical analysis of a classical problem, that of accurate solving of high order SL eigenproblems. It seems that quantum mechanics is the richest source of self-adjoint problems, while non-self-adjoint problems arise in hydrodynamic and magnetohydrodynamic stability theory (see for instance [1] and the vast literature quoted there). The need to compute accurately and efficiently a large set of eigenvalues and eigenfunctions, including those of high index, is now an utmost task.

Our main interest here is to evaluate the capabilities of the new Chebfun package as well as those of conventional spectral methods in meeting these requirements. The latter work in the classical mode, i.e., "discretize-then-solve". On the contrary, the Chebfun spirit consists in the continuous mode, i.e., "solve-then-discretize" (see [2] p. 302).

The effort expended by both classes of methods is also of real interest. It can be assessed in terms of the ease of implementation of the methods as well as in terms of computer resources required to achieve a specified accuracy.

Some FORTRAN software packages have been designed over time to solve various regular and singular SL problems. These seem to be the first attempts to solve numerically (automatically) eigenvalue problems.The most important would be SLEDGE [3,4], the NAG's code SL02F [5,6], SLEIGN and SLEIGN2 [7,8], and later MATSLISE. The SLDRIVER interactive package supports exploration of a set of SL problems with the first four previously mentioned packages. The SLEDGE, SL02F, SLEIGN2, and NAG's D02KDF are "automatic" for eigenvalues and not for eigenfunctions. They have built in *error estimation* and from that they achieve *error control*. They adjust the accuracy of the discretization so that the delivered eigenvalue has estimated error below a user-supplied tolerance.

Essentially, the numerical method used in these software packages replaces the coefficients in the equation by a step function approximation. Their most important drawback remains the impossibility to compute the eigenfunctions and a slow convergence in case of some singular eigenproblems.

The MATSLISE code introduced in [9] can solve some Schrödinger eigenvalue problems by a constant perturbation method of a higher order. Very recently, this code has been improved (see [10]) but it remains for Schrödinger issues which are outside the scope of this paper.

There is also a class of semi-analytical methods which includes the variational iteration method, the homotopy perturbation method, homotopy analysis method, and Adomian decomposition (see for instance [11]) for solving eigenvalue problems. Their accuracy is far from what spectral collocation methods can provide.

In [12], the authors set up an ambitious method based on the Lie group method along with the Magnus expansion in order to solve any order of SL problem with arbitrary boundary conditions.

We believe that spectral collocation methods can contribute to the systematic clarification of some still open issues related to the numeric aspects of SL problems. The most important aspect is how many computed eigenpairs (eigenvalues and eigenfunctions) can we trust when solving a high order SL? This is the outstanding, not completely resolved research issue, we want to address in this paper.

Thus, we will argue that generally Chebfun would provide a greater flexibility in solving various differential problems than the classical spectral methods. This fact is fully true for regular problems. A Chebfun code contains a few lines in which the differential operator is defined along with the boundary conditions and then a subroutine to solve the algebraic eigenproblem. It provides useful information on the optimal order of approximation of eigenvectors and the degree to which the boundary conditions have been satisfied.

Unfortunately, in the presence of various singularities or for problems of higher order than 4, the maximum order of approximation of the unknowns can be reached ($N \geq 4000$) and then Chebfun issues a message that warns about the possible inaccuracy of the results provided.

Alternative use of conventional spectral collocation methods generally helps to overcome this difficulty.

As a matter of fact, in order to resolve a singularity on one end of the integration interval, Chebfun uses only the truncation of the domain. Classical spectral methods can also use this method, but it is not recommended because much more sophisticated methods are at hand in this case. For singular points at finite distances (mainly origin) we will use the so-called *removing technique of independent boundary conditions* (see for a review of this technique our monograph [13] p. 91). The boundary conditions at infinity can be enforced using basis functions that asymptotically satisfy these conditions (Laguerre, Hermite, sinc).

A Chebfun code and two MATLAB codes, one for ChC and another for the SiC method, are provided in order to exemplify. With minor modifications they could be fairly useful for various numerical experiments. These codes are very easy to implement, efficient, and reliable. All our numerical experiments have been carried out using MATLAB R2020a on an Intel (R) Xeon (R) CPU E5-1650 0 @ 3.20 GHz.

The main purpose of this paper was to argue that Chebfun, along with the spectral collocation methods, can be a very feasible alternative to the above software packages regarding accuracy, robustness as well as simplicity of implementation. In addition, these methods can calculate exactly the " whole" set of eigenvectors approximating eigenfunctions and provide *automatic estimation and control of the eigenvalue error*. For self-adjoint problems, checking the orthonormality of computed eigenvectors gives us valuable information on the accuracy of the calculation of these vectors.

The structure of this work is as follows. In Section 2, we recall some specific issues for the regular as well as singular Sturm–Liouville eigenproblems. In Section 3, we review briefly the conventional ChC method as well as Chebfun, the relative drift of a set of eigenvalues and the preconditioning of Chebyshev differentiation matrices. Section 4 is the core part of our study. By analyzing one set of hinged problems and another one of clamped problems, we want to evaluate the applicability of the two classes of methods as well as their performances in terms of the accuracy of the outcomes they produce. There is also a subsection that contains problems equipped with boundary conditions that are a mixture of these two types, clamped and hinged. We end up with Section 5 devoted to conclusions and open problems.

2. 2nd-Order Sturm–Liouville Eigenproblems

The 2nd-order SL equation reads

$$(-1)^n \left(p_n(x) u^{(n)} \right)^{(n)} + (-1)^{n-1} \left(p_{n-1}(x) u^{(n-1)} \right)^{(n-1)} + \ldots$$
$$+ (p_2(x) u'')'' - (p_1(x) u')' = \lambda u(x),\ a < x < b,\ n \in \mathbb{N},\ n \geq 2,$$

along with separated, (self-adjoint) boundary conditions. We shall assume that all coefficient functions are real valued. The technical conditions for the problem to be non singular are: the interval $(a; b)$ is finite; the coefficient functions p_k, $0 < k < n - 1$, the weight w and $1/p_n$ are in $L^1(a,b)$; and the essential infima of p_n and w are both positive. Under these assumptions, the eigenvalues are bounded below (see for instance [14]).

The eigenvalues can be ordered in the usual form: $\lambda_0 \leq \lambda_1 \leq \lambda_2 \leq \ldots$, such that $\lim_{k \to \infty} \lambda_k = +\infty$. In this sequence, each eigenvalue has multiplicity at most n (so $k + n > k$ for all k). The restriction on the multiplicity arises from the fact that for each λ there are at most n linearly independent solutions of the differential equation satisfying either of the endpoint conditions which we shall consider below.

Some of the problems we deal with are also found in the monographic paper [15]. It contains over 50 challenging examples from mathematical physics and applied mathematics along with a summary of SL theory, differential operators, Hilbert function spaces, classification of interval endpoints, and boundary condition functions.

3. Chebfun vs. Conventional Spectral Collocation

3.1. Chebfun

For details on Chebfun we refer to [2,16–18]. The Chebfun system, in object-oriented MATLAB, contains algorithms which amount to spectral collocation methods on Chebyshev grids of *automatically determined resolution*. This is the main difference compared to conventional spectral methods in which the resolution (order of approximation) is imposed almost arbitrarily. Its properties are briefly summarized in [17]. In [16] the authors explain that chebops are the fundamental Chebfun tools for solving ordinary, partial differential or integral equations.

The implementation of chebops combines the numerical analysis idea of spectral collocation with the computer science idea of *lazy or delayed evaluation of the associated spectral discretization matrices*. The grammar of chebops along with a lot of illustrative examples is displayed in the above quoted papers as well as in the text [2]. Thus, one can get a suggestive image of what they can do working with Chebfun.

Moreover, in ([16] p. 12) the authors explain clearly how the Chebfun works, i.e., it solves the eigenproblem for two different orders of approximation, automatically chooses a reference eigenvalue and checks the convergence of the process. At the same time, it warns about the possible failures due to the high non-normality of the analyzed operator (matrix).

Actually, we want to show in this paper that Chebfun along with chebops can do much more, i.e., can accurately solve high order SL problems.

3.2. Spectral Collocation Methods

Spectral methods have been shown to provide exponential convergence for a large variety of problems, generally with smooth solutions, and are often preferred [19]. In all spectral collocation methods designed so far, we have used the collocation differentiation matrices from the seminal paper [20]. We preferred this MATLAB differentiation suite for the accuracy, efficiency as well as for the ingenious way of introducing various boundary conditions.

In order to impose (enforce) the boundary conditions we have used the *boundary bordering*, which is a simplified variant of the above mentioned removing technique of independent boundary conditions, as well as the *basis recombination*. We have used the first technique in the large majority of our papers except [21] where the latter technique has been employed. In the last quoted paper a modified ChT method based on basis recombination has been used in order to solve an Orr-Sommerfeld problem with an eigenparameter dependent boundary condition.

Once eigenvectors are calculated in physical space they are transposed into the space of coefficients using FCT. In this way, it is possible to estimate the way in which their coefficients decrease.

3.3. The Drift of Eigenvalues

Two techniques are used in order to eliminate the "bad" eigenvalues as well as to estimate the stability (accuracy) of ChC or Chebfun computations. The first one is the *drift*, with respect to the order of approximation or the scaling factor, of a set of eigenvalues of interest. The second one is based on the check of the *eigenvectors' orthogonality*.

In other words, we want to separate the "good" eigenvalues from the "bad" ones, i.e., inaccurate eigenvalues. An obvious way to achieve this goal is to compare the eigenvalues computed for different orders of some parameters such as the approximation order (cut-off parameter) N or the scaling factor (length of integration interval). Only those whose difference or "resolution-dependent drift" is "small" can be believed. In this connection, in the paper [22], the so called *absolute (ordinal) drift* with respect to the order of approximation has been introduced. We extend this definition in our recent paper [23] and will use it without repeating it here.

Whenever the exact eigenvalues of a problem are known, the relative drift is reduced to the relative error.

At this point, the following observation is extremely important. In the highly cited monograph [24], the author makes a subtle analysis of spectral methods in solving linear eigenproblems. Among others, he states the so called **Boyd's Eigenvalues Rule-of-Thumb** in which he notices that in solving such a problem with a spectral method using $(N+1)$ terms in the truncated spectral series, the lowest $N/2$ eigenvalues are usually accurate to within a few percent, while the larger $N/2$ numerical eigenvalues differ from those of the differential equation by such large amounts as to be useless.

3.4. Preconditioning

To simplify the introduction of a preconditioner, we use the differential operator

$$L^{(n)}(u) := \frac{d^n u}{dx^n}, \quad x \in (-1,1), \tag{1}$$

subject to *clamped* boundary conditions

$$u^{(\mu)}(-1) = 0, \quad 0 \leq \mu \leq l_n, \tag{2}$$

and

$$u^{(\nu)}(1) = 0, \quad 0 \leq \nu \leq r_n, \tag{3}$$

where $n > 1$, l_n, and r_n are positive integers such that $r_n + l_n + 2 = n$.

It is well known that for general collocation points the first order differentiation matrix has a condition number of order N^2 and the second order differentiation matrix has a condition number of order N^4 as $N \to \infty$. We comment on the preconditioner introduced in [25]. These authors show that the *preconditioning matrix*

$$\mathbf{D} := diag\left((1+x_k)^{l_k+1}(1-x_k)^{r_k+1}\right), \; 2 \leq k \leq N-1,$$

applied to ChC as well as to Chebfun discretization $\mathbf{L}^{(n)}_{ChCor\ Chebfun}$ of differential operator $L^{(n)}$, produces matrices $\mathbf{DL}^{(n)}_{ChCor\ Chebfun}$ of an inferior condition number, namely N^n.

4. Numerical Experiments

4.1. Hinged Ends or Simply Supported Boundary Conditions

4.1.1. The Viola's Eigenproblem-Revisited

Let us consider now the so called *Viola's eigenproblem*. It is encountered in porous stability problems (see [1], Chapter 9) and reads

$$\begin{array}{c} \frac{d^2}{dx^2}\left[(1-\theta x)^3 \frac{d^2 u}{dx^2}\right] = \lambda(1-\theta x)u, \; x \in (0,1), \; 0 \leq \theta < 1, \\ u(0) = u''(0) = u(1) = u''(1). \end{array} \quad (4)$$

It is *singular* as $\theta \to 1^-$ in accordance with the definition introduced in Section 2.

By straightforward variational arguments, we have shown in ([13] p. 50) that the lowest eigenvalue is positive. In this text, we have solved the above problem by ChC using the so called D^2 *strategy* which involves the change of variables

$$v := (1-\theta x)^3 \frac{d^2 u}{dx^2}.$$

The main deficiency of this strategy is the fact that it produces a lot of numerical spurious eigenvalues (at infinity).

In spite of this, we succeeded in stating *the conjecture according to which* $\lambda_1(\theta)$, *the lowest eigenvalue of the problem* (4), *approaches* 1 *as* $\theta \to 1^-$.

Now, taking advantage of Chebfun we solve directly problem (4). Thus, the dependence of the lowest eigenvalue of the problem (4), computed by Chebfun, on the parameter θ is depicted in Figure 1. Actually, we have obtained

$$\lambda_1(0.98765) = 8.775218471808549e - 01,$$

which only partly confirms the above conjecture.

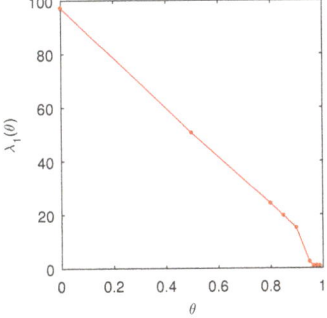

Figure 1. The dependence of the lowest eigenvalue of the Viola's eigenproblem (4), computed by Chebfun, on the parameter θ.

As a validation issue for our computations we have obtained the known value

$$\lambda_1(0.0) = \pi^4,$$

within an approximation of a thousand.

For the highest computable value of parameter θ, we display in Figure 2 the first four eigenvectors of problem (4) and in Figure 3 the Chebyshev coefficients of these eigenvectors.

We have to mention that the singularity in the right end $x = 1$ becomes more prominent as the θ tends to 1. This is confirmed by increasing the degree of the Chebfun approximation. For instance, when $\theta := 0$ only a 25 degree Chebyshev polynomial uses it and when $\theta := 0.98765$ the degree of approximation grows to more than 80 (see Figure 3). It is also worth mentioning that only for θ growing very close to 1 a truncation of the domain along with the use of the option `splitting` have been necessary when Chebfun has been used.

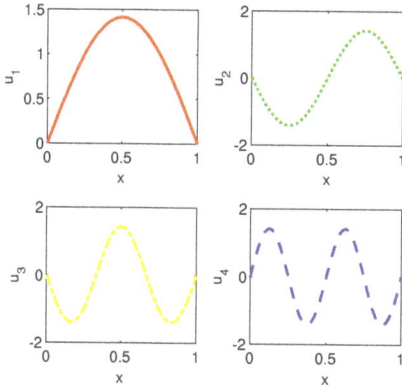

Figure 2. From upper left to lower right we display the first four eigenvectors of the Viola's eigenproblem (4) computed by Chebfun with $\theta = 0.98765$.

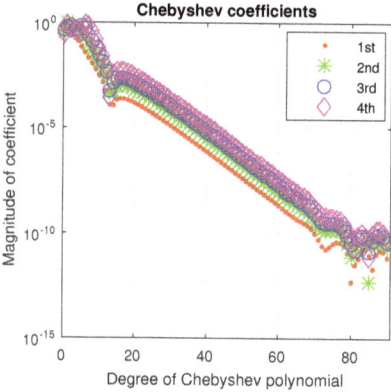

Figure 3. The Chebyshev coefficients of the first four eigenvectors of the Viola's eigenproblem (4) computed by Chebfun with $\theta = 0.98765$. A very narrow rounding-off plateau can be seen.

We have to observe that the problem (4) has been solved by compound matrix method in [26] for $\theta < 0.9$. The author asserts that other methods have to be used in order to resolve the singularity in this problem. We hope that the above analysis sheds some light in this direction.

4.1.2. The Bénard Stability Problem

A simplified form of the Bénard stability problem supplied with self-adjoint boundary conditions reads (see for instance [14,27])

$$u^{(vi)} - (2\nu+3)u^{(iv)} + (\nu^2+4\nu+3)u'' - (\nu^2+2\nu+2)u = \lambda u(x), \; x \in (0,1), \quad (5)$$
$$u(0) = u(1) = u''(0) = u''(1) = u^{(iv)}(0) = u^{(iv)}(1) = 0.$$

The constant ν is regarded as a parameter which typically can take the values

$$\nu_j^{\pm} := -\left(1+j^2\pi^2\right) \pm \left(1+j^2\pi^2\right)^{-1/2}, j = 1,2,\ldots.$$

All our attempts to solve this problem using Chebfun have failed, so we have resorted to the old D^2 *strategy*.

Thus, we rewrite problem (5) as a homogeneous Dirichlet one attached to a second order differential system, namely

$$\begin{aligned} u'' &= v(x), \; x \in (0,1), \\ v'' &= w(x), \\ w'' - (2\nu+3)w + (\nu^2+4\nu+3)v - (\nu^2+2\nu+2)u &= \lambda u(x), \\ u = v = w &= 0 \text{ in } x = 0 \text{ and } x = 1. \end{aligned} \quad (6)$$

Now we apply to each line the ChC discretization. It leads to the generalized and singular *eigenpencil*

$$(\mathbf{A}, \mathbf{B}), \quad (7)$$

where the block matrices are defined by

$$\mathbf{A} := \begin{pmatrix} 4\widetilde{D}^{(2)} & I & Z \\ Z & 4\widetilde{D}^{(2)} & I \\ -(\nu^2+2\nu+2)I & (\nu^2+4\nu+3)I & 4\widetilde{D}^{(2)}-(2\nu+3)I \end{pmatrix},$$

and

$$\mathbf{B} := \begin{pmatrix} Z & Z & Z \\ Z & Z & Z \\ I & Z & Z \end{pmatrix}.$$

The factor 4 in front of $\widetilde{D}^{(2)}$ comes from the shift of interval $(0,1)$ to the canonical Chebyshev interval $[-1,1]$ and the matrix $\widetilde{D}^{(2)}$ signifies the second order Chebyshev differentiation matrix with the homogeneous Dirichlet boundary conditions enforced. The matrices I and Z stand respectively for the identity and zeros matrices of the same dimension as $\widetilde{D}^{(2)}$.

The following short MATLAB code has been used to solve (6):

```
N=256;                                  % order of approximation
nu=-(1+4*(pi^2))-sqrt(1+4*(pi^2));      % parameter \nu
[x,D]=chebdif(N,2); D2=D(2:N-1,2:N-1,2); % differentiation matrices
I=eye(size(D2)); Z=zeros(size(D2));
A=[ 4*D2 -I Z; Z 4*D2 -I; -(nu^2+2*nu+2)*I (nu^2+4*nu+3)*I 4*D2-(2*nu+3)*I];
B=[Z Z Z; Z Z Z; I Z Z];
k = 8 ;                                 % number of computed eigs
E=eigs( @(x)(A\(B*x)), size(A,1),k, 'SM') % Arnoldi method
```

When the order of approximation is N, both matrices in (7) have order $3 \times (N-1)$. This tripling of the dimensions of the matrices involved is not a major disadvantage. On the contrary, if we use Henrici's number as a measure of normality (see for instance our text [13] pp. 22–23), we see from the inequality

$$Henrici(\mathbf{A}) = 0.300293 < Henrici\left(\widetilde{D}^{(2)}\right) = 0.395205,$$

that matrix $\mathbf{A}$ is more normal than $\widetilde{D}^{(2)}$.

In our previous paper [28], we have analyzed various methods to solve singular eigenproblems attached to pencils of the form (7). For the problem at hand we have used the Arnoldi method with the MATLAB sequence `eigs(A^{-1}B)` and with the above code obtained the eigenvalue reported in Table 1. It is very clear that for the values of ν considered, the block matrix $\mathbf{A}$ is non singular and the block matrix $\mathbf{B}$ is singular and independent of ν.

It is extremely important to point out that for the first two values of the parameter ν in Table 1 our results are very close to those reported in [14]. For the other parameter values this no longer happens. A similar situation occurs even for the second eigenvalue. Then, to decide over the accuracy of our outcomes, we resorted to drift. The relative drift, with respect to the order of approximation N, of the first forty eigenvalues when $\nu := -(1+\pi^2) - (1+\pi^2)^{-1/2}$ is displayed in Figure 4. It suggests that the first two eigenvalues are computed with an accuracy of at least 10^{-12} and the first forty with an accuracy of at least 10^{-2}. This leads us to believe that we have produced much better approximations for these eigenvalues than those reported in [14] as well as [27]. Actually, in [14] the authors use the SLEUTH code and accept that "it is clear that the code is not very accurate on this problem".

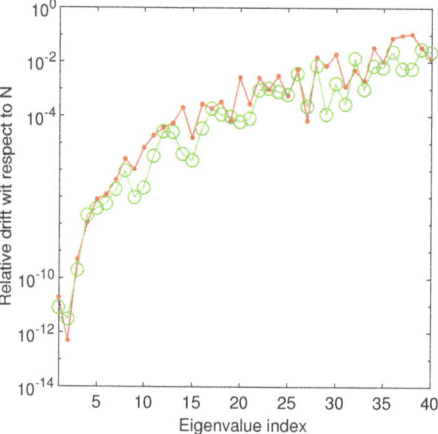

Figure 4. The relative drift of the first forty eigenvalues of Bénard problem is displayed when $\nu := -(1+\pi^2) - (1+\pi^2)^{-1/2}$-red dotted line when $N_1 := 256$ and $N_2 := 128$ and green circled line when $N_1 := 64$ and $N_2 := 128$.

Table 1. The first two eigenvalues of Bénard problem (5) for various ν computed by D^2 strategy along with ChC.

ν	$\lambda_0(\nu)$	$\lambda_1(\nu)$	$\lambda_0(\nu)$ according to [14]
$-(1+\pi^2)$	$-1.000000000102923e+00$	$-3.548769279033568e+04$	-1.000005
$-(1+4\pi^2)$	$-1.000000009534114e+00$	$-9.530184561696226e+03$	-1.0001
$-(1+\pi^2)-(1+\pi^2)^{-1/2}$	$-1.191482998363510e+02$	$-2.802486989002433e+04$	-1×10^{-7}
$-(1+4\pi^2)-(1+4\pi^2)^{-1/2}$	$-1.639502291744172e+03$	$-1.406538196754713e+04$	-3×10^{-5}

4.1.3. A Self-Adjoint Eighth-Order Problem

The eigenproblem of the highest order we consider in this paper is the following

$$u^{(8)}(x) = \lambda u(x), \ 0 < x < 1, \\ u(0) = u''(0) = u^{(4)}(0) = u^{(6)}(0) = u(1) = u''(1) = u^{(4)}(1) = u^{(6)}(1) = 0, \quad (8)$$

with exact eigenvalues $\lambda_k = (k\pi)^8, k = 1, 2, \ldots$.

All our attempts to solve this problem with Chebfun have failed. The abortion message referred to the extremely small conditioning of the eight order Chebyshev collocation differentiation matrix (of the order 10^{-40}).

Instead, the D^2 strategy along with ChC worked well and produced vectors from Figure 5.

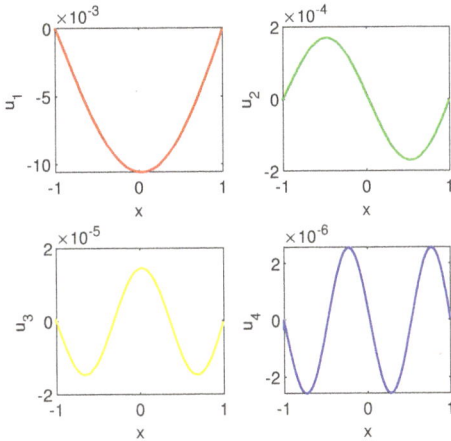

Figure 5. From upper left to lower right we display the first four eigenvectors of problem (8) computed by ChC along with D^2 method when the order of approximation has been $N := 256$.

In order to estimate the error with which the eigenvalues were calculated, we display in Figure 6b the relative drift of the first twelve eigenvalues for different approximation orders. As a result that we know the exact eigenvalues, we also display the relative errors. It is very clear that the first eigenvalue is computed with better accuracy than 10^{-12}. Unfortunately, this means a lower performance by three decimals than that of Magnus expansion reported in Table 10 from [12].

Moreover, this means that we cannot trust more than twelve eigenvalues for this problem.

It is clear that ChC along with the D^2 method have the potential to find the first eigenvalues of an SL problem of arbitrary (even) order with good accuracy. In addition to the Magnus method, this strategy calculates its eigenfunctions (eigenvectors) with reasonable accuracy as can be seen in Figure 6a. We use FCT to compute the Chebyshev coefficients of the eigenvectors.

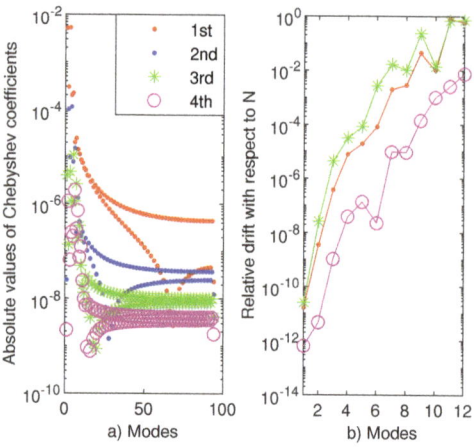

Figure 6. (a) The Chebyshev coefficients of the first four vectors of the problem (8) computed by FCT (fast Chebyshev transform). (b) The relative drift of the first twelve eigenvalues to problem (8), red dotted line $N_1 := 96$, $N_2 := 200$, green stared line $N_1 := 128$, $N_2 := 200$, and magenta circled line $N_1 := 200$, $N_2 := $ exact.

4.2. Clamped Boundary Conditions

4.2.1. A Fourth Order Problem with a Third Derivative Term

With this first example we have to highlight the importance of preconditioning in improving the accuracy of Chebfun. In the papers [25,29], as well as in the monograph [30], the following eigenproblem is carefully studied. It consists of the fourth order differential equation

$$u^{(iv)} + Ru''' = su'', \ x \in (-1,1), \ R \in \mathbb{R}, \quad (9)$$

supplied with the clamped boundary conditions

$$u(\pm 1) = u'(\pm 1) = 0. \quad (10)$$

The eigencondition for this problem is

$$\left(R^2 + 4s\right)^{1/2}\left[\cosh(R) - \cosh\left(R^2 + 4s\right)^{1/2}\right] + 2s \sinh\left(R^2 + 4s\right)^{1/2} = 0. \quad (11)$$

Problems similar to this appear, for example, in linearized stability analysis in fluid dynamics. In [29], the authors noticed spurious eigenvalues when the problem (9) and (10) is solved by ChT method. These spurious eigenvalues appears in the right-half plane suggesting physical instabilities that do not exist.

We have solved the problems (9) and (10) by Chebfun with and without preconditioning. The numerical outcomes are displayed in Table 2. Boldfaced digits in the computed eigenvalues show the extent of agreement with the exact values. Thus, it is very clear that preconditioning Chebfun can considerably improve its accuracy.

Table 2. First two eigenvalues of problems (9) and (10).

i	λ_i Chebfun	λ_i Preconditioned Chebfun	λ_i Solution to (11)
1	$-\mathbf{9.8701}54876048822e+00$	$-\mathbf{9.869604528925013}e+00$	-9.8696044
2	$-\mathbf{2.0192}16607051227e+01$	$-\mathbf{2.019072837497370}e+01$	-20.1907286

4.2.2. A Fourth Order Eigenproblem from Spherical Geometry

In [29], the authors consider the eigenproblem

$$D(D-s)u = 0, \ 0 < r_1 < r_2, \ r_1 > 0, \\ u(r_1) = u(r_2) = u'(r_1) = u'(r_2) = 0, \quad (12)$$

where the operator D is defined by

$$D(u) := u'' + \frac{2}{r}u' - \frac{l(l+1)}{r^2},$$

and l is a positive integer. They solve this problem by a modified ChT method in order to avoid spurious eigenvalues. We have solved this problem by ChC and obtained for the fist two eigenvalues the numerical values

$$s_1 = -3.947819275687863e+01, \quad s_2 = -8.076297512888706e+01,$$

which agree up to the fourth decimal with the the true values (determined from the eigencondition.

The first four vectors to problem (12) computed by Chebfun are depicted in Figure 7a. It is visible that they satisfy the boundary conditions. Their Chebyshev coefficients are displayed Figure 7b. About the first twenty coefficients of the first four eigenvectors decrease just as abruptly and smoothly.

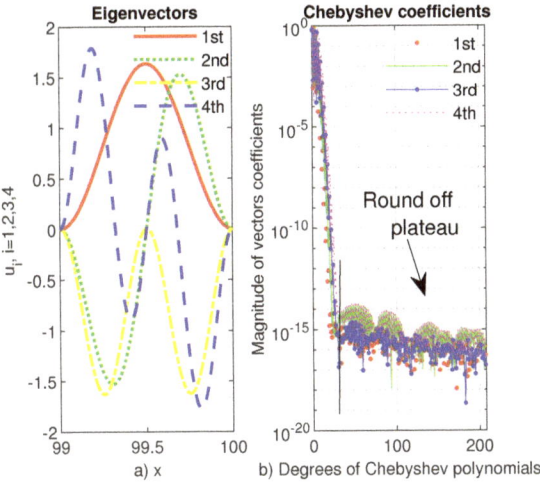

Figure 7. (a) The first four eigenvectors to problem (12) computed by Chebfun. (b) The coefficients of first four eigenvectors to problem (12).

4.2.3. A Set of Sixth Order Eigenproblems

In [31], the authors consider the following sixth order eigenproblems

$$u^{(vi)}(x) = \lambda u^{(j)}(x), \ u(\pm 1) = u'(\pm 1) = u''(\pm 1) = 0, j = 0, 2, 4, \quad (13)$$

and introduce an extremely simple modification to the ChT method which eliminates the spurious eigenvalues when such high order eigenproblems are solved.

We have tried to solve problem (13) with $j := 4$ by Chebfun but all our attempts failed due to the very small conditioning of matrix involved, i.e., around $O(10^{-40})$. The situation became much better with the preconditioner introduced in Section 3.4.

Thus, the first four eigenvectors along with their Chebyshev coefficients are depicted in Figure 8. As it is apparent from the lower panel of this figure the coefficients of degree up to 30 drop sharply to an absolute value below 10^{-10} and then slowly decrease to machine accuracy. This happens at an degree around 120.

The eigenvalues computed by Chebfun agree up to the first three digits with those provided in ([25] p. 405).

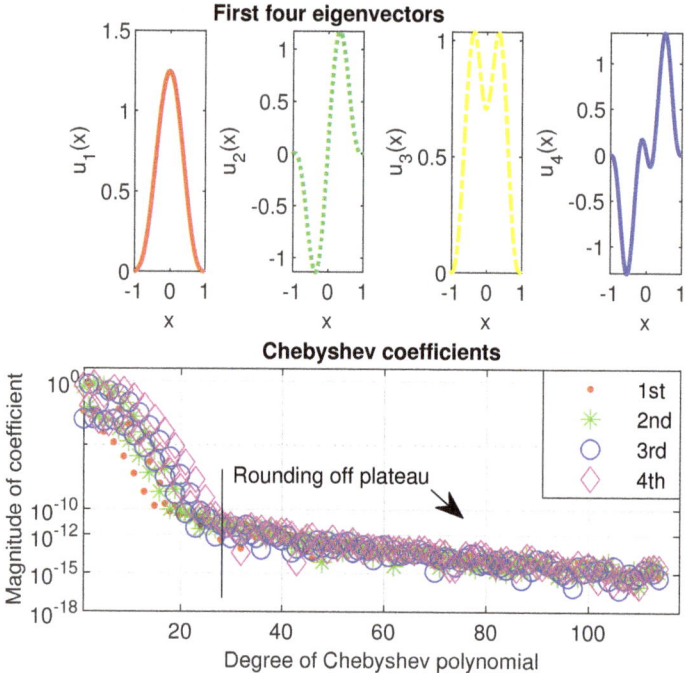

Figure 8. The first four eigenvectors of problem (13) with $j := 4$, computed by Chebfun, are reported in the the upper panels and their Chebyshev coefficients are displayed in the lower panel.

4.3. Problems with Mixed Boundary Conditions

4.3.1. The Free Lateral Vibration of a Uniform Clamped–Hinged Beam

The fourth order eigenproblem

$$u^{(iv)}(x) = \lambda u(x), \ 0 < x < 1, \ u(0) = u'(0) = u(1) = u''(1) = 0, \quad (14)$$

is considered in [32] and is solved by a non conventional spectral collocation method. In this paper, the author shows that the eigenvalues satisfy the transcendental equation

$$\tanh \sqrt[4]{\lambda} = \tan \sqrt[4]{\lambda}. \quad (15)$$

It is extremely important to observe that neither preconditioning nor D^2 strategy can handle the mixture of boundary conditions in (14). Thus, this problem tests how well the Chebfun can cope with various boundary conditions.

As the eigenvalues computed from (15) are compared with those obtained by Magnus expansion in [12], we report in Table 3 the latter eigenvalues compared with those provided by Chebfun. A coincidence of at least three decimals can be observed.

Table 3. The first five eigenvalues to problem (14) computed by Chebfun and compared with those provided by Magnus expansion.

j	λ_j Chebfun	λ_j According to [12]
1	$2.377373239875730e+02$	$2.377210675300e+02$
2	$2.496524908617440e+03$	$2.496487437860e+03$
3	$1.086783642364734e+04$	$1.086758221697e+04$
4	$3.177977410414838e+04$	$3.178009645380e+04$
5	$7.400167551416633e+04$	$7.400084934040e+04$

The first four eigenvectors to problem (14) computed by Chebfun are displayed in Figure 9. It is perfectly visible that they satisfy the boundary conditions assumed in this problem.

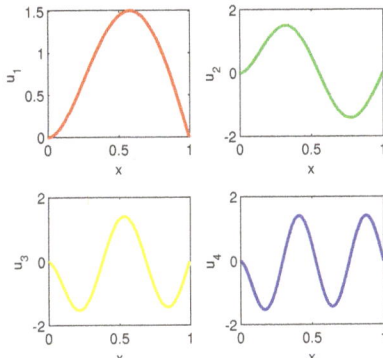

Figure 9. From upper left to lower right we display the first four eigenvectors to problem (14) computed by Chebfun.

The Chebyshev coefficients of the first four eigenvectors to problem (14) computed by Chebfun are displayed in Figure 10. They decrease smoothly to somewhere around 10^{-12} which is an argument in favor of the accuracy of numerical results.

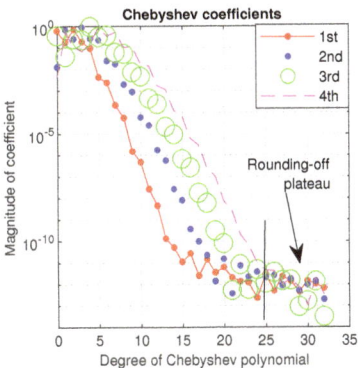

Figure 10. In a log-linear plot we display the Chebyshev coefficients of the first four eigenvectors to problem (14) computed by Chebfun.

4.3.2. A Fourth Order Eigenproblem with Higher Order Boundary Conditions

In order to show again the Chebfun versatility in introducing boundary conditions, we consider the following problem called the cantilevered beam in Euler–Bernouilli theory (see for instance [33]). The equation simply reads

$$u^{(iv)} = \beta^4 u, \; x \in (0, \pi), \tag{16}$$

and is equipped with the following boundary conditions

$$\begin{aligned} u(0) &= 0, & u'(0) &= 0, \\ u''(\pi) &= 0, & u'''(\pi) &= 0. \end{aligned} \tag{17}$$

The first two boundary conditions state that the beam is clamped in 0 and the last two state that the bean is free in the right hand end. The eigenvalues β satisfy the eigencondition

$$\cosh(\beta\pi)\cos(\beta\pi) + 1 = 0. \tag{18}$$

Actually, the problems (16) and (17) are self-adjoint. Without going into details, we will notice that the first eigenvalue of this eigenproblem is the solution of the minimization problem

$$\beta^4 = \min_{v \in V} \frac{\int_0^\pi (v'')^2 dx}{\int_0^\pi v^2 dx},$$

where, roughly, V is a space of continuous functions satisfying the boundary conditions in (17). This Ritz formulation, as well as a weak (variational) formulation can be obtained by multiplying the equation with a function v from V and a double integration by parts.

Again, these boundary conditions are not treatable by preconditioning or D^2 strategy. The following simple and short Chebfun code solves the problems (16) and (17).

```
% Cantilevered beam in Euler-Bernouilli theory
dom=[0,pi];x=chebfun('x',dom);    % the domain
L = chebop(dom);
L.op = @(x,y) diff(y,4);          % the operator
L.lbc = @(y)[y; diff(y,1)];       % fixed b. c.
L.rbc = @(y)[diff(y,2); diff(y,3)];% free b. c.
[U,D]=eigs(L,40,'SM');            % first six eigs.
% Sorted eigenpairs (eigenvalues and eigenvectors)
D=diag(D); [t,o]=sort(D); D=D(o); disp((D.^(1/4)))
U=U(:,o);
```

In Table 4, the first four eigenvalues computed by Chebfun and by Magnus expansion are reported. A satisfactory agreement is observed.

Table 4. The first four eigenvalues of problems (16) and (17) computed by Chebfun compared with numerical solutions to Equation (18).

j	β_j by Chebfun	β_j Exact Solutions of (18)
1	$5.967718563107258e-01$	0.59686
2	$1.494163617547652e+00$	1.49418
3	$2.500244462376521e+00$	2.50025
4	$3.499990154542449e+00$	3.49999

The first four eigenvectors are displayed in Figure 11a and their Chebyshev coefficients are displayed in the same figure panel b. It is clear that Chebfun uses Chebyshev polynomials of slightly lower degree than 20 and from this level only a sharply decreasing rounding-off plateau follows.

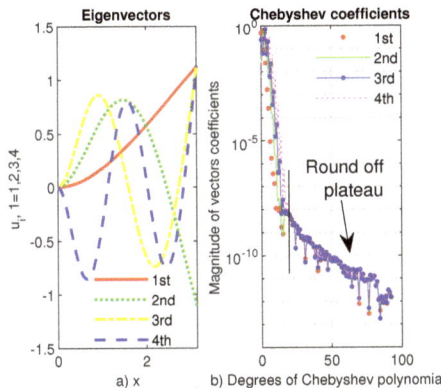

Figure 11. (a) The first four eigenvectors of problem (16) and (17) computed by Chebfun. (b) The absolute values of Chebyshev coefficients of these vectors are displayed in a log-linear plot.

The first four eigenvectors approximating the eigenfunctions are in very good agreement with those exposed in literature. Using the definition of the scalar product of two vectors u and v, namely $u' * v$, we can easily check the orthonormality of eigenvectors.

The curves in Figure 12 clearly show that the eigenvectors of this problem computed by Chebfun are orthonormal.

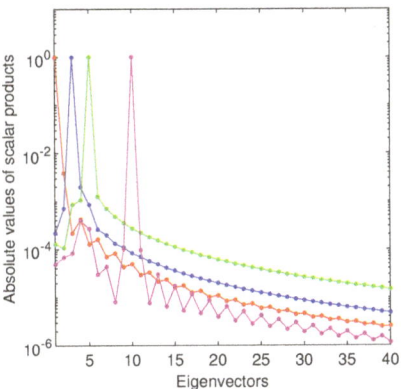

Figure 12. In a log-linear plot we display the scalar products $u'_1 * u_j$—red dotted line, $u'_3 * u_j$—blue dotted line, $u'_5 * u_j$—green dotted line and $u'_{10} * u_j$—magenta dotted line, $j := 1, 2, \ldots, 50$ when the eigenproblem (16) and (17) is solved by Chebfun.

4.3.3. The Harmonic Oscillator and Its Second and Third Powers

We wanted to test our strategy on a problem whose differential equation exhibits *stiffness* in at least part of the range. Thus, along with the well known harmonic oscillator operator

$$h(u) := -u'' + x^2 u, \; x \in (-\infty, \infty)$$

we will consider its second and third powers, namely

$$h^2(u) = u^{(iv)} - 2\left(x^2 u'\right)' + \left(x^4 - 2\right)u, \; x \in (-\infty, \infty),$$

and

$$h^3(u) = -u^{(vi)} + \left(3x^2 u''\right)'' + \left(\left(8 - 3x^4\right)u'\right)' + \left(x^6 - 14x^2\right)u, \; x \in (-\infty, \infty).$$

Actually we want to solve the fourth order eigenvalue problem for $h^2(u)$, namely

$$h^2(u) = \lambda u, \tag{19}$$

and the sixth order eigenvalue problem

$$h^3(u) = \lambda u, \tag{20}$$

corresponding to the cube of the harmonic oscillator operator. The eigenvalues of the harmonic oscillator are $\lambda_k = (2k+1)$, $k = 0, 1, 2, \ldots$, and those of h^2 and h^3 are the second and the third powers, respectively of λ_k. According to the definition for classification of SL problems, given in this Section 2, the eigenproblems (19) and (20) are singular.

We have to observe that no boundary conditions are needed because the problem is of *limit-point type* [15]: the requirement that the eigenfunctions be square integrable suffices as a boundary condition. In [14] the problem (20) is solved by a SLEUTH code along with domain truncation. Actually the authors truncate this problem to the interval $(-100, 100)$, and impose the simplest boundary conditions $u = u' = u'' = 0$ at $x = \pm 100$. Along with these boundary conditions the eigenproblem becomes self-adjoint.

The first four eigenvectors of the cube of harmonic oscillator computed by SiC are displayed in the upper panels of Figure 13. Their sinc coefficients are displayed in the lower panel of the same figure. Roughly speaking, it guarantees us an accuracy of at least 10^{-12} in the computation of these eigenvectors.

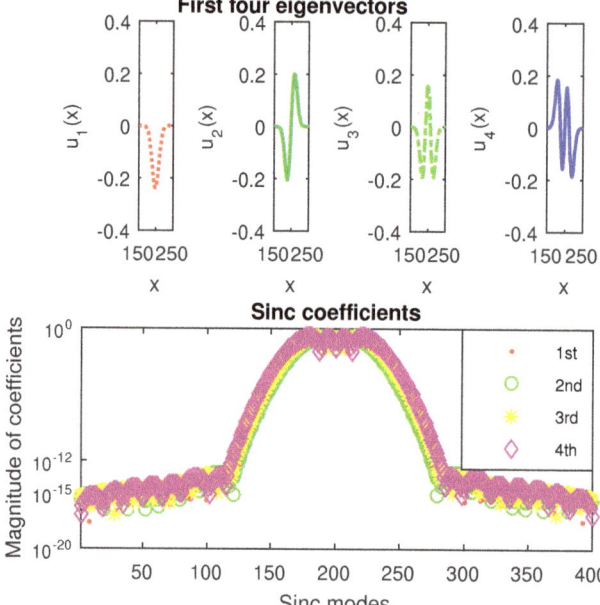

Figure 13. A zoom in on the first four eigenvectors of the cube of harmonic oscillator (20) computed by SiC is displayed in the upper panels and the sinc coefficients of eigenvectors are reported in the lower panel.

SiC computes the integers λ_k with an accuracy of at least six digits.

The relative drift, with respect to N, of the first 250 eigenvalues of the cube of harmonic oscillator computed by SiC are displayed in Figure 14a). It tells us that the first 200 eigenvalues are "good" within an accuracy of approximately 10^{-2}. It also means the "highest" confirmation of Boyd's Eigenvalues Rule-of-Thumb (see Section 3.3).

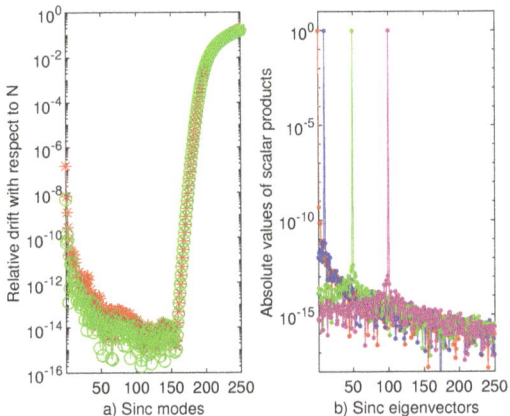

Figure 14. (a)The relative drift of the first 250 eigenvalues of the cube of harmonic oscillator computed by SiC. Red stared line compares the exact values with the eigenvalues computed with $N := 400$ and the circled green line compares the latter eigenvalues with those computed when in SiC $N := 500$. In both cases, the scaling factor h equals 0.1. (b) The orthonormality of the first 250 eigenvectors, i.e., the scalar products, $u'_1 * u_j$ red dotted line, $u'_{10} * u_j$ blue dotted line, $u'_{50} * u_j$ green dotted line and $u'_{100} * u_j$ magenta dotted line, $j := 1, 2, \ldots, 250$.

Trying to explain this spectacular phenomenon we cannot forget the fact that the derivation matrices of SiC are symmetric. This leads to normal matrices (operators) whose eigenpairs are properly computable.

If we compare this result with Table 10 from [14], where the best accuracy in computing of the first eigenvalue is 10^{-2}, we can speak of a total superiority of SiC method over SLEUTH.

In Figure 14b, we display the scalar product of some eigenvectors. They prove that the SiC computed eigenvectors are orthonormal. This means that we can trust at least the first 200 eigenpairs computed by SiC. The following few lines of MATLAB compute the above eigenpairs:

```
% The sinc differentiation matrices [Weideman & Reddy]
N=400;M=6;h=0.1;
%Orders of approximation and differentiation and scaling factor
[x, D] = sincdif(N, M, h); D1=D(:,:,1);D2=D(:,:,2);D6=D(:,:,6);
% The cube of the "harmonic oscillator" operator
L=-D6+D2*(3*diag(x.^2)*D2)+D1*(diag(8-3*(x.^4))*D1)+diag(x.^6-14*(x.^2));
% Finding eigenpairs of L
[U,S]=eigs(L,250,0); S=diag(S); [t,o]=sort(S); S=S(o);
```

Unfortunately, Chebfun along with domain truncation fails in solving the sixth order problem (20) with or without preconditioning. Actually, a warning concerning the very bad conditioning of the matrix is issued.

However, Chebfun behaves fairly well in solving the fourth order problem (19), i.e., computes the corresponding integers with the same accuracy as SiC. We have solved the Equation (19) on the truncated interval $(-X, X)$ for various X along with the boundary conditions $u(\pm X) = u'(\pm X) = 0$. The Chebyshev coefficients of the first four eigenvectors of eigenproblem (19) computed by Chebfun are displayed in Figure 15a). An important aspect must be highlighted, namely the first about 1000 polynomial coefficients decrease steeply and smoothly to about 10^{-14} after which up to the order of 2500 follows a wide rounding-off plateau. This is the polynomial of the highest degree that Chebfun has used in our numerical experiments. The curves in Figure 15b) show the orthonormality of Chebfun eigenvectors.

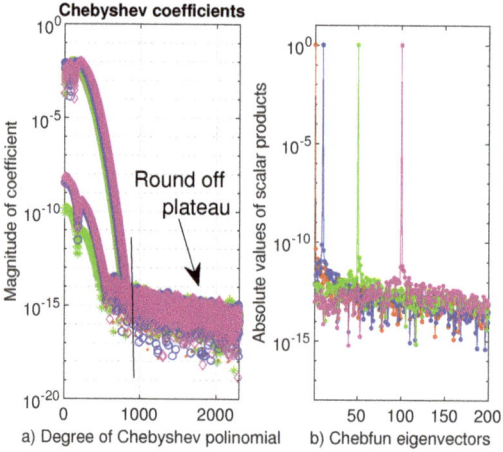

Figure 15. (**a**) We display the Chebyshev coefficients of the first four eigenvectors of eigenproblem (19); red dotted line-first vector, green stared line-second, blue circles-third and magenta diamonds-fourth vector. (**b**) In a log-linear plot we display the scalar products $u'_1 * u_j$—red dotted line, $u'_3 * u_j$—blue dotted line, $u'_5 * u_j$—green dotted line and $u'_{10} * u_j$—magenta dotted line, $j := 1, 2, \ldots, 200$ when the eigenproblem (19) is solved by Chebfun.

The relative drift with respect to the length of integration interval X of the first 250 eigenvalues to problem (19), when it is solved by Chebfun, is displayed in Figure 16. It means that the numerical stability is lost for larger X than 100 and a set of small eigenvalues can be computed with an accuracy better than 10^{-9}.

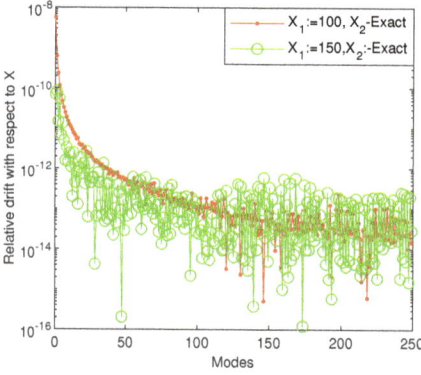

Figure 16. The relative drift (errors) with respect to X of the first 250 eigenvalues of second order harmonic oscillator operator h^2.

5. Conclusions and Open Problems

After analyzing these challenging problems, some firm conclusions can be drawn.

First of all, Chebfun can easily handle any type of boundary condition. This is a significant advantage. Thus, for fourth order eigenproblems, the direct application of Chebfun is versatile in handling various high order boundary conditions and produces reliable outcomes. Furthermore, for problems with clamped boundary conditions both methods, Chebfun as well as ChC, improve their results with two, three decimals by preconditioning.

For sixth order eigenproblems, the Chebfun situation is not so encouraging. Its direct application is very uncertain. Matrices whose conditioning order drops to 10^{-40} appear, which most often lead to inaccurate results. For problems of this order or more, subjected to hinged boundary conditions, the reduction to second-order systems and then the application of the ChC method is the best strategy. In this way, we managed to establish a conjecture for the Viola's problem regarding its lowest eigenvalue.

For fourth order problems on the real line Chebfun along with the truncation of the domain worked fairly well as was the case with the second power harmonic oscillator. As an absolute novelty, we have established in this case the numerical stability with respect to the length of the integration interval. Instead, for the sixth order eigenproblems on the real line, the SiC method remains the unique feasible alternative.

In fact, this method is the best in the sense that we can trust the first half of computed eigenpairs. To our knowledge, no software package has reached this performance so far.

An open problem remains for finding preconditioning methods for the case of hinged boundary conditions or some other types of boundary conditions.

This paper comes shortly after when in another one (see [23]) we have approached, with the same two classes of methods, singular Schröedinger eigenproblems. In this situation we can appreciate that ChC along with Chebfun are a better alternative in some respects to other existing methods for a very wide range of eigenproblems. Both compute eigenvectors (approximating eigenfunctions) and by drift estimation demonstrate numerical stability. In addition, the drift with respect to N shows the degree of accuracy up to which a set of eigenvalues is computed. The situation when both types of methods can be applied to the same problem is the ideal one and the one that produces the safest results.

Funding: This research received no external funding.

Acknowledgments: The author would like to thank the reviewers for their thoughtful comments and efforts towards improving our manuscript.

Conflicts of Interest: The author declares no conflict of interest.

Abbreviations

The following abbreviations are used in this manuscript:

ChC	Chebyshev collocation method
ChT	Chebyshev tau method
D^2	strategy to reduce a 2nd-order equation to a second order system
FCT	fast Chebyshev transform
FD	finite difference method
FE	finite element method
MATSLISE	a MATLAB package for the numerical solution of SL and Schröedinger equations
SiC	sinc spectral collocation
SL	Sturm–Liouville
SLEDGE	Sturm–Liouville estimates determined by global errors
SLEUTH	Sturm–Liouville Eigenvalues using Theta Matrices

References

1. Straughan, B. Stability of Wave Motion in Porous Media; Springer Science+Business Media: New York, NY, USA, 2008.
2. Trefethen, L.N.; Birkisson, A.; Driscoll, T.A. *Exploring ODEs*; SIAM: Philadelphia, PA, USA, 2018.
3. Pruess, S.; Fulton, C.T. Mathematical Software for Sturm-Liouville Problem. *ACM Trans. Math. Softw.* **1993**, *19*, 360–376. [CrossRef]
4. Pruess, S.; Fulton, C.T.; Xie, Y. An Asymptotic Numerical Method for a Class of Singular Sturm-Liouville Problems. *ACM SIAM J. Numer. Anal.* **1995**, *32*, 1658–1676. [CrossRef]

5. Marletta, M.; Pryce, J.D. LCNO Sturm-Liouville problems. Computational difficulties and examples. *Numer. Math.* **1995**, *69*, 303–320. [CrossRef]
6. Pryce, J.D.; Marletta, M. A new multi-purpose software package for Schrödinger and Sturm–Liouville computations. *Comput. Phys. Comm.* **1991**, *62*, 42–54. [CrossRef]
7. Bailey, P.B.; Everitt, W.N.; Zettl, A. Computing Eigenvalues of Singular Sturm-Liouville Problems. *Results Math.* **1991**, *20*, 391–423. [CrossRef]
8. Bailey, P.B.; Garbow, B.; Kaper, H.; Zettl, A. Algorithm 700: A FORTRAN software package for Sturm-Liouville problems. *ACM Trans. Math. Softw.* **1991**, *17*, 500–501. [CrossRef]
9. Ledoux, V.; Van Daele, M.; Vanden Berghe, G. MATSLISE: A MATLAB Package for the Numerical Solution of Sturm-Liouville and Schrödinger Equations. *ACM Trans. Math. Softw.* **2005**, *31*, 532–554. [CrossRef]
10. Baeyens, T.; Van Daele, M. The Fast and Accurate Computation of Eigenvalues and Eigenfunctions of Time-Independent One-Dimensional Schrödinger Equations. *Comput. Phys. Commun.* **2021**, *258*, 107568.. [CrossRef]
11. Abbasb, Y.S.; Shirzadi, A. A new application of the homotopy analysis method: Solving the Sturm—Liouville problems. *Commun. Nonlinear. Sci. Numer. Simulat.* **2011**, *16*, 112—126. [CrossRef]
12. Perera, U.; Böckmann, C. Solutions of Direct and Inverse Even-Order Sturm-Liouville Problems Using Magnus Expansion. *Mathematics* **2019**, *7*, 544. [CrossRef]
13. Gheorghiu, C.I. *Spectral Methods for Non-Standard Eigenvalue Problems: Fluid and Structural Mechanics and Beyond*; Springer: Heidelberg, Germany, 2014.
14. Greenberg, G.; Marletta, M. Oscillation Theory and Numerical Solution of Sixth Order Sturm-Liouville Problems. *SIAM J. Numer. Anal.* **1998** *35*, 2070–2098. [CrossRef]
15. Everitt, W.N. A Catalogue of Sturm-Liouville Differential Equations. In *Sturm-Liouville theory: Past and Present*; Amrein, W.O., Hinz, A.M., Hinz, D.B., Eds.; Birkhäuser Verlag: Basel, Switzerland, 2005; pp. 271–331.
16. Driscoll, T.A.; Bornemann, F.; Trefethen, L.N. The CHEBOP System for Automatic Solution of Differential Equations. *BIT* **2008**, *48*, 701–723. [CrossRef]
17. Driscoll, T.A.; Hale, N.; Trefethen, L.N. *Chebfun Guide*; Pafnuty Publications: Oxford, UK, 2014.
18. Driscoll, T.A.; Hale, N.; Trefethen, L.N. Chebfun-Numerical Computing with Functions. Available online: http://www.chebfun.org (accessed on 15 November 2019).
19. Gheorghiu, C.I. *Spectral Collocation Solutions to Problems on Unbounded Domains*; Casa Cărţii de Ştiinţă Publishing House: Cluj-Napoca, Romania, 2018.
20. Weideman, J.A.C.; Reddy, S.C. A MATLAB Differentiation Matrix Suite. *ACM Trans. Math. Softw.* **2000**, *26*, 465–519. [CrossRef]
21. Gheorghiu, C.I.; Pop, I.S. A Modified Chebyshev-Tau Method for a Hydrodynamic Stability Problem. In *Proceedings of the International Conference on Approximation and Optimization*; Stancu, D.D., Coman, G., Breckner, W.W., Blaga, P., Eds.; Transilvania Press: Cluj-Napoca, Romania, 1996; Volume II, pp. 119–126.
22. Boyd, J.P. Traps and Snares in Eigenvalue Calculations with Application to Pseudospectral Computations of Ocean Tides in a Basin Bounded by Meridians. *J. Comput. Phys.* **1996**, *126*, 11–20. [CrossRef]
23. Gheorghiu, C.I. Accurate Spectral Collocation Computation of High Order Eigenvalues for Singular Schrödinger Equations. *Computation* **2021**, *9*, 2. [CrossRef]
24. Boyd, J.P. *Chebyshev and Fourier Spectral Methods*; Dover Publications: New York, NY, USA, 2000; pp. 127–158.
25. Huang, W.; Sloan, D.M. The Pseudospectral Method for Solving Differential Eigenvalue Problems. *J. Comput. Phys.* **1994**, *111*, 399–409. [CrossRef]
26. Straughan, B. *The Energy Method, Stability, and Nonlinear Convection*; Springer: New York, NY, USA, 1992; pp. 218–222.
27. Lesnic, D.; Attili, S. An Efficient Method for Sixth-order Sturm-Liouville Problems. *Int. J. Sci. Technol.* **2007**, *2*, 109–114.
28. Gheorghiu, C.I.; Rommes, J. Application of the Jacobi-Davidson method to accurate analysis of singular linear hydrodynamic stability problems. *Int. J. Numer. Meth. Fluids* **2013**, *71*, 358–369. [CrossRef]
29. Gardner, D.R.; Trogdon, S.A.; Douglass, R.W. A Modified Spectral Tau Method That Eliminates Spurious Eigenvalues. *J. Comput. Phys.* **1989**, *80*, 137–167. [CrossRef]
30. Fornberg, B. *A Practical Guide to Pseudospectral Methods*; Cambridge University Press: Cambridge, UK, 1998; pp. 89–90.
31. McFadden, G.B.; Murray, B.T.; Boisvert, R.F. Elimination of Spurious Eigenvalues in the Chebyshev Tau Spectral Method. *J. Comput. Phys.* **1990**, *91*, 228–239. [CrossRef]
32. Mai-Duy, N. An effective spectral collocation method for the direct solution of high-order ODEs. *Commun. Numer. Methods Eng.* **2006**, *22*, 627–642. [CrossRef]
33. Zhao, S.; Wei, G.W.; Xiang, Y. DSC analysis of free-edged beams by an iteratively matched boundary method. *J. Sound. Vib.* **2005**, *284*, 487–493. [CrossRef]

Article

Lagrangian Curve Flows on Symplectic Spaces

Chuu-Lian Terng [1,*] and Zhiwei Wu [2]

[1] Department of Mathematics, University of California at Irvine, Irvine, CA 92697, USA
[2] School of Mathematics (Zhuhai), Sun Yat-Sen University, Zhuhai 519082, China; wuzhiwei3@mail.sysu.edu.cn
* Correspondence: cterng@math.uci.edu

Abstract: A smooth map γ in the symplectic space $\mathbb{R}^{2n}$ is *Lagrangian* if $\gamma, \gamma_x, \ldots, \gamma_x^{(2n-1)}$ are linearly independent and the span of $\gamma, \gamma_x, \ldots, \gamma_x^{(n-1)}$ is a Lagrangian subspace of $\mathbb{R}^{2n}$. In this paper, we (i) construct a complete set of differential invariants for Lagrangian curves in $\mathbb{R}^{2n}$ with respect to the symplectic group $Sp(2n)$, (ii) construct two hierarchies of commuting Hamiltonian Lagrangian curve flows of C-type and A-type, (iii) show that the differential invariants of solutions of Lagrangian curve flows of C-type and A-type are solutions of the Drinfeld-Sokolov's $\hat{C}_n^{(1)}$-KdV flows and $\hat{A}_{2n-1}^{(2)}$-KdV flows respectively, (iv) construct Darboux transforms, Permutability formulas, and scaling transforms, and give an algorithm to construct explicit soliton solutions, (v) give bi-Hamiltonian structures and commuting conservation laws for these curve flows.

Keywords: Lagrangian curve flows; KdV type hierarchies; Darboux transforms

Citation: Terng, C.L.; Wu, Z. Lagrangian Curve Flows on Symplectic Spaces. *Symmetry* **2021**, *13*, 298. https://doi.org/10.3390/sym13020298

Academic Editor: Valentin Lychagin
Received: 10 January 2021
Accepted: 3 February 2021
Published: 9 February 2021

Publisher's Note: MDPI stays neutral with regard to jurisdictional claims in published maps and institutional affiliations.

Copyright: © 2021 by the authors. Licensee MDPI, Basel, Switzerland. This article is an open access article distributed under the terms and conditions of the Creative Commons Attribution (CC BY) license (https://creativecommons.org/licenses/by/4.0/).

1. Introduction

The modern theory of soliton equations dates from the famous numerical computation of the interaction of solitary waves of the Korteweg-de Vries (KdV) equation by Zabusky and Kruskal [1] in 1965. In 1967, Gardner, Green, Kruskal, and Miura [2] applied the Gelfand-Levitan's inverse scattering transform of the one-dimensional linear Schrödinger operator to solve the Cauchy problem for rapidly decaying initial data for the KdV equation. In 1968, Lax [3] introduced the Lax-pair for KdV. Zakharov and Faddeev [4] gave a Hamiltonian formulation of KdV, and proved that KdV is completely integrable by finding action-angle variables. Zakharov and Shabat [5] found a Lax pair of 2×2 first order differential operators for the non-linear Schrödinger equation (NLS), Adler-Kostant-Symes gave a method to construct completely integrable Hamiltonian systems using splitting of Lie algebras (cf. [6–9]), Kupershmidt-Wilson [10] constructed $n \times n$ modified KdV (mKdV) using a loop algebra, and finally Drinfeld-Sokolov [11] gave a general method to construct soliton hierarchies from affine Kac-Moody algebras. In particular, soliton equations have many remarkable properties including: a Lax pair, infinite families of explicit soliton solutions, Bäcklund and Darboux transformations that generate new solutions from a given one by solving a first order system, a permutability formula to superpose solutions, a rational loop group action, a scattering theory and an inverse scattering transform to solve the Cauchy problem, a bi-Hamiltonian structure, and infinitely many commuting Hamiltonians. For more detail and references, we refer readers to the following books and survey articles: [11–18].

Soliton equations are also found in classical differential geometry: the sine-Gordon equation (SGE) arose first through the theory of surfaces of negative constant Gauss curvature in $\mathbb{R}^3$, and the reduced 3-wave equation can be found in Darboux's work [19] on triply orthogonal coordinate systems of $\mathbb{R}^3$. These equations were rediscovered later independently of their geometric history. The main contribution of the classical geometers lies in their methods for constructing explicit solutions of these equations from geometric transformations.

There are many classes of submanifolds in space forms and symmetric spaces whose Gauss-Codazzi equations are soliton equations. For example, the Gauss-Codazzi equations for the following classes of submanifolds are soliton equations: n-dimensional submanifolds of constant sectional curvature -1 in in $\mathbb{R}^{2n-1}$ (cf. [20,21]), isometric immersions of space forms in space forms (cf. [22,23]), flat Lagrangian submanifolds in $\mathbb{R}^{2n}$ [24], conformally flat submanifolds in spheres [25], and isothermic submanifolds in $\mathbb{R}^n$ (cf. [26–28]). For a survey of submanifold geometry and related soliton equations see [29].

Next we discuss how curve flows appeared in soliton theory. In 1906, da Rios, a student of Levi-Civita, wrote a master's thesis, in which he modeled the movement of a thin vortex by the motion of a curve propagating in $\mathbb{R}^3$ along its binormal with curvature as speed, i.e.,

$$\gamma_t = k\mathbf{b}.$$

This is the vortex filament equation (VFE). It was much later, in 1971, that Hasimoto showed in [30] the equivalence of VFE with the NLS,

$$q_t = i(q_{xx} + 2|q|^2 q).$$

In fact, if $\gamma(x,t)$ is a solution of VFE, then there exists a function $\theta(t)$ such that

$$q(x,t) = k(x,t)\exp(i(\theta(t) - \int_{-\infty}^{x} \tau(s,t)\,\mathrm{d}s))$$

is a solution of the NLS, where k, τ are the curvature and torsion of the curve. This correspondence between the VFE and NLS given above uses the Frenet frame. If we use the parallel normal frame, then the correspondence can be stated as follows: If γ is a solution of the VFE, then there exists an orthonormal moving frame $g = (e_1, e_2, e_3) : \mathbb{R}^2 \to SO(3)$ such that

$$g^{-1}g_x = \begin{pmatrix} 0 & -k_1 & -k_2 \\ k_1 & 0 & 0 \\ k_2 & 0 & 0 \end{pmatrix},$$

and $q = k_1 + ik_2$ is a solution of the NLS, where $e_1(\cdot, t)$ is tangent to the curve $\gamma(\cdot, t)$, $e_2(\cdot, t)$ and $e_3(\cdot, t)$ are parallel normal fields along $\gamma(\cdot, t)$, and $k_1(\cdot, t)$ and $k_2(\cdot, t)$ are the principal curvatures along $e_2(\cdot, t)$ and $e_3(\cdot, t)$ respectively. Since the NLS is a soliton equation, we can use techniques in soliton theory to study geometric and Hamiltonian aspects of the VFE.

The NLS admits an $so(3)$ valued Lax pair with phase space $C^\infty(\mathbb{R}, V)$, where

$$V = \left\{ \begin{pmatrix} 0 & -k_1 & -k_2 \\ k_1 & 0 & 0 \\ k_2 & 0 & 0 \end{pmatrix} \middle| k_1, k_2 \in \mathbb{R} \right\}.$$

Please note that the differential invariants constructed from the parallel frames for curves in $\mathbb{R}^3$ lie in $C^\infty(\mathbb{R}, V)$. Hence a good way to construct integrable curve flows on a homogeneous space $M = G \cdot p_0 = G/H$ is to find a class of curves in G/H, which has a moving frame $g : \mathbb{R} \to G$ so that $\gamma = g \cdot p_0$, $g^{-1}g_x$ gives a complete set of differential invariants, and $g^{-1}g_x$ lies in the phase space of a soliton equation. A more detailed discussion of how to use this scheme to construct integrable curve flows can be found in [31].

There are many recent works on integrable geometric curve flows in homogeneous spaces. For example, Langer-Perline studied Poisson structures and local geometric invariants of the VFE in [32,33], and constructed curve flows that relate to Fordy-Kulish NLS type hierarchies associated with Hermitian symmetric spaces in [34]. Doliwa-Santini constructed curve flows in $\mathbb{R}^2$ and $\mathbb{R}^3$ that give the mKdV and NLS respectively in [35]. Ferapontov gave hydro-dynamic type curve flows on homogeneous isoparametric hypersurfaces in sphere in [36]. Yasui-Sasaki studied the integrability of the VFE in [37]. Chou-Qu constructed integrable curve flows in affine plane in [38] and integrable curve flows in the plane for all Klein geometries in [39]. Anco constructed integrable curve flows on the symmetric space $\frac{U}{K}$ in [40]. Sanders-Wang studied curve flows in $\mathbb{R}^n$ whose curvatures are solutions of the vector mKdV in [41]. Terng-Thorbergsson constructed curve flows on Adjoint orbits of a compact Lie group G that relate to the n-wave equation associated with G in [42], Terng-Uhlenbeck explained the relation between the Schrödinger flow on compact Hermitian symmetric space and the Fordy-Kulish NLS system and wrote down a bi-Hamiltonian structure, geometric conservation laws, and commuting curve flows in [43] for the Schrödinger flows. Terng constructed Darboux transforms and explicit soliton solutions of the Airy curve flow in $\mathbb{R}^n$ in [44]. Mari Beffa gave natural Poisson structures on semi-simple homogeneous spaces and discussed their relations to integrable curve flows in [45,46]. Readers are referred to these papers for more references.

Drinfeld and Sokolov in [11] associated with each affine Kac-Moody algebra $\hat{\mathcal{G}}$ a hierarchy of soliton equations of KdV type, which will be called the $\hat{\mathcal{G}}$-*KdV hierarchy*. It was proved in [11] that the KdV hierarchy is the $\hat{A}_1^{(1)}$-KdV hierarchy and the Gelfand-Dickey hierarchy is the $\hat{A}_{n-1}^{(1)}$-KdV hierarchy.

There are recent works on integrable curve flows on flat spaces whose differential invariants satisfy the $\hat{\mathcal{G}}$-KdV hierarchies. The first example was given by Pinkall, who in [47] constructed a hierarchy of central affine curve flows on $\mathbb{R}^2$ invariant under the group $SL(2,\mathbb{R})$ and showed that their differential invariant (the central affine curvature) satisfies the KdV hierarchy. Calini-Ivey-Mari Beffa in [48] (for $n=3$) and Terng and Wu in [49] (for general n) constructed a hierarchy of curve flows on the affine space $\mathbb{R}^n$ invariant under $SL(n,\mathbb{R})$ whose differential invariants satisfy the $\hat{A}_{n-1}^{(1)}$-KdV hierarchies. Terng and Wu also constructed in [50] two hierarchies of curve flows on $\mathbb{R}^{n+1,n}$, whose differential invariants under the group $O(n+1,n)$ are solutions of the $\hat{B}_n^{(1)}$-KdV and $\hat{A}_{2n}^{(2)}$-KdV hierarchies respectively. In this paper, we construct two hierarchies of curve flows on the symplectic space $\mathbb{R}^{2n}$ whose differential invariants under the symplectic group are solutions of the $\hat{C}_n^{(1)}$-KdV and the $\hat{A}_{2n-1}^{(2)}$-KdV hierarchies respectively.

We need to set up some more notations before we explain our results. Let $\mathbb{R}^{2n}$ be the symplectic space with the symplectic form

$$\omega(X,Y) = X^t S_n Y, \quad \text{where } S_n = \sum_{i=1}^{2n}(-1)^{i+1} e_{i,2n+1-i}, \qquad (1)$$

$Sp(2n) = \{g \in GL(2n,\mathbb{R}) \mid g^t S_n g = S_n\}$ the group of linear isomorphisms of $\mathbb{R}^{2n}$ that preserves ω, and

$$sp(2n) = \{A \in sl(2n) \mid A^t S_n + S_n A = 0\}$$

the Lie algebra of $Sp(2n)$. A linear subspace V of $\mathbb{R}^{2n}$ is *isotropic* if $\omega(x,y) = 0$ for all $x,y \in V$. A maximal isotropic subspace has dimension n, and is called *Lagrangian*. The action of $Sp(2n)$ on the space of Lagrangian subspaces of $\mathbb{R}^{2n}$ defined by $g \cdot V = gV$ is transitive.

Definition 1. *A smooth map $\gamma : \mathbb{R} \to \mathbb{R}^{2n}$ is a Lagrangian curve if*

(i) $\gamma(s), \gamma_s(s), \ldots, \gamma_s^{(2n-1)}(s)$ *are linearly independent for all $s \in \mathbb{R}$,*

(1) *the span of $\gamma(s), \ldots, \gamma_s^{(n-1)}(s)$ is a Lagrangian subspace of $\mathbb{R}^{2n}$ for all $s \in \mathbb{R}$,*

where $\gamma_s^{(i)} = \frac{d^i \gamma}{ds^i}$.

We show that if $\gamma : \mathbb{R} \to \mathbb{R}^{2n}$ is Lagrangian then there exists a unique orientation preserving parameter $x = x(s)$ such that $\omega(\gamma_x^{(n)}, \gamma_x^{(n-1)}) = (-1)^n$. We call such parameter the *Lagrangian parameter* for γ.

Let

$$\mathcal{M}_{2n} = \left\{ \gamma \in \mathbb{R}^{2n} \mid \gamma \text{ is Lagrangian}, \omega(\gamma_s^{(n)}, \gamma_s^{(n-1)}) = (-1)^n \right\}.$$

$V_n = \oplus_{i=1}^n \mathbb{R} e_{n+1-i,n+i}$, where $\oplus$ is the direct sum.

We prove that given $\gamma \in \mathcal{M}_{2n}$, there exists a unique $g = (g_1, \ldots, g_{2n}) : \mathbb{R} \to Sp(2n)$ such that $g_i = \gamma_x^{(i-1)}$ for $1 \leq i \leq n+1$ and

$$g^{-1}g_x = b + u$$

for some $u = \sum_{i=1}^{n} u_i e_{n+1-i,n+i} \in C^\infty(\mathbb{R}, V_n)$, where

$$b = \sum_{i=1}^{n-1} e_{i+1,i}. \tag{2}$$

We call this g the *Lagrangian moving frame* and $u = \sum_{i=1}^{n} u_i e_{n+1-i,n+i}$ the *Lagrangian curvature* along γ.

It is easy to see that

$$\gamma(x) = (1, x, \frac{x^2}{2!}, \cdots, \frac{x^{2n-1}}{(2n-1)!})^t$$

is in $\mathcal{M}_{2n}$ with Lagrangian frame $g(x) = \exp(bx)$ and zero Lagrangian curvature.

Definition 2. *The Lagrangian curvature map*

$$\Psi : \mathcal{M}_{2n} \to C^\infty(\mathbb{R}, V_n),$$

is defined by $\Psi(\gamma) = u$, *where u is the Lagrangian curvature of $\gamma \in \mathcal{M}_{2n}$.*

It follows from the theory of existence and uniqueness of solutions of ordinary differential equations that the Lagrangian curvatures form a complete set of differential invariants for curves in $\mathcal{M}_{2n}$.

A *Lagrangian curve flow* is an evolution equation on $\mathcal{M}_{2n}$, i.e., the flow preserves the Lagrangian parameter. Such flow can be written in the form $\gamma_t = g\xi(u)$ so that $g\xi(u)$ is tangent to $\mathcal{M}_{2n}$ at γ, where $g(\cdot, t)$ and $u(\cdot, t)$ are the Lagrangian moving frame and Lagrangian curvature along $\gamma(\cdot, t)$ and $\xi(u)$ is a $\mathbb{R}^{2n \times 1}$ valued differential polynomial of u in x variable.

Please note that when $n = 1$, we have $sp(2) = sl(2, \mathbb{R})$, $\omega(X, Y) = \det(X, Y)$, the Lagrangian parameter, frame, curvature are the central affine parameter, frame, central affine curvature on $\mathbb{R}^2$ under the group $SL(2, \mathbb{R})$, and the Lagrangian curve flows on $\mathbb{R}^2$ are the central affine curve flows studied in [47] (see also in [51,52]). For example,

$$\gamma_t = \frac{u_x}{4}\gamma - \frac{u}{2}\gamma_x$$

is a Lagrangian flow on $\mathbb{R}^2$ and its Lagrangian curvature u satisfies the KdV,

$$u_t = \frac{1}{4}(u_{xxx} - 6uu_x).$$

In this paper, we construct two hierarchies of Lagrangian curve flows on $\mathbb{R}^{2n}$ whose Lagrangian curvatures are solutions of the $\hat{C}_n^{(1)}$-KdV and $\hat{A}_{2n-1}^{(2)}$-KdV hierarchies respectively. In particular, we obtain the following results:

(1) We construct a sequence of commuting Lagrangian curve flows of C-type and A-type respectively on $\mathcal{M}_{2n}$ such that the third flows are

$$\gamma_t = -\frac{3}{4n}(u_1)_x \gamma - \frac{3}{2n} u_1 \gamma_x + \gamma_{xxx}, \tag{3}$$

$$\gamma_t = -\frac{3}{2n-1} u_1 \gamma_x + \gamma_{xxx} \tag{4}$$

respectively, where u_1 is the first Lagrangian curvature.

(2) The Lagrangian curvature map Ψ maps the space of solutions of Lagrangian curve flows of C-type (A-type resp.) modulo $Sp(2n)$ bijectively onto the space of solutions of $\hat{C}_n^{(1)}$-KdV ($\hat{A}_{2n-1}^{(2)}$-KdV resp.) flows. For example, the Lagrangian curvatures u_1, u_2 of a solution γ of (3) and (4) satisfy the third $\hat{C}_2^{(1)}$-KdV flow

$$\begin{cases} (u_1)_t = -\frac{5}{4}u_1^{(3)} + 3u_2' + \frac{3}{4}u_1 u_1', \\ (u_2)_t = -\frac{3}{8}u_1^{(5)} + u_2^{(3)} + \frac{3}{8}(u_1 u_1^{(3)} + u_1' u_1'') - \frac{3}{4}u_1 u_2'. \end{cases} \quad (5)$$

and the third $\hat{A}_3^{(2)}$-KdV flow

$$\begin{cases} (u_1)_t = 3(u_2)_x, \\ (u_2)_t = (u_2)_{xxx} - (u_1 u_2)_x \end{cases} \quad (6)$$

respectively.

(3) A bi-Hamiltonian structure and commuting conservation laws for Lagrangian curve flows of C- and A-types are given. For example, the curve flows (3) and (4) are Hamiltonian flows for functionals

$$\hat{F}_3(\gamma) = \oint u_2 + \frac{2n-3}{4n} u_1^2 dx$$

$$\hat{H}_3(\gamma) = \oint u_2 + \frac{n-2}{2n-1} u_1^2 dx$$

respectively on $\mathcal{M}_{2n}$ with respect to the second Hamiltonian structure, where u is the Lagrangian curvature of γ.

(4) We construct Darboux transforms (DTs), Permutability formulas, scaling transforms, and give an algorithm to compute explicit soliton solutions of these flows.

This paper is organized as follows: We construct Lagrangian moving frames in Section 2, and review the constructions of the $\hat{C}_n^{(1)}$-KdV and $\hat{A}_{2n-1}^{(2)}$-KdV hierarchies in Section 3. Lagrangian curve flows of C- and A- types and the evolutions of their Lagrangian curvatures are given in Section 4. In Section 5, we construct Darboux transforms (DTs) and a Permutability formula for the $\hat{C}_n^{(1)}$-KdV and for the Lagrangian curve flows of C-type. DTs for the A case and its Permutability formula are given in Section 6. The scaling transforms are given in Section 7. Bi-Hamiltonian structures and commuting conserved functionals are given in Section 8. We give an outline of a method for constructing integrable curve flows whose differential invariants satisfy the $\hat{\mathcal{G}}^{(1)}$-KdV hierarchy for general simple real non-compact Lie algebra $\mathcal{G}$ and give some open problems in the last section.

2. Lagrangian Moving Frame

In this section, we prove the existence of Lagrangian parameter and construct the Lagrangian moving frame and curvatures for Lagrangian curves (cf. Definition 1).

Proposition 1. *If $\gamma : \mathbb{R} \to \mathbb{R}^{2n}$ is a Lagrangian curve, then there exists a unique Lagrangian parameter $x = x(s)$, i.e., $\omega(\gamma_x^{(n)}, \gamma_x^{(n-1)}) = (-1)^n$.*

Proof. If $\omega(\gamma_s^{(n)}, \gamma_s^{(n-1)})$ is zero at s_0, then it follows from $\omega(\gamma_s^{(i)}, \gamma_s^{(j)}) = 0$ for all $0 \leq i, j \leq n-1$ that $\omega(\gamma_s^{(n)}, \gamma_s^{(i)}) = 0$ at s_0. Hence $\gamma(s_0), \gamma_s(s_0), \ldots, \gamma_s^{(n)}(s_0)$ span an $(n+1)$-dimension isotropic subspace. However, the maximal dimension of an isotropic subspace is n, a contradiction. Hence $\omega(\gamma_s^{(n)}, \gamma_s^{(n-1)})$ never vanishes. Choose $x = x(s)$ such that $(\frac{dx}{ds})^{2n-1} = (-1)^n \omega(\gamma_s^{(n)}, \gamma_s^{(n-1)})$. □

Theorem 1. *If $\gamma \in \mathcal{M}_{2n}$, then there exists a unique Lagrangian moving frame g along γ, i.e., $g^{-1}g_x = b + \sum_{i=1}^{n} u_i e_{n+1-i,n+i}$ for some $u_1, \ldots, u_n$, where b is defined by (2).*

Proof. Let $u_1 = (-1)^{n-1}\omega(\gamma_x^{(n+1)}, \gamma_x^{(n)})$, and $g_{n+2} = \gamma_x^{(n+1)} - u_1\gamma_x^{(n-1)}$. We derive g_i's and u_i's by the recursive formula:

$$u_j = (-1)^{n-j}\omega((g_{n+j})_x, g_{n+j}) = (-1)^{n-j}\omega(d_x^{n+j}\gamma, g_{n+j}), \quad 2 \le j \le n-1,$$
$$g_{n+j+1} = d_x g_{n+j} - u_j \gamma_x^{(n-j)}, \quad 2 \le j \le n-1,$$
$$u_n = \omega((g_{2n})_x, g_{2n}).$$

Then $g = (\gamma, \ldots, \gamma_x^{(n)}, g_{n+2}, \ldots, g_{2n})$ satisfies $g^{-1}g_x = b + u$, i.e., g is a Lagrangian moving frame along γ. □

Example 1. *For $n = 1$, we have $\omega(X,Y) = \det(X,Y)$, thus $\gamma \in \mathcal{M}_2$ if and only if $\det(\gamma, \gamma_x) = 1$. So the Lagrangian parameter is the central affine parameter, the Lagrangian frame along γ is $g = (\gamma, \gamma_x)$ is the central affine moving frame along γ, and the Lagrangian curvature is the central affine curvature. Moreover,*

$$g^{-1}g_x = \begin{pmatrix} 0 & u_1 \\ 1 & 0 \end{pmatrix}.$$

Example 2. *The Lagrangian frame $g = (\gamma, \gamma_x, \gamma_{xx}, g_4)$ along $\gamma \in \mathcal{M}_4$ satisfies*

$$g^{-1}g_x = \begin{pmatrix} 0 & 0 & 0 & u_2 \\ 1 & 0 & u_1 & 0 \\ 0 & 1 & 0 & 0 \\ 0 & 0 & 1 & 0 \end{pmatrix},$$

where

$$u_1 = -\omega(\gamma_x^{(3)}, \gamma_{xx}), \quad u_2 = \omega((g_4)_x, g_4) = \omega(\gamma_x^{(4)}, \gamma), \quad g_4 = \gamma_x^{(3)} - u_1 \gamma_x.$$

It follows from the Existence and Uniqueness of ordinary differential equations that $\{u_1, \cdots, u_n\}$ forms a complete set of local differential invariants for $\gamma \in \mathcal{M}_{2n}$ under the $Sp(2n)$-action. So we have the following:

Proposition 2. *The Lagrangian curvature map $\Psi : \mathcal{M}_{2n} \to C^{\infty}(\mathbb{R}, V_n)$ defined by Definition 2 is onto and $\Psi^{-1}(u)$ is a $Sp(2n)$-orbit.*

Example 3. *A Lagrangian curve in $\mathbb{R}^{2n}$ with zero Lagrangian curvature is of the form:*

$$\gamma = c_0(1, x, \frac{x^2}{2}, \cdots, \frac{x^{2n-1}}{(2n-1)!})^t, \quad c_0 \in Sp(2n).$$

3. The $\hat{C}_n^{(1)}$-KdV and the $\hat{A}_{2n-1}^{(2)}$-KdV Hierarchies

In this section, we review the constructions of the $\hat{C}_n^{(1)}$-, $\hat{A}_{2n-1}^{(2)}$-, $\hat{C}_n^{(1)}$-KdV, and $\hat{A}_{2n-1}^{(2)}$-KdV hierarchies and derive some elementary properties of these hierarchies (cf. [11,53]).

3.1. The $\hat{C}_n^{(1)}$-KdV Hierarchy

A *splitting* of a Lie algebra $\mathcal{L}$ is a pair of Lie subalgebras $\mathcal{L}_+, \mathcal{L}_-$ such that $\mathcal{L} = \mathcal{L}_+ \oplus \mathcal{L}_-$ as linear subspaces (but not as subalgebras). For $\xi \in \mathcal{L}$, we write

$$\xi = \xi_+ + \xi_-, \quad \text{where} \quad \xi_+ \in \mathcal{L}_+, \xi_- \in \mathcal{L}_-.$$

A *vacuum sequence* is a linearly independent, commuting sequence $\{J_j \mid j \geq 1\}$ in $\mathcal{L}_+$.
Let
$$Sp(2n,\mathbb{C}) = \{A \in GL(2n,\mathbb{C}) \mid A^t S_n A = S_n\},$$
and $sp(2n,\mathbb{C})$ its Lie algebra. Then $sp(2n)$ is a real form of $sp(2n,\mathbb{C})$ defined by the involution $\tau(A) = \bar{A}$.

Let
$$\hat{\mathcal{C}}_n^{(1)} := \left\{ A = \sum_i A_i \lambda^i \mid A_i \in sp(2n) \right\},$$
$$(\hat{\mathcal{C}}_n^{(1)})_+ = \left\{ \sum_{i \geq 0} A_i \lambda^i \in \hat{\mathcal{C}}_n^{(1)} \right\}, \quad (\hat{\mathcal{C}}_n^{(1)})_- = \left\{ \sum_{i < 0} A_i \lambda^i \in \hat{\mathcal{C}}_n^{(1)} \right\}.$$

Then $((\hat{\mathcal{C}}_n^{(1)})_+, (\hat{\mathcal{C}}_n^{(1)})_-)$ is a splitting of $\hat{\mathcal{C}}_n^{(1)}$.

Please note that $\xi(\lambda) = \sum_i \xi_i \lambda^i$ is in $\hat{\mathcal{C}}_n^{(1)}$ if and only if ξ satisfy the $sp(2n)$-reality condition, i.e.,
$$\xi(\lambda)^t S_n + S_n \xi(\lambda) = 0, \quad \overline{\xi(\bar{\lambda})} = \xi(\lambda).$$

A meromorphic map $f : \mathbb{C} \to SL(2n,\mathbb{C})$ is said to satisfy the $Sp(2n)$-reality condition if
$$f(\lambda)^t S_n f(\lambda) = S_n, \quad \overline{f(\bar{\lambda})} = f(\lambda). \tag{7}$$

For $\xi(\lambda) = \sum_i \xi_i \lambda^i$, we have
$$\xi_+(\lambda) = \sum_{i \geq 0} \xi_i \lambda^i, \quad \xi_-(\lambda) = \sum_{i < 0} \xi_i \lambda^i.$$

Let B_n^+ and N_n^+ denote the subgroups of upper, strictly upper triangular matrices in $Sp(2n)$ respectively, and $\mathcal{B}_n^+, \mathcal{N}_n^+$ the corresponding Lie subalgebras of $sp(2n)$.

Set
$$J = \sum_{i=1}^{2n-1} e_{i+1,i} + e_{1,2n} \lambda = b + e_{1,2n} \lambda \in (\hat{\mathcal{C}}_n^{(1)})_+.$$

Then
$$J^i = (b^t)^{n-i} \lambda + b^i, \quad 1 \leq i \leq 2n-1, \tag{8}$$
$$J^{2n} = \lambda I_{2n}. \tag{9}$$

It is easy to check that J^{2j-1} is in $(\hat{\mathcal{C}}_n^{(1)})_+$, but J^{2j} is not. So $\{J^{2j-1} \mid j \geq 1\}$ is a vacuum sequence. Note that
$$[J, (\hat{\mathcal{C}}_n^{(1)})_-]_+ = \mathcal{B}_n^+.$$

Next we use the general method given in [53] to construct the $\hat{\mathcal{C}}_n^{(1)}$-hierarchy generated by the vacuum sequence $\{J^{2j-1} \mid j \geq 1\}$. First a direct computation gives the following known results:

Theorem 2 ([49,53]). *Given $q \in C^\infty(\mathbb{R}, \mathcal{B}_n^+)$, then there exists a unique*
$$P(q,\lambda) = \sum_{i \leq 1} P_{1,i}(q) \lambda^i$$

in $\hat{\mathcal{C}}_n^{(1)}$ satisfying
$$\begin{cases} [\partial_x + J + q, P(q,\lambda)] = 0, \\ P^{2n}(q,\lambda) = \lambda I_{2n}. \end{cases} \tag{10}$$

Moreover, $P_{1,i}(q)$ can be computed recursively by equating the coefficients of λ^i in (10) and they are polynomials in u and x-derivatives of u (i.e., a differential polynomial in u).

Please note that if operators A, B commute, then A and B^j also commute. Hence it follows from the first equation of (10) that we have

$$[\partial_x + J + q, P^{2j-1}(q,\lambda)] = 0. \tag{11}$$

Write the power series

$$P^{2j-1}(q,\lambda) = \sum_i P_{2j-1,i}(q)\lambda^i. \tag{12}$$

We compare coefficient of λ^i of (11) to obtain

$$[\partial_x + b + q, P_{2j-1,i}(q)] = [P_{2j-1,i-1}(q), e_{1,2n}], \tag{13}$$

which implies that the left hand side lies in $\mathcal{B}_n^+$. So

$$q_{t_{2j-1}} = [\partial_x + b + q, P_{2j-1,0}(q)], \quad j \geq 1. \tag{14}$$

defines a flow on $C^\infty(\mathbb{R}, \mathcal{B}_n^+)$. We call (14) the $(2j-1)$-th $\hat{\mathcal{C}}_n^{(1)}$-flow.

We need the following well-known elementary result to explain the Lax pair:

Proposition 3. *Let $\mathcal{G}$ be the Lie algebra of G, and $A, B : \mathbb{R}^2 \to \mathcal{G}$ smooth maps. Then the following statements are equivalent:*

(1) *the linear system*

$$\begin{cases} g_x = gA, \\ g_t = gB \end{cases}$$

is solvable for $g : \mathbb{R}^2 \to G$,

(2) *A, B satisfy*

$$A_t = B_x + [A, B] = [\partial_x + A, B],$$

(3) *$[\partial_x + A, \partial_t + B] = 0.$*

Proposition 4. *The following statements are equivalent for smooth $q : \mathbb{R}^2 \to \mathcal{B}_n^+$:*

(1) *q is a solution of (14),*

(2) *the following linear system is solvable for $h : \mathbb{R}^2 \to Sp(2n)$,*

$$\begin{cases} h^{-1}h_x = b + q, \\ h^{-1}h_t = P_{2j-1,0}(q). \end{cases} \tag{15}$$

(3) *the following linear system is solvable for $F(x,t,\lambda) \in SL(2n,\mathbb{C})$,*

$$\begin{cases} F_x = F(J+q), \\ F_t = F(P^{2j-1}(q,\lambda))_+, \\ F(x,t,\lambda)^t S_n F(x,t,\lambda) = S_n, \quad \overline{F(x,t,\bar\lambda)} = F(x,t,\lambda). \end{cases} \tag{16}$$

The last equation says that $F(x,t,\lambda)$ satisfies the $Sp(2n)$-reality condition (7) in λ.

Proof. Equation (13) implies that the coefficients of λ^i for $i > 0$ of

$$[\partial_x + J + q, \partial_t + (P^{2j-1}(q,\lambda))_+]$$

are zero. The constant term is $[\partial_x + b + q, \partial_t + P_{2j-1,0}(q)]$. This proves that $[\partial_x + J + q, \partial_t + (P^{2j-1}(q,\lambda))_+] = 0$ is equivalent to $[\partial_x + b + q, \partial_t + P_{2j-1,0}(q)] = 0$. It follows from Proposition 3 that (2) and (3) are equivalent.

Equation (14) can be written as

$$(b+q)_t = (P_{2j-1,0}(q))_x + [b+q, P_{2j-1,0}(q)].$$

It follows from Proposition 3 that (1) and (2) are equivalent. □

The group $C^\infty(\mathbb{R}, N_n^+)$ acts on $C^\infty(\mathbb{R}, \mathcal{B}_n^+)$ by gauge transformation,

$$f(\partial_x + b + q)f^{-1} = \partial_x + b + f * q \tag{17}$$

for $f \in C^\infty(\mathbb{R}, N_n^+)$ and $q \in C^\infty(\mathbb{R}, \mathcal{B}_n^+)$, where

$$f * q = f(b+q)f^{-1} - f_x f^{-1} - b. \tag{18}$$

The following Proposition shows that $C^\infty(\mathbb{R}, V_n)$ is a cross-section of this gauge action.

Proposition 5. *Given $q \in C^\infty(\mathbb{R}, \mathcal{B}_n^+)$, then there exist a unique $\triangle \in C^\infty(\mathbb{R}, N_n^+)$ and $u = \sum_{i=1}^n u_i e_{n+1-i, n+i}$ in $C^\infty(\mathbb{R}, V_n)$ such that*

$$\triangle(\partial_x + J + q)\triangle^{-1} = \partial_x + J + u. \tag{19}$$

In particular, $u = \triangle * q$.

Proof. Let $\mathcal{G}_j = \oplus_{i=1}^{2n-j} \mathbb{R} e_{i, i+j}$, $\mathcal{G}_{-j} = \oplus_{i=1}^{2n-j} \mathbb{R} e_{i+j, i}$ for $0 \leq j \leq 2n - 1$. Equation (19) implies that

$$\triangle(J + q) - \triangle_x = (J + u)\triangle. \tag{20}$$

Proposition is proved by equating components of $\mathcal{G}_j$ of (20) for $|j| \leq 2n - 1$. □

It can be checked by the same method for the $\hat{A}_n^{(1)}$-hierarchy (cf. [53]) that flow (14) is invariant under the $C^\infty(\mathbb{R}, N_n^+)$-action. So given $u \in C^\infty(\mathbb{R}, V_n)$ and $j \geq 1$, there exists a unique $\mathcal{N}_n^+$-valued differential polynomial $\eta_j(u)$ satisfying

$$[\partial_x + J + u, (P(u)^{2j-1})_+ - \eta_j(u)] \in C^\infty(\mathbb{R}, V_n). \tag{21}$$

The induced quotient flow of (14) on the cross-section $C^\infty(\mathbb{R}, V_n)$ is obtained by projecting (14) down along gauge orbits. So the induced quotient flow on $C^\infty(\mathbb{R}, V_n)$ is

$$u_{t_{2j-1}} = [\partial + J + u, P_{2j-1,0}(u) - \eta_j(u)]. \tag{22}$$

The above equation is the $(2j-1)$-th $\hat{C}_n^{(1)}$-KdV flow.

As a consequence of the construction, we have the following.

Proposition 6. *The following statements are equivalent for smooth $u : \mathbb{R}^2 \to V_n$:*

(i) u *is a solution of* (22),
(ii) $[\partial_x + J + u, \partial_t + (P^{2j-1}(u,\lambda))_+ - \eta_j(u)] = 0$.
(iii) *The following linear system is solvable for $g : \mathbb{R}^2 \to Sp(2n)$,*

$$\begin{cases} g^{-1} g_x = b + u, \\ g^{-1} g_t = P_{2j-1,0}(u) - \eta_j(u). \end{cases} \tag{23}$$

(iv) The following linear system is solvable for $E(x,t,\lambda) \in SL(2n,\mathbb{C})$ for all parameter $\lambda \in \mathbb{C}$,

$$\begin{cases} E_x = E(J+u), \\ E_t = E((P^{2j-1}(u,\lambda))_+ - \eta_j(u)), \\ E(x,t,\lambda)^t S_n E(x,t,\lambda) = S_n, \quad \overline{E(x,t,\bar\lambda)} = E(x,t,\lambda). \end{cases} \quad (24)$$

Example 4. *The third $\hat{C}_1^{(1)}$-KdV flow is the KdV for $q = u_1$:*

$$q_t = \frac{1}{4}(q_{xxx} - 6qq_x).$$

Example 5. The third $\hat{C}_2^{(1)}$-KdV flow

The $\hat{C}_2^{(1)}$-KdV flows are for $u = u_1 e_{23} + u_2 e_{14}$. We compare coefficients of λ^i in (10) to compute $P_{1,i}(q)$. Then $P_{2j-1,i}(q)$ can be computed from $P_{1,k}(q)$'s. We obtain the first few terms of $P^3(q,\lambda)$:

$$P_{3,1}(u) = \begin{pmatrix} 0 & 1 & 0 & -\frac{3}{4}u_1 \\ 0 & 0 & 1 & 0 \\ 0 & 0 & 0 & 1 \\ 0 & 0 & 0 & 0 \end{pmatrix},$$

$$P_{3,0}(u) = \begin{pmatrix} -\frac{3}{8}u_1' & u_2 - \frac{3}{8}u_1'' & u_2' - \frac{3}{8}(u_1)_x^{(3)} & \xi \\ \frac{1}{4}u_1 & -\frac{1}{8}u_1' & -\frac{1}{2}u_1'' + u_2 + \frac{1}{4}u_1^2 & -u_2' + \frac{3}{8}(u_1)_x^{(3)} \\ 0 & \frac{1}{4}u_1 & \frac{1}{8}u_1' & u_2 - \frac{3}{8}u_1'' \\ 1 & 0 & \frac{1}{4}u_1 & \frac{3}{8}u_1' \end{pmatrix}, \text{ where}$$

$$\xi = -\frac{3}{8}(u_1)_x^{(4)} + u_2'' + \frac{3}{8}u_1 u_1'' - \frac{3}{4}u_1 u_2.$$

Therefore the third $\hat{C}_2^{(1)}$-KdV flow is (5).

3.2. The $\hat{A}_{2n-1}^{(2)}$-KdV Hierarchy ($n \geq 2$)

Let κ be the involution of $sl(2n,\mathbb{C})$ defined by

$$\kappa(X) = -S_n X^t S_n^{-1},$$

where S_n is as in (1). Then the fixed point set of κ is $sp(2n,\mathbb{C})$ and the -1 eigen-space of κ in $sl(2n)$ is

$$\mathcal{P}_{2n} = \{\xi \in sl(2n) \mid \xi = S_n \xi^t S_n^{-1}\}.$$

Let

$$\hat{A}_{2n-1}^{(2)} = \left\{ A(\lambda) = \sum_{i \leq m_0} A_i \lambda^i \mid A_i \in sl(2n,\mathbb{R}), \kappa(A(-\lambda)) = A(\lambda) \right\},$$

and

$$(\hat{A}_{2n-1}^{(2)})_+ = \left\{ \sum_{i \geq 0} A_i \lambda^i \in \hat{A}_{2n-1}^{(2)} \right\}, \quad (\hat{A}_{2n-1}^{(2)})_- = \left\{ \sum_{i < 0} A_i \lambda^i \in \hat{A}_{2n-1}^{(2)} \right\}.$$

Then $((\hat{A}_{2n-1}^{(2)})_+, (\hat{A}_{2n-1}^{(2)})_-)$ is a splitting of $\hat{A}_{2n-1}^{(2)}$.

Please note that the following are equivalent for $A(\lambda) = \sum_i A_i \lambda^i$:

(1) $A \in \hat{A}_{2n-1}^{(2)}$,
(2) $A_{2i} \in sp(2n)$ and $A_{2i+1} \in \mathcal{P}_{2n}$ for all i,

(3) A satisfies
$$-S_n A(-\lambda)^t S_n^{-1} = A(\lambda), \quad \overline{A(\bar{\lambda})} = A(\lambda).$$

Set
$$\beta = \frac{1}{2}(e_{1,2n-1} + e_{2,2n}). \tag{25}$$

$$J_B = \frac{1}{2}(e_{1,2n-1} + e_{2,2n})\lambda + \sum_{i=1}^{2n-1} e_{i+1,i} = \beta\lambda + b.$$

Please note that $J_B^{2i-1} \in (\hat{\mathcal{A}}_{2n-1}^{(2)})_+$ and
$$J_B^{2n} = \lambda J_B.$$

Then $\{J_B^{2j-1} \mid j \geq 1\}$ is a vacuum sequence in $(\hat{\mathcal{A}}_{2n-1}^{(2)})_+$.

Next we use the general method given in [53] to construct the $\hat{\mathcal{A}}_{2n-1}^{(2)}$-hierarchy generated by $\{J_B^{2j-1} \mid j \geq 1\}$. Similarly, we have the following:

Theorem 3. *Given $q \in C^\infty(\mathbb{R}, \mathcal{B}_n^+)$ and $j \in \mathbb{Z}$, then there exists a unique*
$$Q(q,\lambda) = \sum_{i \leq 1} Q_{1,i}(q) \lambda^i \in \hat{\mathcal{A}}_{2n-1}^{(2)}$$

satisfying
$$\begin{cases} [\partial_x + J_B + q, Q(q,\lambda)] = 0, \\ Q^{2n}(q,\lambda) = \lambda Q(q,\lambda). \end{cases} \tag{26}$$

Moreover, $Q_{1,i}(q)$'s are polynomial differentials in q and derivatives of q and can be computed recursively by equating the coefficient of λ^i of (26).

Proof. It was proved in [11] that given any $\xi = \sum_{i \leq i_0} \xi_i \lambda^i \in sl(2n, \mathbb{R})$, there exists unique diagonal matrices h_i such that
$$\xi = \sum_{j \leq 2n(i_0+1)} h_i J^j,$$

where $J = b + e_{1,2n}\lambda$.

Given permutation s in S_{2n} and $h = \text{diag}(h_1, \ldots, h_{2n})$, let
$$h^s = \text{diag}(h_{s(1)}, \ldots, h_{s(2n)}).$$

Let $\theta \in S_{2n}$ be the cyclic permutation defined by $\theta(1) = 2n$, and $\theta(i) = i - 1$ for $2 \leq i \leq 2n$. A simple computation implies that
$$Jh = h^\theta J, \quad J^i h = h^{\theta^i} J^i. \tag{27}$$

Please note that
$$\frac{1}{2}(e_{1,2n-1} + e_{2n}) = \text{diag}(\frac{1}{2}, \frac{1}{2}, 0, \ldots, 0) J^2.$$

$$q = \sum_{i=1}^{2n} k_i J^{i-2n},$$

where $q = (q_{ij})$ and k_i are diagonal matrices defined by
$$k_i = \text{diag}(q_{1,2n-i+1}, q_{2,2n-i+2}, \ldots, q_{i,2n}, 0, \ldots, 0).$$

Write
$$Q(q,\lambda) = \mathrm{diag}(\frac{1}{2},\frac{1}{2},0,\ldots,0)J^2 + \mathrm{diag}(0,1,\ldots,1)J + \sum_{j\leq 0} h_j J^j,$$

We compare the coefficients of J^i's of both sides of each equation in (26) and use (27) to solve h_j uniquely as differential polynomial of q. This gives the formula for $Q(q,\lambda)$. We plug in Formulas (8) and (9) to obtain $Q_{1,j}(q)$'s. □

The first equation of (26) implies that
$$[\partial_x + J_B + q, Q^{2j-1}(q,\lambda)] = 0. \tag{28}$$

Write $Q^{2j-1}(q,\lambda)$ as a power series in λ,
$$Q^{2j-1}(q,\lambda) = \sum_i Q_{2j-1,i}(q)\lambda^i. \tag{29}$$

We compare the coefficient of λ^i of (28) to obtain
$$[\partial_x + b + q, Q_{2j-1,i}(q)] = [Q_{2j-1,i-1}(q), \beta], \tag{30}$$

where β is defined by (25). So the left hand side of (30) is $\mathcal{B}_n^+$-valued and
$$q_{t_{2j-1}} = [\partial_x + b + q, Q_{2j-1,0}(q)], \quad j \geq 1, \tag{31}$$

is a flow on $C^\infty(\mathbb{R}, \mathcal{B}_n^+)$. This is the $(2j-1)$-th flow in the $\hat{A}^{(2)}_{2n-1}$-hierarchy.

We use the same proof of Proposition 4 to obtain the following:

Proposition 7. *The following statements are equivalent for smooth $q : \mathbb{R}^2 \to \mathcal{B}_n^+$:*
(i) *q is a solution of (31).*
(ii) *The following linear system is solvable for smooth $g : \mathbb{R}^2 \to Sp(2n)$,*
$$\begin{cases} g^{-1}g_x = b + q, \\ g^{-1}g_t = Q_{2j-1,0}(u). \end{cases} \tag{32}$$

(iii) *The following linear system is solvable for $F(x,t,\lambda) \in SL(2n,\mathbb{C})$ for all parameter $\lambda \in \mathbb{C}$,*
$$\begin{cases} F_x = F(J_B + q), \\ F_t = F((Q_{2j-1}(u,\lambda))_+, \\ F(x,t,\lambda)^t S_n F(x,t,-\lambda) = S_n, \quad \overline{F(x,t,\bar{\lambda})} = F(x,t,\lambda). \end{cases}$$

It follows from Proposition 5 that there exist a unique $\triangle \in C^\infty(\mathbb{R}, \mathcal{N}_n^+)$ and $u \in C^\infty(\mathbb{R}, V_n)$ such that $u = \triangle * q$. So given $u \in C^\infty(\mathbb{R}, V_n)$ and $j \geq 1$, there exists a unique $\zeta_j(u) \in C^\infty(\mathbb{R}, \mathcal{N}_n^+)$ such that
$$[\partial_x + b + u, Q_{2j-1,0}(u) - \zeta_j(u)] \in C^\infty(\mathbb{R}, V_n). \tag{33}$$

The $(2j-1)$-th $\hat{A}^{(2)}_{2n-1}$-KdV flow is the following flow on $C^\infty(\mathbb{R}, V_n)$:
$$u_{t_{2j-1}} = [\partial_x + b + u, Q_{2j-1,0}(u) - \zeta_j(u)]. \tag{34}$$

Proposition 8. *The following statements are equivalent for smooth $u : \mathbb{R}^2 \to V_n$:*
(i) *u is a solution of (34).*

(ii) *The following linear system is solvable for* $g : \mathbb{R}^2 \to Sp(2n)$,

$$\begin{cases} g^{-1}g_x = b + u, \\ g^{-1}g_t = Q_{2j-1,0}(u) - \xi_j(u). \end{cases} \tag{35}$$

(iii) *The following linear system is solvable for* $E(x, t, \lambda) \in SL(2n, \mathbb{C})$ *for all* $\lambda \in \mathbb{C}$,

$$\begin{cases} E_x = E(J_B + u), \\ E_t = E(Q_{2j-1}(u, \lambda))_+ - \xi_j(u)), \\ E(x, t, \lambda)^t S_n E(x, t, -\lambda) = S_n, \quad \overline{E(x, t, \bar\lambda)} = E(x, t, \lambda). \end{cases}$$

Example 6. *We use* (26) *to compute* $Q_{1,i}(u)$, *then use these to compute* $Q_{3,0}(u)$. *A direct computation implies that the third* $\hat{A}_3^{(2)}$-*KdV flow is* (6).

Definition 3. $F(x,t,\lambda)$ ($E(x,t,\lambda)$ *resp.*) *is a frame of a solution* $q : \mathbb{R}^2 \to \mathcal{B}_n^+$ *of* (14) ($u : \mathbb{R}^2 \to V_n$ *of* (22) *resp.*) *if* $F(x,t,\lambda)$ ($E(x,t,\lambda)$ *resp.*) *is holomorphic for all* $\lambda \in \mathbb{C}$ *and satisfies the linear system* (16) ((24) *resp.*). *Frames for solutions of* (31) *and* (34) *are defined similarly.*

It follows from the constructions of the $\hat{C}_n^{(1)}$-KdV and $\hat{A}_{2n-1}^{(2)}$-KdV flows that we have the following.

Proposition 9.

(1) *Let* $F(x, t, \lambda)$ *be a frame of a solution* $q : \mathbb{R}^2 \to \mathcal{B}_n^+$ *of* (14) ((31) *resp.*), *and the unique* $\triangle : \mathbb{R}^2 \to N_n^+$ *such that* $u := \triangle * q$ *is* V_n-*valued (as in Proposition* 5*). Then* u *is a solution of* (22) ((34) *resp.*) *and* $E(x, t, \lambda) = F(x, t, \lambda)\triangle^{-1}(x, t)$ *is a frame of* u, *where* $*$ *is the gauge action defined by* (18) *or equivalently* (17).

(2) *Let* E *be a solution* $u : \mathbb{R}^2 \to V_n$ *of* (22) ((34) *resp.*), *and* $\triangle : \mathbb{R}^2 \to N_n^+$ *satisfying* $\triangle_t \triangle^{-1} = \eta_j(u)$ ($\triangle_t \triangle^{-1} = \xi_j(u)$ *resp.*), *where* $\eta_j(u)$ ($\xi_j(u)$ *resp.*) *is defined by* (21) ((33) *resp.*). *Then* $q := \triangle^{-1} * u$ *is a solution of* (14) ((31) *resp.*) *and* $F = E\triangle$ *is a frame of* q.

4. Lagrangian Curve Flows on $\mathbb{R}^{2n}$

In this section, we

(i) give a description of the tangent space of $\mathcal{M}_{2n}$ at γ and show that it is isomorphic to $C^\infty(\mathbb{R}, \mathbb{R}^n)$,

(ii) construct two hierarchies of Lagrangian curve flows whose curvatures satisfy the $\hat{C}_n^{(1)}$-KdV and the $\hat{A}_{2n-1}^{(2)}$-KdV flows respectively.

Henceforth in this paper we set

$$e_1 = (1, 0, \ldots, 0)^t \in \mathbb{R}^{2n}.$$

Theorem 4. *Let* g *and* u *denote the Lagrangian frame and Lagrangian curvature along* $\gamma \in \mathcal{M}_{2n}$, *and* $\Psi : \mathcal{M}_{2n} \to C^\infty(\mathbb{R}, V_n)$ *the Lagrangian curvature map. Then*

$$\mathrm{d}\Psi_\gamma(\delta\gamma) = [\partial_x + b + u, g^{-1}\delta g], \tag{36}$$

where $b = \sum_{i=1}^{2n-1} e_{i+1,i}$. *Moreover,*

(1) $C : \mathbb{R} \to sp(2n)$ *satisfies*

$$[\partial_x + b + u, C] \in C^\infty(\mathbb{R}, V_n). \tag{37}$$

if and only if gCe_1 *is tangent to* $\mathcal{M}_{2n}$ *at* γ,

(2) *if* ξ *is tangent to* $\mathcal{M}_{2n}$ *at* γ *then there exists a unique smooth* $C : \mathbb{R} \to sp(2n)$ *satisfying* (37) *such that* $\xi = gCe_1$.

Proof. A direct computation gives (36) (cf. [49]).

Suppose $\delta\gamma$ is tangent to $\mathcal{M}_{2n}$ at γ. By (36), $d\Psi_\gamma(\delta\gamma) = [\partial + b + u, g^{-1}\delta g]$ is in V_n. So $C := g^{-1}\delta g$ satisfies (37).

Suppose C satisfies (37). Let η_i denote the i-th column of gC. Please note that ξ is tangent to $\mathcal{M}_{2n}$ at γ if and only if

$$\begin{cases} \omega(\xi_x^{(i)}, \gamma_x^{(j)}) + \omega(\gamma_x^{(i)}, \xi_x^{(j)}) = 0, & 1 \leq i,j \leq n-1, \\ \omega(\xi_x^{(n-1)}, \gamma_x^{(n)}) + \omega(\gamma_x^{(n-1)}, \xi_x^{(n)}) = 0. \end{cases} \tag{38}$$

To prove η_1 satisfies (38), we let $\rho = [\partial_x + b + u, C]$. Then

$$(gC)_x = g_x C + g C_x = gC(b+u) + g\rho.$$

Since $\rho \in V_n$, $\eta_i = (\eta_1)_x^{(i-1)}$ for $1 \leq i \leq n+1$. By

$$(gC)^t S_n g + g^t S_n gC = C^t g^t S_n g + g^t S_n gC = C^t S_n + S_n C = 0,$$

so η_1 satisfies (38). □

By (14) and (33), we see that both $P_{2j-1,0}(u) - \eta_j(u)$ and $Q_{2j-1,0}(u) - \xi_j(u)$ satisfy (37). So it follows from Theorem 4 that

$$\gamma_t = g(P_{2j-1,0}(u) - \eta_j(u))e_1, \quad \gamma_t = g(Q_{2j-1,0}(u) - \xi_j(u))e_1$$

are flows on $\mathcal{M}_{2n}$. Since $\eta_j(u)$ and $\xi_j(u)$ are all strictly upper triangular, we have

$$\eta_j(u)e_1 = \xi_j(u)e_1 = \zeta_j(u)e_1 = 0.$$

Hence we have the following.

Proposition 10. *Let g and u be the Lagrangian moving frame and Lagrangian curvature along $\gamma \in \mathcal{M}_{2n}$ respectively. Then*

$$\gamma_t = g(P_{2j-1,0}(u) - \eta_j(u))e_1 = gP_{2j-1,0}(u)e_1, \tag{39}$$

$$\gamma_t = g(Q_{2j-1,0}(u) - \xi_j(u))e_1 = gQ_{2j-1,0}(u)e_1, \tag{40}$$

are Lagrangian curve flows on $\mathcal{M}_{2n}$, where $P_{2j-1,0}(u), Q_{2j-1,0}(u), \eta_j(u)$, and $\xi_j(u)$ are given by (12), (29), (21), and (33) respectively.

We call (39) and (40) the $(2j-1)$-th Lagrangian curve flow on $\mathcal{M}_{2n}$ of C-type and A-type respectively.

Example 7. Lagrangian curve flows of C-type

(i) When $n = 1$, $sp(2) = sl(2, \mathbb{R})$, the symplectic form $\omega(\xi, \eta)$ defined by (1) is $\det(\xi, \eta)$, $\gamma \in \mathcal{M}_2$ if and only if γ satisfies $\det(\gamma, \gamma_x) = 1$, and the Lagrangian parameter, moving frame, and curvature for $\gamma \in \mathcal{M}_2$ are the central affine parameter, moving frame and curvature respectively. The third $\hat{C}_1^{(1)}$-KdV is the KdV,

$$q_t = \frac{1}{4}(q_{xxx} - 6qq_x). \tag{41}$$

The third Lagrangian curve flow of C-type on $\mathcal{M}_2$ is

$$\gamma_t = \frac{1}{4}(u_1)_x \gamma - \frac{1}{2} u_1 \gamma_x, \tag{42}$$

which is the third central affine curve flow on the affine plane (cf. [47]). Moreover, if γ is a solution of (42), then its Lagrangian curvature is a solution of the KdV (41).

(ii) Let $g = (\gamma, \gamma_x, \gamma_{xx}, g_4)$ be the Lagrangian moving frame of $\gamma \in \mathcal{M}_4$, and u_1, u_2 the Lagrangian curvatures as in Example 2. From Example 5, we see that the first column of $P_{3,0}(u)$ is

$$(-\frac{3}{8}(u_1)_x, \frac{1}{4}u_1, 0, 1)^t.$$

So the third Lagrangian curve flow of C-type on $\mathcal{M}_4$ is

$$\gamma_t = -\frac{3}{8}(u_1)_x \gamma + \frac{1}{4}u_1 \gamma_x + g_4,$$

where g_4 is the fourth column of the Lagrangian frame of γ. This is the curve flow (3) for $n = 2$ because $g_4 = \gamma_{xxx} - u_1 \gamma_x$ (given in Example 2).

Similar computation implies that the first column of $P_{5,0}(u)$ is

$$(-\frac{5}{32}(u_1)_x^{(3)} + \frac{3}{8}(u_2)_x + \frac{5}{32}u_1(u_1)_x, \frac{1}{16}(u_1)_x^{(2)} - \frac{1}{4}u_2 - \frac{3}{32}u_1^2, \frac{1}{8}(u_1)_x, -\frac{1}{4}u_1)^t.$$

Hence the fifth Lagrangian curve flow of C-type on $\mathcal{M}_4$ is

$$\gamma_x = (-\frac{5}{32}(u_1)_x^{(3)} + \frac{3}{8}(u_2)_x + \frac{5}{32}u_1(u_1)_x)\gamma$$
$$+ (\frac{1}{16}(u_1)_x^{(2)} - \frac{1}{4}u_2 - \frac{3}{32}u_1^2)\gamma_x + \frac{1}{8}(u_1)_x \gamma_{xx} - \frac{1}{4}u_1 g_4.$$

(iii) We use Equation (10) to compute $P_{1,i}(u)$ and the first column of $P_{3,0}(u)$ for general n. Then we see that the third Lagrangian curve flow of C-type on $\mathcal{M}_{2n}$ for $n \geq 3$ is (3).

Example 8. Lagrangian curve flows of A-type

We use the algorithm given in Theorem 3 to compute $Q_{1,i}(u)$. Then we use these $Q_{1,i}(u)$'s to compute $Q_{i,0}(u)$. Then we obtain the following:

(i) The third Lagrangian curve flow of A-type on $\mathcal{M}_4$ is

$$\gamma_t = \gamma_x^{(3)} - u_1 \gamma_x. \tag{43}$$

The fifth Lagrangian curve flow of A-type on $\mathcal{M}_4$ is

$$\gamma_t = (-\frac{5}{3}(u_2)_x - \frac{1}{9}(u_1)_x^{(3)} + \frac{1}{6}u_1(u_1)_x)\gamma - \frac{1}{9}(6u_2 + (u_1)_x^{(2)} + u_1^2)\gamma_x$$
$$+ \frac{1}{3}(u_1)_x \gamma_{xx} - \frac{2}{3}u_1 g_4.$$

(ii) The third Lagrangian curve flow of A-type on $\mathcal{M}_{2n}$ ($n \geq 2$) is (4). Since $\gamma_x^{(3)} = u_1 \gamma_x + g_4$, (4) becomes (43) when $n = 2$.

Theorem 4 (1) states that $g\xi$ is tangent to $\mathcal{M}_{2n}$ at γ if and only if there is a C satisfying (37) and $\xi = Ce_1$. So to get a better description of the tangent space of $\mathcal{M}_{2n}$ at γ, we need to understand properties of C that satisfies (37).

Theorem 5. *Let $u \in C^\infty(\mathbb{R}, V_n)$ and $v = \sum_{i=1}^n v_i e_{n+i,n+1-i} : \mathbb{R} \to V_n^t$ a smooth map. Let $\pi_0 : sp(2n) \to V_n^t$ be the linear projection onto V_n^t defined by*

$$\pi_0(y) = \sum_{i=1}^n y_{n+i,n+1-i} e_{n+i,n+1-i}, \quad y = (y_{ij}). \tag{44}$$

If $C = (C_{ij}) : \mathbb{R} \to sp(2n)$ satisfies
$$[\partial_x + b + q, C] \in C^\infty(\mathbb{R}, V_n), \quad \pi_0(C) = v, \tag{45}$$
then we have the following:

(i) There exists differential polynomial $\phi_{ij}(u,v)$ that is linear in v such that $C_{ij} = \phi_{ij}(u,v)$ for all $1 \le i, j \le 2n$, and $\phi_{n+i, n+1-i}(u,v) = v_i$ for $1 \le i \le 2n$.

(ii) $\phi_{2i,1}(u,v) = v_i + \phi_i$ for $1 \le i \le n$, where ϕ_i's are differential polynomials in $u, v_{i+1}, \cdots, v_n$.

(iii) There exist differential polynomials h_{2i+1} for $0 \le i \le n-1$ such that
$$C_{2i+1,1} = h_{2i+1}(u, C_{2i+2,1}, \cdots, C_{2n,1}).$$

(iv) $C_{i,j}$'s are differential polynomials of $u, C_{21}, \cdots, C_{2n,1}$.

Conversely, given $u \in C^\infty(\mathbb{R}, V_n)$ and $v \in C^\infty(\mathbb{R}, V_n^t)$, define $C = (C_{ij})$ by $C_{ij} = \phi_{ij}(u,v)$ for $1 \le i, j \le 2n$. Then C satisfies (45).

Proof. Let $\mathcal{G}_i = \text{span}\{e_{j,i+j} \mid 1 \le i+j \le 2n\}$. For $\xi \in sp(2n)$, we use $\xi_{\mathcal{G}_i}$ to denote the $\mathcal{G}_i$-component of ξ with respect to $sp(2n) = \oplus_{i=1-2n}^{2n-1} \mathcal{G}_i$, and write $C = \sum_{i=1-2n}^{2n-1} C_i, C_i \in \mathcal{G}_i$. Set $[\partial_x + b + u, C] = \sum_{i=1}^n \eta_i e_{n+1-i, n+i}$. Then

$$(C_i)_x + [b, C_{i+1}] + [u, C]_{\mathcal{G}_i} = \begin{cases} \eta_j e_{n+1-j, n+j}, & i = 2j-1, \\ 0, & \text{else.} \end{cases} \tag{46}$$

We prove (i) by induction. When $i = 1 - 2n$, we have $C_{2n,1} = v_n$. From $(C_{1-2n})_x + [b, C_{2-2n}] = 0$, we get $C_{2n-1,1} = -C_{2n,2} = -\frac{1}{2}(v_n)_x$. For $j < 0$, $ad(b) : \mathcal{G}_{2j} \to \mathcal{G}_{2j-1}$ is a bijection, and $\dim(ad(b)(\mathcal{G}_{2j+1})) = \dim(\mathcal{G}_{2j}) = \dim(\mathcal{G}_{2j+1}) - 1$. Then by (46) and induction, C_j ($j < 0$) are differential polynomials in u, v_i and the linear system (46) implies (ii).

Please note that $ad(b) : \mathcal{G}_0 \to \mathcal{G}_{-1}$ is bijection, and $[u, C]_{\mathcal{G}_{-1}}$ depends only in $u, v_1, \cdots, v_n$. Hence C_0 can be solved uniquely from $C_i, i < 0$. This proves (iii).

For $j > 0$, $ad(b) : \mathcal{G}_{2j+1} \to \mathcal{G}_{2j}$ is a bijection. Hence $\mathcal{G}_{2j+1}$ is a differential polynomial in $\mathcal{G}_{2j}$. In addition, $ad(b) : \mathcal{G}_{2j+2} \to \mathcal{G}_{2j+1}$ is an injection and $\dim(ad(b)(\mathcal{G}_{2j+2})) = \dim(\mathcal{G}_{2j+1}) - 1$. Then by induction, C_j ($j > 0$) are differential polynomials in $u, v_1, \cdots, v_n$. This proves (i). Moreover, from the argument, we see that η_i's are differential polynomials in u and $v_1, \cdots, v_n$.

Statement (iv) is a consequence of (i) and (ii).

The proof of (i) to (iv) implies that the converse is also true. □

Corollary 1. *Let u and g be the Lagrangian curvature and frame of $\gamma \in \mathcal{M}_{2n}$. Then $T_\gamma \mathcal{M}_{2n}$ is the set of all $g\xi$, where $\xi = (\xi_1, \ldots, \xi_{2n})^t$ satisfies $\xi_{2i+1} = h_{2i+1}(u, \xi_{2i+2}, \cdots, \xi_{2n})$ and h_{2i+1} is given in Proposition 5. In particular, the tangent space of $\mathcal{M}_{2n}$ at $\gamma \in \mathcal{M}_{2n}$ is isomorphic to $C^\infty(\mathbb{R}, \mathbb{R}^n)$.*

Proof. It follows from Theorem 4 (1) and Proposition 5 (iv). □

Corollary 2. *Given $C_1, C_2 : \mathbb{R} \to sp(2n)$ satisfying (37), then we have the following:*

(1) *If the first columns of C_1 and C_2 are the same, then $C_1 = C_2$.*
(2) *If $\pi_0(C_1) = \pi_0(C_2)$, then $C_1 = C_2$, where π_0 is the projection defined by (44).*

Proof. (1) follows from Proposition 5 (iv), and (2) follows from Proposition 5 (i). □

It follows from Proposition 5 (i) that we have the following:

Corollary 3. *Given smooth $u : \mathbb{R} \to V_n$ and $v : \mathbb{R} \to V_n^t$, there exists a unique $C : \mathbb{R} \to sp(2n)$ satisfying (45) and entries of C are polynomial differentials of u, v and linear in v.*

The above Corollary leads us to define a natural linear differential operator P_u defined below.

Definition 4. *Given $u \in C^\infty(\mathbb{R}, V_n)$, let*

$$P_u : C^\infty(\mathbb{R}, V_n^t) \to C^\infty(\mathbb{R}, sp(2n))$$

be the linear differential operator defined by $P_u(v) = $ the unique $C \in C^\infty(\mathbb{R}, sp(2n))$ satisfies (45).

It follows from the definition of P_u and Theorem 5 that we have the following:

Proposition 11. *Let $u \in C^\infty(\mathbb{R}, V_n)$. Then*
(i) *C satisfies (37) if and only if $C = P_u(v)$ for some $v \in C^\infty(\mathbb{R}, V_n^t)$.*
(ii) *If C satisfies (37), then $C = P_u(\pi_0(C))$.*

Example 9. *Let $u = u_1 e_{23} + u_2 e_{14} \in C^\infty(S^1, V_2)$, and $\xi = \xi_1 e_{32} + \xi_2 e_{41} \in C^\infty(S^1, V_2^t)$. We use the algorithm given in the proof of Proposition 5 to obtain:*

$$P_u(\xi) = \begin{pmatrix} -\frac{1}{2}(\xi_2''' + 3\xi_1' - u_1\xi_2') & C_{12} & C_{13} & C_{14} \\ \frac{1}{2}\xi_2'' + \xi_1 & -\frac{1}{2}\xi_1' & C_{23} & -C_{13} \\ -\frac{1}{2}\xi_2' & \xi_1 & \frac{1}{2}\xi_1' & C_{12} \\ \xi_2 & \frac{1}{2}\xi_2' & \frac{1}{2}\xi_2'' + \xi_1 & \frac{1}{2}(\xi_2''' + 3\xi_1' - u_1\xi_2') \end{pmatrix},$$

and

$$C_{12} = -\frac{1}{2}\xi_2^{(4)} - \frac{3}{2}\xi_1'' + \frac{1}{2}(u_1\xi_2')' + u_2\xi_2,$$

$$C_{23} = -\frac{1}{2}\xi_2^{(4)} - 2\xi_1'' + \frac{1}{2}(u_1\xi_2')' + u_2\xi_2 + u_1\xi_1,$$

$$C_{13} = -\frac{1}{2}\xi_2^{(5)} - \frac{3}{2}\xi_1^{(3)} + \frac{1}{2}(u_1\xi_2')'' + (u_2\xi_2)' + \frac{1}{2}u_2\xi_2',$$

$$C_{14} = -\frac{1}{2}\xi_2^{(6)} - \frac{3}{2}\xi_1^{(4)} + \frac{1}{2}(u_1\xi_2')^{(3)} + \frac{1}{2}u_1\xi_2^{(4)} + (u_2\xi_2)'' + u_2\xi_2'' + \frac{1}{2}u_2'\xi_2'$$

$$+ u_2\xi_1 + \frac{3}{2}u_1\xi_1'' - \frac{1}{2}u_1(u_1\xi_2')' - u_1u_2\xi_2.$$

Example 10 (Tangent space of $\mathcal{M}_4$ at γ)**.** *Let $u = u_1 e_{23} + u_2 e_{14}$ and $g = (\gamma, \gamma', \gamma'', g_4)$ be the Lagrangian curvature and frame along γ as in Example 2, where $g_4 = \gamma''' - u_1\gamma'$. It follows from Corollary 1 and Proposition 11 that the tangent space of $\mathcal{M}_4$ at γ is*

$$\{gP_u(\xi)e_1 \mid \xi \in C^\infty(\mathbb{R}, V_n^t)\}.$$

We use the formula of $P_u(\xi)$ given in Example 9. Set $\eta_1 = \frac{1}{2}\xi_2'' + \xi_1$ and $\eta_2 = \xi_2$. Then we have $\xi_1 = \eta_1 - \frac{1}{2}\eta_2''$ and $\xi_2 = \eta_2$. So the first column of $P_u(\xi)$ is

$$(\frac{1}{4}\eta_2''' - \frac{3}{2}\eta_1' + \frac{1}{2}u_1\eta_2', \eta_1, -\frac{1}{2}\eta_2', \eta_2)^t.$$

Hence the tangent space of $\mathcal{M}_4$ at γ is the space of

$$(\frac{1}{4}\eta_2''' - \frac{3}{2}\eta_1' + \frac{1}{2}u_1\eta_2')\gamma + \eta_1\gamma_x - \frac{1}{2}\eta_2'\gamma_{xx} + \eta_2 g_4,$$

where η_1, η_2 are smooth functions.

Proposition 12. *Let $u \in C^\infty(\mathbb{R}, V_n)$, $P_{2j-1,0}(u)$, $Q_{2j-1,0}(u)$, $\xi_j(u)$, $\eta_j(u)$ as in (12), (29), (21), (33) respectively. Then*

$$P_u(\pi_0(P_{2j-1,0}(u))) = P_{2j-1,0}(u) - \eta_j(u),$$
$$P_u(\pi_0(Q_{2j-1,0}(u))) = Q_{2j-1,0}(u) - \xi_j(u),$$

and the $(2j-1)$-th $\hat{C}_n^{(1)}$-KdV and $\hat{A}_{2n-1}^{(2)}$-KdV flows can be written respectively as

$$u_t = [\partial_x + b + u, P_u(\pi_0(P_{2j-1,0}(u)))],$$
$$u_t = [\partial_x + b + u, P_u(\pi_0(Q_{2j-1,0}(u)))].$$

Proof. It follows from (21) and (34) that both $P_{2j-1,0}(u) - \eta_j(u)$ and $Q_{2j-1,0}(u) - \xi_j(u)$ satisfies $[\partial_x + b + q, C]$ is V_n-valued. Proposition follows from Proposition 11 (ii). □

Theorem 6.

(i) If $\gamma \in \mathcal{M}_{2n}$ is a solution of the $(2j-1)$-th Lagrangian curve flow (39) of C-type ((40) of A-type resp.), then its Lagrangian curvature u is a solution of the $(2j-1)$-th $\hat{C}_n^{(1)}$-KdV flow (22) ($\hat{A}_{2n-1}^{(2)}$-KdV flow (34) resp.).

(ii) Let $u \in C^\infty(\mathbb{R}^2, V_n)$ be a solution of (22) ((34) resp.), and $g : \mathbb{R}^2 \to Sp(2n)$ a solution of (23) ((35) resp.). Then $\gamma(x,t) := g(x,t)e_1$ is solution of the $(2j-1)$-th isotropic curve flow of C-type (39) (A-type (40) resp.) with Lagrangian curvature $u(\cdot, t)$ and Lagrangian moving frame $g(\cdot, t)$.

(iii) Let q be a solution of the $(2j-1)$-th $\hat{C}_n^{(1)}$-flow (14) ($\hat{A}_{2n-1}^{(2)}$-flow (31) resp.), and $g : \mathbb{R}^2 \to Sp(2n)$ a solution of (15) ((32) resp.). Then $\gamma(x,t) = g(x,t)e_1$ is a solution of the $(2j-1)$-th Lagrangian curve flow (39) of C-type ((4) of A-type resp.).

Proof. We claim that $Z := g^{-1}g_t$ satisfies (37). Since $g^{-1}g_x = b + u$ with $u \in V_n$, a direct computation implies that

$$y := (g^{-1}g_x)_t = -Z(b+u) - g^{-1}g_{xt} \tag{47}$$

is V_n-valued. By (47), we obtain

$$Z_x + [b + u, Z] = -g^{-1}g_x g^{-1}g_t + g^{-1}g_{xt} = y,$$

which is V_n-valued. So Z satisfies (37). By definition of $\eta_j(u)$, $P_{2j-1}(u) - \eta_j(u)$ also satisfies (37). The first column of gZe_1 is γ_t, which is $P_{2j-1}(u)e_1$. Since $\eta_j(u)$ is strictly upper triangular, the first column of $P_{2j-1}(u) - \eta_j(u)$ is also $P_{2j-1}(u)e_1$. It follows from Corollary 2 that $Z = P_{2j-1}(u) - \eta_j(u)$. Hence we have proved g satisfies (23). By Proposition 6, u is a solution of (22). This proves (i).

Since $g^{-1}g_x = b + u$, $g(\cdot, t)$ is the Lagrangian frame along $\gamma = ge_1$. So $\gamma_t = g_t e_1 = g(P_{2j-1}(u) - \eta_j(u))e_1 = gP_{2j-1}(u)e_1$. This proves (ii). (iii) is proved similarly. □

Remark 1. *We use the same proof as in [49] for the n-dimensional central affine curve flow to show that solutions of the Cauchy problem of (22) give solutions of the Cauchy problem for Lagrangian curve flow (39) with both rapidly decaying and periodic initial data. Similar results hold for the Lagrangian curve flows (34) and (40).*

5. Darboux Transforms for the $\hat{C}_n^{(1)}$-Hierarchy

In this section, we use the loop group factorization method given in [54] to construct Darboux transformations for the $\hat{C}_n^{(1)}$-, $\hat{C}_n^{(1)}$-KdV, and the Lagrangian curve flows of C-type. We also give a Permutability formula for these Darboux transforms. To use this method, we need to identify the loop groups, find simple rational elements, and write down formulas for the factorizations.

Let $(\hat{\mathbb{C}}_n^{(1)})_+$ denote the group of holomorphic maps $f : \mathbb{C} \to GL(2n, \mathbb{C})$ satisfying the $Sp(2n)$-reality condition (7), i.e.,

$$\overline{f(\bar{\lambda})} = f(\lambda), \quad f(\lambda)^t S_n f(\lambda) = S_n,$$

and $\mathbb{R}\hat{\mathbb{C}}_n^{(1)}$ the group of rational maps $f : \mathbb{C} \to GL(2n, \mathbb{C})$ satisfying (7) and $f(\infty) = I$. Then the Lie algebras of $(\hat{\mathbb{C}}_n^{(1)})_+$ and $\mathbb{R}\hat{\mathbb{C}}_n^{(1)}$ are contained in $(\hat{\mathcal{C}}_n^{(1)})_+$ and $(\hat{\mathcal{C}}_n^{(1)})_-$ respectively.

Next Proposition gives the uniqueness of factorization.

Proposition 13. *Let $f_1, g_1 \in (\hat{\mathbb{C}}_n^{(1)})_+$, and $f_2, g_2 \in \mathbb{R}\hat{\mathbb{C}}_n^{(1)}$. If $f_1 f_2 = g_1 g_2$, then $f_1 = g_1$ and $f_2 = g_2$.*

Proof. Let $h := g_1^{-1} f_1 = g_2 f_2^{-1}$. Then h is both holomorphic for $\lambda \in \mathbb{C}$ and at $\lambda = \infty$. So h is constant. However, at $\lambda = \infty$, $h = I$. Therefore, $h \equiv I$. □

The following result was proved in [54] for soliton hierarchies constructed from a splitting of loop algebras. So it works for both the $\hat{\mathcal{C}}_n^{(1)}$- and $\hat{A}_{2n-1}^{(2)}$-hierarchies given in Section 3.

Theorem 7 ([54]). *Let $F(x, t, \lambda)$ be a frame of a solution q of (14) ((31) resp.) and $g \in \mathbb{R}\hat{\mathbb{C}}_n^{(1)}$. Then there exists an open neighborhood $\mathcal{O}$ of $(0,0)$ such that we can factor*

$$g(\lambda) F(x, t, \lambda) = \tilde{F}(x, t, \lambda) \tilde{g}(x, t, \lambda)$$

with $\tilde{F}(x, t, \cdot)$ in $(\hat{\mathbb{C}}_n^{(1)})_+$ and $\tilde{g}(x, t, \cdot) \in \mathbb{R}\hat{\mathbb{C}}_n^{(1)}$ for all $(x, t) \in \mathcal{O}$. Moreover, write

$$\tilde{g}(x, t, \lambda) = I + g_{-1}(x, t) \lambda^{-1} + g_{-2}(x, t) \lambda^{-2} + \cdots.$$

Then $\tilde{q} = q + [g_{-1}, \beta]$ is a new solution of (14) ((31) resp.) and $\tilde{F}$ is a frame of $\tilde{q}$, where $\beta = e_{1,2n}$ ($\beta = \frac{1}{2}(e_{1,2n-1} + e_{2,2n})$ resp.).

Theorem 8. *Let $g \bullet q$ denote the solution of (14) ((31) resp.) constructed from the frame $F(x, t, \lambda)$ of solution q of (14) ((31) resp.) satisfying $F(0, 0, \lambda) = I$. Then $g \bullet q$ defines an action of $\mathbb{R}\hat{\mathbb{C}}_n^{(1)}$ on the space of solutions of (14) ((31) resp.).*

Proof. It suffices to prove that $(gf) \bullet q = g \bullet (f \bullet q)$ for $f, g \in \mathbb{R}\hat{\mathbb{C}}_n^{(1)}$. Assume that

$$f(\lambda) F(x, t, \lambda) = F_1(x, t, \lambda) \tilde{f}(x, t, \lambda),$$
$$g(\lambda) F_1(x, t, \lambda) = \tilde{F}(x, t, \lambda) \tilde{g}(x, t, \lambda),$$

where $\tilde{f}(x, t, \cdot), \tilde{g}(x, t, \cdot)$ are in $\mathbb{R}\hat{\mathbb{C}}_n^{(1)}$ and $F_1(x, t, \lambda), \tilde{F}(x, t, \lambda)$ are holomorphic for $\lambda \in \mathbb{C}$. It follows from Theorem 7 that we have

$$f \bullet q = q + [\tilde{f}_{-1}, \beta],$$
$$g \bullet (f \bullet q) = f \bullet q + [\tilde{g}_{-1}, \beta] = q + [\tilde{f}_{-1} + \tilde{g}_{-1}, \beta]$$

are solutions of (14), where

$$\tilde{f}(x, t, \lambda) = I + \tilde{f}_{-1}(x, t) \lambda^{-1} + \tilde{f}_{-2}(x, t) \lambda^{-2} + \cdots,$$
$$\tilde{g}(x, t, \lambda) = I + \tilde{g}_{-1}(x, t) \lambda^{-1} + \tilde{g}_{-2}(x, t) \lambda^{-2} + \cdots.$$

To obtain $(gf) \bullet q$, we factor

$$(gf)F = g(fF) = g(F_1 \tilde{f}) = (gF_1)\tilde{f} = (\tilde{F}\tilde{g})\tilde{f} = \tilde{F}(\tilde{g}\tilde{f}).$$

Since $\tilde{f}(x,t,\cdot)$ and $\tilde{g}(x,t,\cdot)$ are in $\mathbb{R}\hat{\mathbb{C}}_n^{(1)}$, $\tilde{g}\tilde{f}$ is in $\mathbb{R}\hat{\mathbb{C}}_n^{(1)}$. Please note that the coefficient of λ^{-1} of $\tilde{g}\tilde{f}$ is $\tilde{f}_{-1} + \tilde{g}_{-1}$. Hence it follows from Theorem 7 that we have

$$(gf) \bullet q = q + [\beta, \tilde{f}_{-1} + \tilde{g}_{-1}].$$

So we have proved that $(gf) \bullet q = g \bullet (f \bullet q)$. □

Given a linear subspace V of $\mathbb{R}^{2n}$, let

$$V^\perp = \{y \in \mathbb{R}^{2n} \mid \omega(\xi, y) = 0 \quad \text{for all } \xi \in V\}.$$

Lemma 1. *Let $\mathbb{R}^{2n} = V_1 \oplus V_2$ be a direct sum of linear subspaces, and π the projection of $\mathbb{R}^{2n}$ onto V_1 along V_2. Then we have $\omega(\pi X, Y) = \omega(X, \pi_s Y)$, where*

$$\pi_s = S_n^{-1} \pi^t S_n$$

is a projection of $\mathbb{R}^{2n}$ onto $V_2^\perp$ regarding $\mathbb{R}^{2n} = V_2^\perp \oplus V_1^\perp$.

Proof. Please note that

$$\omega(\pi(X), Y) = X^t \pi^t S_n Y = \omega(X, S_n^{-1}\pi^t S_n(Y)) = \omega(X, \pi_s Y),$$

where ω is the symplectic form defined by (1).

If $\omega(\pi_s X, Y) = 0$ for all $X \in \mathbb{R}^{2n}$, then $\omega(X, \pi Y) = 0$. Hence $(\text{Im}(\pi_s))^\perp \subset \text{Ker}(\pi) = V_2$, which implies $\text{Im}(\pi_s) = V_2^\perp$.

On the other hand, if $Y \in \text{Ker}(\pi_s)$, then $\omega(\pi X, Y) = \omega(X, \pi_s Y) = 0$ for any $X \in \mathbb{R}^{2n}$. So $Y \in V_1^\perp$, which implies $\text{Ker}(\pi_s) = V_1^\perp$. □

We use Lemma 1 and a direct computation to get:

Lemma 2.

(1) *A linear subspace of $\mathbb{R}^{2n}$ is Lagrangian if and only if $V^\perp = V$.*
(2) *Let π be a projection of $\mathbb{R}^{2n}$. Then*

$$\text{Im}\pi \text{ and } \text{Ker}\pi \text{ are Lagrangian subspaces,} \tag{48}$$

if and only if

$$\pi_s = I_{2n} - \pi. \tag{49}$$

Given $\alpha \in \mathbb{R} \setminus 0$ and a projection π of $\mathbb{R}^{2n}$, let

$$k_{\alpha,\pi}(\lambda) = I + \frac{2\alpha}{\lambda - \alpha}(I - \pi). \tag{50}$$

A direct computation implies that

$$k_{\alpha,\pi}^{-1}(\lambda) = I - \frac{2\alpha}{\lambda + \alpha}(I - \pi). \tag{51}$$

Lemma 3. *Given $\alpha \in \mathbb{R} \setminus 0$, if π is a projection of $\mathbb{R}^{2n}$ satisfying (48) then*

$$k_{\alpha,\pi}^t(\lambda) S_n k_{\alpha,\pi}(\lambda) = \frac{\lambda + \alpha}{\lambda - \alpha} S_n. \tag{52}$$

Proof. Lemmas 1 and 2 (2) implies that $S_n^{-1}\pi^t S_n = I - \pi$. So we have $I - \pi^t = S_n \pi S_n^{-1}$. Then a direct computation gives (52). □

Lemma 4. *Let $\alpha \in \mathbb{R} \setminus 0$, π a projection satisfying (48), and $f : \mathbb{C} \to GL(2n, \mathbb{C})$ a meromorphic map, holomorphic at $\lambda = \alpha, -\alpha$, and f satisfying the $Sp(2n)$-reality condition (7). Let $\tilde{V}_1 = f(\alpha)^{-1}(\operatorname{Im}\pi)$, and $\tilde{V}_2 = f(-\alpha)^{-1}(\operatorname{Ker}\pi)$. Assume that $\tilde{V}_1 \cap \tilde{V}_2 = \{0\}$. Let $\tilde{\pi}$ be the projection onto $\tilde{V}_1$ along $\tilde{V}_2$. Then*
(1) *$\tilde{V}_1$ and $\tilde{V}_2$ are Lagrangian subspaces,*
(2) *$\tilde{f} := k_{\alpha,\pi} f k_{\alpha,\tilde{\pi}}^{-1}$ is holomorphic at $\lambda = \alpha, -\alpha$ and satisfies the $Sp(2n)$-reality condition (7).*

Proof. Since f satisfies the $Sp(2n)$-reality condition, $f(r) \in Sp(2n)$ for all $r \in \mathbb{R}$. Hence $f(r)^{-1}(V_i)$ is again a Lagrangian subspace. This proves (i).

By (51), we have

$$\tilde{f}(\lambda) = (I + \frac{2\alpha}{\lambda - \alpha}(I - \pi))f(\lambda)(I - \frac{2\alpha}{\lambda + \alpha}(I - \tilde{\pi})).$$

Please note that $\tilde{f}$ has a simple pole at $\lambda = \alpha$ and the residue of $\tilde{f}$ at $\lambda = \alpha$ is $2\alpha(I - \pi)f(\alpha)\tilde{\pi}$, which is zero because

$$(I - \pi)f(\alpha)\operatorname{Im}\tilde{\pi} = (I - \pi)f(\alpha)f(\alpha)^{-1}(\operatorname{Im}\pi) = (I - \pi)(\operatorname{Im}\pi) = 0.$$

Similarly, $\tilde{f}$ has a simple pole at $\lambda = -\alpha$ and its residue is $2\alpha\pi f(-\alpha)(I - \tilde{\pi})$, which is again zero because its image is

$$\pi f(-\alpha)\operatorname{Ker}\tilde{\pi} = \pi f(-\alpha)f(-\alpha)^{-1}(\operatorname{Ker}\pi) = 0.$$

This proves $\tilde{f}$ is holomorphic at $\lambda = \alpha, -\alpha$.

It follows from (52) that $k_{\alpha,\pi}^{-1} = \frac{\lambda - \alpha}{\lambda + \alpha}(k_{\alpha,\pi})_s$. Since f satisfies $f(\lambda)^{-1} = (f(\lambda))_s$, a direct computation shows that $\tilde{f}^{-1} = \tilde{f}_s$. Hence $\tilde{f}$ satisfies (7). □

Theorem 9 (Darboux transform for the $\hat{C}_n^{(1)}$-flow (14)).

Let $c \in Sp(2n)$ be a constant, $F(x, t, \lambda)$ the frame of a solution q of (14) satisfying $F(0,0,\lambda) = c$, $\alpha \in \mathbb{R}\setminus\{0\}$, and π a projection of $\mathbb{R}^{2n}$ satisfying (48). Let

$$\tilde{V}_1(x,t) = F^{-1}(x,t,\alpha)(\operatorname{Im}\pi), \quad \tilde{V}_2(x,t) = F^{-1}(x,t,-\alpha)(\operatorname{Ker}\pi).$$

Then
(1) *$\tilde{V}_i(x,t)$ are Lagrangian subspaces for all $(x,t) \in \mathbb{R}^2$,*
(2) *there is an open subset $\mathcal{O}$ of $(0,0)$ in $\mathbb{R}^2$ such that $\mathbb{R}^{2n} = \tilde{V}_1(x,t) \oplus \tilde{V}_2(x,t)$ for all $(x,t) \in \mathcal{O}$,*
(3) *let $\tilde{\pi}(x,t)$ be the projection of $\mathbb{R}^{2n}$ onto $\tilde{V}_1(x,t)$ along $\tilde{V}_2(x,t)$, and*

$$\tilde{F}(x,t,\lambda) = k_{\alpha,\pi}F(x,t,\lambda)k_{\alpha,\tilde{\pi}(x,t)}^{-1}(\lambda), \qquad (53)$$

then

$$\tilde{q} = q + 2\alpha[e_{1,2n}, \tilde{\pi}] \qquad (54)$$

is a new solution of (14) and $\tilde{F}$ is a frame for $\tilde{q}$.

In particular, we have

$$k_{\alpha,\pi} \bullet q = q + 2\alpha[e_{1,2n}, \tilde{\pi}] \qquad (55)$$

if F is chosen so that $F(0,0,\lambda) = I$.

Proof. Let $V_1 = \operatorname{Im}\pi$, and $V_2 = \operatorname{Ker}\pi$. By assumption, V_1, V_2 are Lagrangian. Since $F(x,t,\lambda)$ satisfies the $Sp(2n)$ reality condition (7) and $\alpha, -\alpha \in \mathbb{R}$, (1) follows.

By assumption, $V_1 \cap V_2 = \{0\}$. Please note that $\tilde{V}_1(0,0) = F(0,0,\alpha)^{-1}(V_1) = c^{-1}(V_1)$ and $\tilde{V}_2(0,0) = c^{-1}V_2$. So $(c^{-1}V_1) \cap (c^{-1}V_2) = \{0\}$. Since $\mathcal{O} = \{(x,t) \mid \tilde{V}_1(x,t) \cap \tilde{V}_2(x,t) = \{0\}\}$ is an open subset of $\mathbb{R}^2$ and $(0,0) \in \mathcal{O}$, (2) follows.

(3) follows from Lemma 4 and Theorem 7. □

Theorem 9 can be reformulated as follows:

Theorem 10. *Let q be a solution of (14), $\lambda \in \mathbb{R} \setminus 0$, and $D_{q,\lambda}$ the following linear system*

$$D_{q,\lambda} \begin{cases} y_x = -(e_{1,2n}\lambda + b + q)y, \\ y_t = -(P^{2j-1}(q,\lambda))_+ y. \end{cases} \tag{56}$$

Then (56) is solvable. Moreover, let $\{v_1, \ldots, v_{2n}\}$ be a basis of $\mathbb{R}^{2n}$ such that the span of $v_1, \ldots, v_n$ and the span of $v_{n+1}, \ldots, v_{2n}$ are Lagrangian subspaces. Let y_i be the solution of $D_{q,\alpha}$ with initial data $y_i(0,0) = v_i$, and y_{n+i} the solution of $D_{q,-\alpha}$ with $y_{n+i}(0,0) = v_{n+i}$ for $1 \le i \le n$. Let $\tilde{V}_1(x,t)$ be the span of $y_1(x,t), \ldots, y_n(x,t)$, and $\tilde{V}_2(x,t)$ the span of $y_{n+1}(x,t), \ldots, y_{2n}(x,t)$. Then

(1) $\tilde{V}_i(x,t)$ *is Lagrangian for all $(x,t) \in \mathbb{R}^2$ and $i = 1, 2$,*
(2) *there is an open subset $\mathcal{O}$ of $(0,0)$ such that $\tilde{V}_1(x,t) \cap \tilde{V}_2(x,t) = 0$ for all $(x,t) \in \mathcal{O}$,*
(3) $\tilde{q}$ *defined by (54) is a solution of (14) defined on $\mathcal{O}$, where $\tilde{\pi}$ is the projection onto $\tilde{V}_1(x,t)$ along $\tilde{V}_2(x,t)$.*

Remark 2 (Bäcklund transformations for the $\hat{C}_n^{(1)}$-flow).

Let $q, \tilde{q}, F, \tilde{F}$ be as in Theorem 9. Then we have

$$\begin{cases} F^{-1}F_x = J + q, \\ F^{-1}F_t = B(q,\lambda), \end{cases} \qquad \begin{cases} \tilde{F}^{-1}\tilde{F}_x = J + \tilde{q}, \\ \tilde{F}^{-1}\tilde{F}_t = B(\tilde{q},\lambda), \end{cases} \tag{57}$$

where $B(q,\lambda) = (P^{2j-1}(q,\lambda))_+$. Let $\tilde{k}(x,t) = k_{\alpha,\tilde{\pi}(x,t)}$. Then it follows from (57) and (53) that we have

$$\begin{cases} \tilde{k}(J+q) - \tilde{k}_x = (J+\tilde{q})\tilde{k}, \\ \tilde{k}B(q,\lambda) - \tilde{k}_t = B(\tilde{q},\lambda)\tilde{k}. \end{cases} \tag{58}$$

Equate the residues of (58) at $\lambda = \alpha$ to get

$$(BT)_{q,\lambda} \begin{cases} \tilde{\pi}_x = [\alpha e_{1,2n} - b - q, \tilde{\pi}] - 2\alpha[e_{1,2n}, \tilde{\pi}]\tilde{\pi}, \\ \tilde{\pi}_t = B(q + 2\alpha[e_{1,2n}, \tilde{\pi}], \alpha)(I - \tilde{\pi}) - (I - \tilde{\pi})B(q,\alpha), \end{cases}$$

which is the Bäcklund transformation for the $\hat{C}_n^{(1)}$-flow. Moreover,

(1) $(BT)_{q,\lambda}$ is solvable for $\tilde{\pi}$ if and only if q is a solution of the $\hat{C}_n^{(1)}$-flow,
(2) if π_0 is a projection satisfies (49), then the solution $\tilde{\pi}$ of $(BT)_{q,\alpha}$ with initial data $\tilde{\pi}(0,0) = \pi_0$ also satisfies (49) and $\tilde{q}$ defined by (54) is also a solution of the $\hat{C}_n^{(1)}$-flow. In fact, $\tilde{q} = k_{\alpha,\pi} \bullet q$.

The following DTs for (22) is a consequence of Proposition 9 and Theorem 9.

Theorem 11 (DT for $\hat{C}_n^{(1)}$-KdV).

Let E be a frame of a solution u of (22), $k_{\alpha,\pi}$ as in Lemma 3, $\triangle : \mathbb{R}^2 \to N_n^+$ satisfying $\triangle_t \triangle^{-1} = \eta_j(u)$, and $q = \triangle^{-1} * u$. Let

$$\tilde{V}_i(x,t) = \triangle^{-1}(x,t)E^{-1}(x,t,\alpha_i)(V_i)$$

for $i = 1, 2$, and $\tilde{\pi}(x,t)$ the projection onto $\tilde{V}_1(x,t)$ along $\tilde{V}_2(x,t)$. Let $\tilde{q}$ be defined by (54), $\tilde{\triangle} : \mathbb{R}^2 \to N_n^+$ the unique map such that $\tilde{u} = \tilde{\triangle} * \tilde{q}$ is V_n valued. Then $\tilde{u}$ is a solution of (22) and

$$\tilde{E}(x,t,\lambda) = k_{\alpha,\pi}(\lambda) E(x,t,\lambda) \triangle(x,t) \tilde{k}_{\alpha,\tilde{\pi}(x,t)}^{-1}(\lambda) \tilde{\triangle}^{-1}(x,t)$$

is a frame of $\tilde{u}$, where $*$ is defined by (18).

As a consequence of Theorems 9 and 6 (iii), we have

Theorem 12 (DT for Lagrangian curve flow of C-type).

Let γ be a solution of the Lagrangian curve flow (39), and $g(\cdot, t)$ and $u(\cdot, t)$ the Lagrangian frame and curvature along $\gamma(\cdot, t)$. Let E be the frame of the solution u of (22) satisfying $E(0, 0, \lambda) = g(0, 0)$. Let $\triangle$, $k_{\alpha,\pi}$, and $\tilde{\pi}$ be as in Theorem 11. Then

$$\tilde{\gamma} = (2\pi - I) g \triangle (2\tilde{\pi} - I) e_1$$

is a new solution of (39).

Example 11. [1-soliton solutions of C-type]

First, we apply Theorem 9 to the trivial solution $q = 0$ of the third $\hat{C}_2^{(1)}$-flow to construct 1-soliton solutions and their corresponding frames. Then we use Theorem 11 to construct solutions of the third $\hat{C}_2^{(1)}$-KdV flow (5). In the end, we apply Theorem 12 to get explicit solutions of the third Lagrangian curve flow of C-type on $\mathbb{R}^4$:

$$\gamma_t = -\frac{3}{8}(u_1)_x \gamma + \frac{1}{4} u_1 \gamma_x + g_4 = -\frac{3}{8}(u_1)_x \gamma - \frac{3}{4} u_x \gamma_x + \gamma_{xxx}.$$

Please note that

$$F(x, t, \lambda) = \exp(Jx + J^3 t)$$

is a frame of the solution $q = 0$ of the third $\hat{C}_2^{(1)}$-flow. We use $\lambda = z^4$ to write down $F(x, t, \lambda)$ in terms of known functions,

$$F(x, t, z^4) = \frac{1}{4} \begin{pmatrix} m_1(x,t,z) & zm_2(x,t,z) & z^2 m_3(x,t,z) & z^3 m_4(x,t,z) \\ \frac{1}{z} m_4(x,t,z) & m_1(x,t,z) & zm_2(x,t,z) & z^2 m_3(x,t,z) \\ \frac{1}{z^2} m_3(x,t,z) & \frac{1}{z} m_4(x,t,z) & m_1(x,t,z) & zm_2(x,t,z) \\ \frac{1}{z^3} m_2(x,t,z) & \frac{1}{z^2} m_3(x,t,z) & \frac{1}{z} m_4(x,t,z) & m_1(x,t,z) \end{pmatrix},$$

where

$$\begin{pmatrix} m_1(x,t,z) \\ m_2(x,t,z) \\ m_3(x,t,z) \\ m_4(x,t,z) \end{pmatrix} = \begin{pmatrix} 1 & 1 & 1 & 1 \\ 1 & i & -1 & -i \\ 1 & -1 & 1 & -1 \\ 1 & -i & -1 & i \end{pmatrix} \begin{pmatrix} e^{zx+z^3 t} \\ e^{i(zx-z^3 t)} \\ e^{-(zx+z^5 t)} \\ e^{-i(zx-z^3 t)} \end{pmatrix}.$$

(Although the entries of $F(x, t, z^3)$ involves z^i in the denominators, use power series expansion and a simple computation to see that they are holomorphic at $z = 0$).

Next we apply DTs for the third $\hat{C}_2^{(1)}$-flow to the trivial solution $q = 0$ and $z = 1$. Let π be the projection onto V_1 along V_2, where

$$V_1 = \left\{ \begin{pmatrix} 1 \\ 0 \\ 1 \\ 0 \end{pmatrix}, \begin{pmatrix} 1 \\ 0 \\ -1 \\ 0 \end{pmatrix} \right\}, \quad V_2 = \left\{ \begin{pmatrix} 0 \\ 1 \\ 0 \\ 1 \end{pmatrix}, \begin{pmatrix} 0 \\ 1 \\ 0 \\ -1 \end{pmatrix} \right\}.$$

Then $\tilde{\pi}$ is the projection onto $\tilde{V}_1$ along $\tilde{V}_2$, where

$$\tilde{V}_1 = E(x,t,1)^{-1}V_1 = \text{span}\{p_1, p_2\}$$

$$= \text{span}\left\{\begin{pmatrix} \cosh(x+t) \\ -\sinh(x+t) \\ \cosh(x+t) \\ -\sinh(x+t) \end{pmatrix}, \begin{pmatrix} \cos(x-t) \\ -\sin(x-t) \\ -\cos(x-t) \\ \sin(x-t) \end{pmatrix}\right\},$$

$$\tilde{V}_2 = E(x,t,1)^{-1}V_1 = \text{span}\{p_3, p_4\}$$

$$= \text{span}\left\{\begin{pmatrix} \sinh(x+t) \\ \cosh(x+t) \\ \sinh(x+t) \\ \cosh(x+t) \end{pmatrix}, \begin{pmatrix} -\sin(x-t) \\ \cos(x-t) \\ \sin(x-t) \\ -\cos(x-t) \end{pmatrix}\right\}.$$

From a direct computation, we have

$$\tilde{\pi}(x,t) = (\tilde{p}_1, \tilde{p}_2, 0, 0)(\tilde{p}_1, \tilde{p}_2, \tilde{p}_3, \tilde{p}_4)^{-1}$$

$$= \frac{1}{2}\left(\frac{1}{\sqrt{y_1}}\tilde{p}_1, (1+\frac{y_2}{y_3})\tilde{p}_2, 0, \frac{y_2}{y_3}\tilde{p}_2\right)\begin{pmatrix} \frac{1}{\sqrt{y_1}}\tilde{p}_1^t \\ \tilde{p}_2^t \\ 0 \\ \tilde{p}_4^t \end{pmatrix},$$

where

$$y_1 = \cosh(2(x+t)),$$
$$y_2 = \sin(2(x-t)),$$
$$y_3 = (1-\sin(2(x-t)))(1+\frac{1}{2}\sin(4(x-t))).$$

Applying (54), we can get a solution of the third $\hat{C}_2^{(1)}$-flow,

$$\tilde{q} = q_1(e_{11} - e_{44}) + q_2(e_{12} + 4_{34}) + q_3(e_{13} - e_{24}) + q_4 e_{14}.$$

Using the algorithm in the proof of Proposition 5, we get a new solution of (5),

$$\tilde{u} = \triangle * q = \tilde{u}_1 e_{23} + u_2 e_{14},$$

where

$$\tilde{u}_1 = 3q_1' + 2q_2 + q_1^2,$$
$$\tilde{u}_2 = (q_1)_x^{(3)} + q_2'' + q_3' + q_4 + 2q_1 q_3 - q_2^2 - q_1' q_2 + q_1 q_2' + q_1 q_1''.$$

We use Theorem 12, and the formula for $\tilde{\pi}$, and a direct computation to see that

$$\tilde{\gamma}(x,t) = \begin{pmatrix} 1 & 0 & 0 & 0 \\ -x & -1 & 0 & 0 \\ \frac{1}{2}x^2 & x & 1 & 0 \\ -(\frac{1}{6}x^3+t) & -\frac{1}{2}x^2 & -x & -1 \end{pmatrix}\begin{pmatrix} \tilde{\zeta}_1 \\ \tilde{\zeta}_2 \\ \tilde{\zeta}_3 \\ \tilde{\zeta}_4 \end{pmatrix}$$

is a solution of the third Lagrangian curve flow of C-type on $\mathcal{M}_4$, where

$$\zeta_1 = \frac{1}{y_1}\cosh^2(x+t) + (1+\frac{y_2}{y_3})\cos^2(x-t) - \frac{y_2}{2y_3}\sin(2(x-t)) - 1,$$

$$\zeta_2 = -\frac{1}{4}(\cosh(2(x+t)) + (1+\frac{y_2}{y_3})\sin(2(x-t)) - 2\frac{y_2}{y_3}\sin^2(x-t)),$$

$$\zeta_3 = \frac{1}{2}(\frac{1}{y_1}\cosh^2(x+t) - (1+\frac{y_2}{y_3})\cos^2(x-t) + \frac{1}{2}\frac{y_2}{y_3}\sin(2(x-t))),$$

$$\zeta_4 = -\frac{1}{4}(\cosh(2(x+t)) - (1+\frac{y_2}{y_3})\sin(2(x-t)) + 2\frac{y_2}{y_3}\sin^2(x-t)).$$

Next we give a Permutability formula for DTs of the $\hat{C}_n^{(1)}$ flows. The following Lemma follows from Lemma 4.

Lemma 5. *Let α_1, α_2 be distinct real constants, π_1, π_2 projections of $\mathbb{R}^{2n}$ satisfying (48), and τ_1, τ_2 projections defined by*

$$\begin{cases} \mathrm{Im}(\tau_1) = k_{\alpha_2,\pi_2}(\alpha_1)\mathrm{Im}(\pi_1), & \mathrm{Ker}(\tau_1) = k_{\alpha_2,\pi_2}(-\alpha_1)\mathrm{Ker}(\pi_1), \\ \mathrm{Im}(\tau_2) = k_{\alpha_1,\pi_1}(\alpha_2)\mathrm{Im}(\pi_2), & \mathrm{Ker}(\tau_2) = k_{\alpha_1,\pi_1}(-\alpha_2)\mathrm{Ker}(\pi_2). \end{cases}$$

Then τ_1, τ_2 satisfy (48) and

$$k_{\alpha_2,\tau_2}k_{\alpha_1,\pi_1} = k_{\alpha_1,\tau_1}k_{\alpha_2,\pi_2}.$$

It follows from Lemma 5 and Theorems 8 and 9 that we have

Theorem 13 (Permutability Formula for the $\hat{C}_n^{(1)}$-flows).

Let $F(x,t,\lambda)$ be the frame of the solution q of the $(2j-1)$-th $\hat{C}_n^{(1)}$-flow (14) with $F(0,0,\lambda) = I_{2n}$, α_i, π_i, τ_i for $i = 1, 2$ as in Lemma 5. Then we have the following:

(1) *Let $\tilde{V}_i = F(x,t,\alpha_i)^{-1}(\mathrm{Im}\pi_i)$, and $\tilde{W}_i = F(x,t,-\alpha_i)^{-1}(\mathrm{Ker}\pi_i)$, $\tilde{\pi}_i$ the projection onto $\tilde{V}_i$ along $\tilde{W}_i$ for $i=1,2$. Then*

$$q_1 := k_{\alpha_1,\pi_1} \bullet q = q + 2\alpha_1[e_{1,2n}, \tilde{\pi}_1],$$
$$q_2 := k_{\alpha_2,\pi_2} \bullet q = q + 2\alpha_2[e_{1,2n}, \tilde{\pi}_2].$$

(2) $k_{\alpha_2,\tau_2} \bullet (k_{\alpha_1,\pi_1} \bullet q) = k_{\alpha_1,\tau_1} \bullet (k_{\alpha_2,\pi_2} \bullet q).$

(3) *Let $\tilde{\tau}_1(x,t), \tilde{\tau}_2(x,t)$ be the projections defined by*

$$\mathrm{Im}(\tilde{\tau}_1) = k_{\alpha_2,\tilde{\pi}_2}(\alpha_1)\mathrm{Im}(\tilde{\pi}_1), \quad \mathrm{Ker}(\tilde{\tau}_1) = k_{\alpha_2,\tilde{\pi}_2}(-\alpha_1)\mathrm{Ker}(\tilde{\pi}_1),$$
$$\mathrm{Im}(\tilde{\tau}_2) = k_{\alpha_1,\tilde{\pi}_1}(\alpha_2)\mathrm{Im}(\tilde{\pi}_2), \quad \mathrm{Ker}(\tilde{\tau}_2) = k_{\alpha_1,\tilde{\pi}_1}(-\alpha_2)\mathrm{Ker}(\tilde{\pi}_2).$$

Then we have

$$q_{12} := k_{\alpha_2,\tau_2} \bullet (k_{\alpha_1,\pi_1} \bullet q) = k_{\alpha_1,\tau_1} \bullet (k_{\alpha_2,\pi_2} \bullet q)$$
$$= q_1 + 2\alpha_2[e_{1,2n}, \tilde{\tau}_2] = q_2 + 2\alpha_1[e_{1,2n}, \tilde{\tau}_1].$$

In particular, q_{12} can be obtained algebraically from $\tilde{\pi}_1$ and $\tilde{\pi}_2$.

The Permutability Theorem 13 gives an algebraic formula for constructing k-solitons and their frames from k 1-solitons for the $\hat{C}_n^{(1)}$-flow. If $\tilde{F}$ is a frame of the k-soliton solution $\tilde{q}$ of $\hat{C}_n^{(1)}$-flow, then $\tilde{\gamma} = \tilde{F}(x,t,0)e_1$ is a k-soliton solution of the Lagrangian curve flow of C-type and its Lagrangian curvature $\tilde{u}$ is a k-soliton of the $\hat{C}_n^{(1)}$-KdV flow.

6. Darboux Transforms for the $\hat{A}^{(2)}_{2n-1}$-Hierarchy

In this section, we construct Darboux transformations for the $\hat{A}^{(2)}_{2n-1}$, $\hat{A}^{(2)}_{2n-1}$-KdV, and the Lagrangian curve flows of A type. We also give a Permutability formula for these Darboux transforms.

Let $(\hat{A}^{(2)}_{2n-1})_+$ denote the group of holomorphic maps $f : \mathbb{C} \to SL(2n+1, \mathbb{C})$ satisfying the reality condition (7), i.e.,

$$\overline{f(\bar{\lambda})} = f(\lambda), \quad f(-\lambda)^t S_n f(\lambda) = S_n, \tag{59}$$

and $\mathbb{R}\hat{A}^{(2)}_{2n-1}$ the group of rational maps $f : \mathbb{C} \to SL(2n+1, \mathbb{C})$ satisfying (7) with $f(\infty) = I$. Then the Lie algebras of $(\hat{A}^{(2)}_{2n-1})_+$ and $\mathbb{R}\hat{A}^{(2)}_{2n-1}$ are subalgebras of $(\hat{A}^{(2)}_{2n-1})_+$ and $(\hat{A}^{(2)}_{2n-1})_-$ respectively.

Please note that the second condition of (59) is equivalent to

$$f^{-1}(\lambda) = f(-\lambda)_s,$$

where $A_s = S_n^{-1} A^t S_n$.

Please note that the restriction of the symplectic form w to a linear subspace V of $\mathbb{R}^{2n}$ is non-degenerate if and only if $\mathbb{R}^{2n} = V \oplus V^\perp$.

Lemma 6. *Let π be a projection. Then $\mathrm{Ker}(\pi) = (\mathrm{Im}(\pi))^\perp$ if and only if*

$$\pi = \pi_s. \tag{60}$$

Lemma 7. *Let π be a projection of $\mathbb{R}^{2n}$ satisfying (60), and $\alpha \in \mathbb{R}\setminus\{0\}$. Then $k_{\alpha,\pi}$ defined by (50) is in $\mathbb{R}\hat{A}^{(2)}_{2n-1}$.*

Lemma 8. *Let $\alpha \in \mathbb{R} \setminus 0$, π a projection satisfying (60), and $f : \mathbb{C} \to GL(2n, \mathbb{C})$ a meromorphic map, holomorphic at $\lambda = \alpha$ and $\lambda = -\alpha$, and satisfying (59). Let $\tilde{V} = f(\alpha)^{-1}(V)$, where $V = \mathrm{Im}\pi$. Then*

(1) $\tilde{V}^\perp = f(-\alpha)^{-1}(V^\perp)$,

(2) *assume that $\tilde{V} \cap \tilde{V}^\perp = 0$, let $\tilde{\pi}$ be the projection onto $\tilde{V}$ along $\tilde{V}^\perp$, then*

$$\tilde{f} = k_{\alpha,\pi} f k_{\alpha,\tilde{\pi}}^{-1}$$

is holomorphic at $\lambda = \alpha, -\alpha$ and satisfies (59).

Proof. Set $V = \mathrm{Im}\pi$. If $Y \in \tilde{V}^\perp$, then

$$0 = \omega(f(\alpha)^{-1}V, Y) = \omega(f(-\alpha)_s V, Y) = \omega(V, f(-\alpha)Y).$$

Hence $f^{-1}(-\alpha)Y \in V^\perp$, which implies that $f^{-1}(-\alpha)(\tilde{V}^\perp) \subset V^\perp$. Since they have the same dimension, $f^{-1}(-\alpha)(\tilde{V}^\perp) = V^\perp$. This proves (1).

By (51), we have

$$\tilde{f}(\lambda) = (I + \frac{2\alpha}{\lambda - \alpha}(I - \pi))f(\lambda)(I - \frac{2\alpha}{\lambda + \alpha}(I - \tilde{\pi})).$$

Please note that $\tilde{f}$ has a simple pole at $\lambda = \alpha$ and $\lambda = -\alpha$. The residue of $\tilde{f}$ at $\lambda = \alpha$ is $2\alpha(I - \pi)f(\alpha)\tilde{\pi}$, which is zero because $\mathrm{Im}(f(\alpha)\tilde{\pi}) = f(\alpha)(\tilde{V}) = V$ and $\mathrm{Ker}(I - \pi) = V$. The residue of $\tilde{f}$ at $\lambda = -\alpha$ is $-2\alpha\pi f(-\alpha)(I - \tilde{\pi})$, which is zero because $\pi f(-\alpha)\tilde{V}^\perp = \pi V^\perp = 0$. Hence $\tilde{f}$ is holomorphic at $\lambda = \alpha, -\alpha$. Since both f and $k_{\alpha,\pi}$ satisfies (59), $\tilde{f}$ satisfies (59). □

Using Lemma 8, Theorem 7 and a proof similar to the proof of Theorem 9, we obtain the following:

Theorem 14 (DT for the $\hat{A}^{(2)}_{2n-1}$-hierarchy).
Let $c \in Sp(2n)$ be a constant, and $F(x,t,\lambda)$ be the frame of a solution q of the $(2j-1)$-th $\hat{A}^{(2)}_{2n-1}$-flow (31) with $F(0,0,\lambda) = c$, and π a projection satisfying (60). Given $\alpha \in \mathbb{R}\backslash\{0\}$, let

$$\tilde{V}(x,t) = F(x,t,\alpha)^{-1}(V), \quad \text{where } V = \operatorname{Im}\pi.$$

Then
(1) there exists an open neighborhood $\mathcal{O}$ of $(0,0)$ in $\mathbb{R}^2$ such that $\mathbb{R}^{2n} = \tilde{V}(x,t) \oplus \tilde{V}(x,t)^\perp$ for all $(x,t) \in \mathbb{R}^2$,
(2) let $\tilde{\pi}(x,t)$ be the projection onto $\tilde{V}(x,t)$ along $\tilde{V}^\perp(x,t)$, then

$$\tilde{q} = q + \alpha[e_{1,2n-1} + e_{2,2n}, \tilde{\pi}] \tag{61}$$

is a solution of (31) defined on $\mathcal{O}$ and

$$\tilde{F}(x,t,\lambda) = k_{\alpha,\pi}(\lambda) F(x,t,\lambda) k^{-1}_{\alpha,\tilde{\pi}(x,t)}(\lambda)$$

is a frame of $\tilde{q}$.
In particular, if F satisfies $F(0,0,\lambda) = I_{2n}$, then we have

$$k_{\alpha,\pi} \bullet q = q + \alpha[e_{1,2n-1} + e_{2,2n}, \tilde{\pi}]. \tag{62}$$

Theorem 14 can be reformulated as follows:

Theorem 15. Let q be a solution of (31), $\lambda \in \mathbb{R} \setminus 0$, and $B_{q,\lambda}$ the following linear system

$$B_{q,\lambda} \begin{cases} y_x = -(\beta\lambda + b + q)y, \\ y_t = -(Q^{2j-1}(q,\lambda))_+ y, \end{cases} \tag{63}$$

where $\beta = \frac{1}{2}(e_{1,2n-1} + e_{2,2n})$. Then we have the following:
(1) (63) is solvable.
(2) Let $\{v_1,\ldots,v_{2n}\}$ be a basis of $\mathbb{R}^{2n}$ such that $w(v_i, v_{n+j}) = 0$ for all $1 \leq i,j \leq n$. Let y_i be the solution of $D_{q,\alpha}$ with initial data $y_i(0,0) = v_i$, and y_{n+i} the solution of $D_{q,-\alpha}$ with $y_{n+i}(0,0) = v_{n+i}$ for $1 \leq i \leq n$. Let $\tilde{V}_1(x,t)$ be the span of $y_1(x,t),\ldots,y_n(x,t)$, and $\tilde{V}_2(x,t)$ the span of $y_{n+1}(x,t),\ldots,y_{2n}(x,t)$. Then
(a) $\tilde{V}_2(x,t) = \tilde{V}_1(x,t)^\perp$ for all $(x,t) \in \mathbb{R}^2$ and $i = 1,2$,
(b) there is an open subset $\mathcal{O}$ of $(0,0)$ such that $\tilde{V}_1(x,t) \cap \tilde{V}_2(x,t) = 0$,
(c) $\tilde{q}$ defined by (61) is a solution of (31) defined on $\mathcal{O}$, where $\tilde{\pi}$ is the projection onto $\tilde{V}_1(x,t)$ along $\tilde{V}_2(x,t)$.

Bäcklund transformations for the $\hat{A}^{(2)}_{2n-1}$-flows are obtained in the similar way as for the $\hat{C}^{(1)}_n$-flows.
As a consequence of Proposition 9 and Theorem 14, we obtain the following:

Theorem 16 (DT for $\hat{A}^{(2)}_{2n-1}$-KdV (33)).
Let E be a frame of a solution u of (34), $\triangle : \mathbb{R}^2 \to N_n^+$ a solution of $\triangle_t \triangle^{-1} = \xi_j(u)$, and $q = \triangle^{-1} * u$, where $\xi_j(u)$ is defined by (33). Let π be a projection satisfying (60), and $k_{\alpha,\pi}$ defined by (50), and $\tilde{V}(x,t) = \triangle^{-1}(x,t) E^{-1}(x,t,\alpha)(\operatorname{Im}\pi)$. Then
(1) there exists an open subset containing $(0,0)$ such that $\mathbb{R}^{2n} = \tilde{V}(x,t) \oplus \tilde{V}^\perp(x,t)$,

(2) let $\tilde{\pi}(x,t)$ denote the projection onto $\tilde{V}(x,t)$ along $\tilde{V}(x,t)^\perp$, $\tilde{q}$ defined by (61), and $\tilde{\triangle}:$ $\mathbb{R}^2 \to N_n^+$ such that $\tilde{\triangle} * \tilde{q}$ is V_n-valued. Then $\tilde{u} = \tilde{\triangle} * \tilde{q}$ is a new solution of (34) and

$$\tilde{E} = k_{\alpha,\pi} E \triangle k_{\alpha,\tilde{\pi}}^{-1} \tilde{\triangle}^{-1}$$

is a frame of $\tilde{u}$.

Theorems 14 and 6 (iii) give the following:

Theorem 17 (DT for Lagrangian curve flows of A-type).
Let γ be a solution of the Lagrangian curve flow (40) of A-type, and $g(\cdot,t)$, $u(\cdot,t)$ the Lagrangian frame and Lagrangian curvature along $\gamma(\cdot,t)$. Let E be the frame of the solution u of (31) satisfying $E(0,0,\lambda) = g(0,0)$. Let $\triangle$, α, π, $\tilde{\pi}$ be as in Theorem 16. Then

$$\tilde{\gamma} = (2\pi - I) g \triangle (2\tilde{\pi} - I) e_1$$

is a new solution of (40) and its Lagrangian curvature $\tilde{u}$ is a solution of (31).

Example 12. 1-soliton solutions of A-type
Please note that $u = 0$ is the trivial solution of the third $\hat{A}^{(2)}_{2n-1}$-flow with frame $F(x,t,\lambda) = \exp(xJ_B(\lambda) + tJ_B^3(\lambda))$. By Theorem 6 (iii),

$$\gamma(x,t) = F(x,t,0)e_1 = \exp(bx + b^3 t)e_1$$

is the Lagrangian curve flow (39) with zero Lagrangian curvature and

$$g(x,t) = \exp(bx + b^3 t)$$

as its Lagrangian frame.
Please note that the linear system $B_{q,\lambda}$ given by (63) for $q = 0$ is

$$B_{0,\lambda} \begin{cases} y_x = -J_B y, \\ y_t = -J_B^3 y. \end{cases}$$

Since

$$J_B^{2n} = \lambda J_B, \quad (J_B^3)^{2n} = \lambda^3 J_B^3,$$

the solution of $B_{0,\lambda}$ for any given initial data can be written down explicitly. Hence Theorem 15 gives an algorithm to compute explicit formula for 1-solitons $\tilde{q}$ and its frame for the third $\hat{A}^{(2)}_{2n-1}$-flow. Theorem 17 gives the corresponding 1-soliton solution $\tilde{\gamma}$ of the third Lagrangian curve flow of A-type and the Lagrangian curvature $\tilde{u}$ of $\tilde{\gamma}$ is a 1-soliton solution of the third $\hat{A}^{(2)}_{2n-1}$-KdV flow.

Next we give the Permutability formula. First it follows from Lemma 8 that we have the following:

Lemma 9. Let $\alpha_1, \alpha_2 \in \mathbb{R}\setminus\{0\}$ such that $|\alpha_1| \neq |\alpha_2|$, and π_i projections of $\mathbb{R}^{2n}$ satisfying $\mathrm{Ker}\,\pi_i = (\mathrm{Im}\,\pi_i)^\perp$. Then $\tilde{V}_1 = k_{\alpha_2,\pi_2}(\alpha_1)(\mathrm{Im}\,\pi_1)$ and $\tilde{V}_2 = k_{\alpha_1,\pi_1}(\alpha_2)(\mathrm{Im}\,\pi_2)$ are non-degenerate, and

$$k_{\alpha_1,\tau_1} k_{\alpha_2,\pi_2} = k_{\alpha_2,\tau_2} k_{\alpha_1,\pi_1},$$

where τ_i is the projection onto $\tilde{V}_i$ along $\tilde{V}_i^\perp$ for $i = 1,2$.

Similarly, Lemma 9, Theorems 8 and 14 give the following:

Theorem 18. [Permutability for DTs of the $\hat{A}^{(2)}_{2n-1}$-flow]

Let α_i, π_i, τ_i be as in Lemma 9 for $i = 1, 2$. Let F be the frame of a solution q of the $(2j-1)$-th $\hat{A}^{(2)}_{2n-1}$-flow with $F(0, 0, \lambda) = I$, $\tilde{V}_i(x,t) = F(x,t,\alpha_i)^{-1}(\operatorname{Im}\pi_i)$, and $\tilde{\pi}_i(x,t)$ the projection onto $\tilde{V}_i(x,t)$ along $\tilde{V}_i(x,t)^{\perp}$. Let $\tilde{W}_1 = k_{\alpha_2, \tilde{\pi}_2}(\alpha_1)(\operatorname{Im}\tilde{\pi}_1)$, $\tilde{W}_2 = k_{\alpha_1, \tilde{\pi}_1}(\alpha_2)(\operatorname{Im}\tilde{\pi}_2)$, and $\tilde{\tau}_i$ be the projection onto $\tilde{W}_i$ along $\tilde{W}_i^{\perp}$. Then we have

$$q_i := k_{\alpha_i, \pi_i} \bullet q = q + \alpha_i[\beta, \tilde{\pi}], \quad i = 1, 2,$$
$$k_{\alpha_1, \tau_1} \bullet (k_{\alpha_2, \pi_2} \bullet q) = k_{\alpha_2, \tau_2} \bullet (k_{\alpha_1, \pi_1} \bullet q),$$
$$q_{12} := k_{\alpha_1, \tau_1} \bullet (k_{\alpha_2, \pi_2} \bullet q) = q_1 + \alpha_2[\beta, \tilde{\tau}_2] = q_2 + \alpha_1[\beta, \tilde{\tau}_1],$$

where $\beta = e_{1, 2n-1} + e_{2, 2n}$.

The Permutability Theorem 18 gives an algebraic formula to construct k-solitons of the $(2j-1)$-th $\hat{A}^{(2)}_{2n-1}$-flow and their frames from k 1-solitons of the $(2j-1)$-th $\hat{A}^{(2)}_{2n-1}$-flow. If $\tilde{F}$ is a frame of the k-soliton solution $\tilde{q}$ of $\hat{A}^{(2)}_{2n-1}$-flow, then $\tilde{\gamma} = \tilde{F}(x,t,0)e_1$ is a k-soliton solution of the Lagrangian curve flow of A-type and its Lagrangian curvature $\tilde{u}$ is a k-soliton of the $\hat{A}^{(2)}_{2n-1}$-KdV flow.

7. Scaling Transforms

In this section, we construct scaling transforms and give relations between DTs and scaling transforms for the $\hat{C}^{(1)}_n$-flows and $\hat{A}^{(2)}_{2n-1}$-flows.

Theorem 19. *Let α_i, π_i, τ_i as in Lemma 5 (9 resp.), and $F(x,t,\lambda)$ the frame of the solution q of the $(2j-1)$-th $\hat{C}^{(1)}_n$-flow (14) ($\hat{A}^{(2)}_{2n-1}$-flow (31) resp.) with $F(0,0,\lambda) = I_{2n+1}$. Let $r \in \mathbb{R} \setminus \{0\}$, and*

$$\Gamma(r) = \operatorname{diag}(1, r, \ldots, r^{2n-1}). \tag{64}$$

Then
(1) $(r \odot q)(x,t) = r\Gamma(r)^{-1}q(rx, r^{2j-1}t)\Gamma(r)$ *is a solution of the $(2j-1)$-th $\hat{C}^{(1)}_n$-flow (the $\hat{A}^{(2)}_{2n-1}$-flow resp.)*,
(2) *for the $\hat{C}^{(1)}_n$ case,*

$$(r \odot F)(x,t,\lambda) := \Gamma(r)^{-1} F(rx, r^{2j-1}t, r^{-2n}\lambda)\Gamma(r)$$

is the frame of the solution $r \odot q$ of the $\hat{C}^{(1)}_n$-flow (14),
(3) *for the $\hat{A}^{(2)}_{2n-1}$-case,*

$$(r \odot F)(x,t,\lambda) := \Gamma(r)^{-1} F(rx, r^{2j-1}t, r^{-(2n-1)}\lambda)\Gamma(r)$$

is the frame of the solution $r \odot q$ of the $\hat{A}^{(2)}_{2n-1}$-flow (31).

Proof. First we prove the Theorem for the $\hat{C}^{(1)}_n$-flows. Set $\hat{F}(x,t,\lambda) = \Gamma(r)^{-1}F(rx, r^{2j-1}t, r^{-2n}\lambda)$. Please note that

$$r\Gamma(r)^{-1}(e_{1,2n}r^{-2n}\lambda + b)\Gamma(r) = e_{1,2n}\lambda + b = J(\lambda). \tag{65}$$

Since F is a frame of q, $F^{-1}F_x = J + q$. direct computation implies that

$$\hat{F}^{-1}\hat{F}_x = J + rq(rx, r^{2j-1}t, r^{-2n}\lambda).$$

Let $P(x,t,\lambda) = P(q(x,t), \lambda)$ be the solution of (10). So $P_x + [J+q, P] = 0$. Set

$$\hat{P}(x,t,\lambda) = r\Gamma(r)^{-1}P(q(rx, r^{2j-1}t), r^{-2n}\lambda)\Gamma(r).$$

49

We use (65) and a direct computation to see that
$$\hat{P}_x + [J + r \odot q, \hat{P}] = 0.$$

This shows that $\hat{P} = P(r \odot q, \lambda)$. A direct computation implies that
$$\hat{F}^{-1}\hat{F}_t = \Gamma(r)^{-1}(r^{2j-1}P^{2j-1}(rx, r^{2j-1}t, r^{-2n}\lambda)_+ \Gamma(r))$$
$$= \Gamma(r)^{-1}(rP(rx, r^{2j-1}t, r^{-2n}\lambda))_+^{2j-1}\Gamma(r) = (\hat{P}^{2j-1})_+$$
$$= (P^{2j-1}(r \odot q, \lambda))_+.$$

It follows from Proposition 4 that $r \odot q$ is a solution of (14) and $\hat{F}$ is a frame of $r \odot q$. This proves (1) and (2) for the $\hat{C}_n^{(1)}$-hierarchy.

Similar proof gives (1) and (3) for the $\hat{A}_{2n-1}^{(2)}$-hierarchy. □

It follows from Theorem 19 (2) and Theorem 6 (iii) that we have the following:

Corollary 4. *Let $c \in \mathbb{R} \setminus 0$, and γ a solution of the $(2j-1)$-th Lagrangian curve flow of C-type or A-type. Then*
$$(c \odot \gamma)(x, t) := \Gamma(c)\gamma(cx, ct)$$
is again a solution, where $\Gamma(c)$ is defined by (64).

In particular, let $\tilde{\gamma}$ be the solution of the third Lagrangian curve flow on $\mathcal{M}_4$ constructed in Example 11. Then $c \odot \tilde{\gamma}$ is also a solution for all $c \in \mathbb{R} \setminus 0$.

Corollary 5. *Let $u = \sum_{i=1}^n u_i e_{n+1-i, n+i}$ be a solution of the $(2j-1)$-th $\hat{C}_n^{(1)}$-KdV flow (22) ($\hat{A}_{2n-1}^{(2)}$-KdV flow (34) resp.), $r \in \mathbb{R}\setminus\{0\}$, $\Gamma(r)$ as in (7). Then we have the following:*

(1) $r \odot u = \sum_{i=1}^n r^{2i} u_i(rx, r^{2j-1}t)e_{n+1-i, n-i}$ *is a solution of the $(2j-1)$-th $\hat{C}_n^{(1)}$-KdV flow (22) ($\hat{A}_{2n-1}^{(2)}$-KdV flow (34) resp.).*

(2) *If $E(x, t, \lambda)$ is a frame of the solution u of (22), then*
$$(r \odot E)(x, t, \lambda) := \Gamma(r)^{-1}E(rx, r^{2j-1}t, r^{-2n}\lambda)\Gamma(r)$$
is a frame of $r \odot u$.

(3) *If $E(x, t, \lambda)$ is a frame of the solution u of (34), then*
$$(r \odot E)(x, t, \lambda) := \Gamma(r)^{-1}E(rx, r^{2j-1}t, r^{-(2n-1)}\lambda)\Gamma(r)$$
is a frame of $r \odot u$.

Corollary 6. *$r \odot u$ defines an action of the multiplicative group $\mathbb{R}^+$ on the space of solutions of (22) ((34) resp.).*

Next we give a relation between the scaling transforms and Darboux transforms. First we need a Lemma.

Lemma 10. *Let $r \in \mathbb{R} \setminus 0$, $\Gamma(r)$ defined by (64), and $A_s = S_n^{-1}A^t S_n$ as before. Then*

(1) $\Gamma(r)_s = r^{2n+1}\Gamma(r)^{-1}$,

(2) *let π be a projection of $\mathbb{R}^{2n}$, and $\hat{\pi} = \Gamma(r)\pi\Gamma(r)^{-1}$, then*

  (a) *if $\pi_s = \pi$, then $\hat{\pi}_s = \hat{\pi}$,*
  (b) *if $\pi_s = I - \pi$, then $\hat{\pi}_s = I - \hat{\pi}$.*

Proof. It is clear that $\Gamma(r)S_n\Gamma(r) = r^{2n+1}S_n$, which gives (1). (2) follows from (1). □

It follows from Lemma 10, the formulas for $r \odot q$ in Theorem 19 and (55), (62) that we have the following.

Theorem 20. *Let $r, \alpha \in \mathbb{R} \setminus 0$, $\Gamma(r)$ as in (64), π a projection of $\mathbb{R}^{2n}$, and $\hat{\pi} = \Gamma(r) \pi \Gamma(r)^{-1}$.*

(1) *If q is a solution of the $(2j-1)$-th $\hat{C}_n^{(1)}$-flow (14) and π satisfies $\pi_s = I - \pi$, then*

$$k_{r^{-2n}, \hat{\pi}} \bullet q = r^{-1} \odot (k_{1,\pi} \bullet (r \odot q)).$$

(2) *If q is a solution of the $(2j-1)$-th $\hat{A}_{2n-1}^{(2)}$-flow (31) and π satisfies $\pi_s = \pi$, then*

$$k_{r^{-(2n-1)}, \hat{\pi}} \bullet q = r^{-1} \odot (k_{1,\pi} \bullet (r \odot q)).$$

8. Bi-Hamiltonian Structure

The existence of a bi-Hamiltonian structure and using it to generate the hierarchy are two of the well-known properties for soliton hierarchies (cf. [11,55,56]). In this section, we use the linear operator P_u defined in Definition 4 to write down the bi-Hamiltonian structure for the $\hat{C}_n^{(1)}$-KdV and $\hat{A}_{2n-1}^{(2)}$-KdV. The pull back of this bi-Hamiltonian structure to $\mathcal{M}_{2n}$ via the Lagrangian curvature map Ψ gives the bi-Hamiltonian structure for the Lagrangian curve flows of C and A-type.

Let

$$\langle \xi, \eta \rangle = \oint \text{tr}(\xi \eta) \, dx$$

denote the standard L^2 inner product on $C^\infty(S^1, sl(2n, \mathbb{R}))$.

The bi-Hamiltonian structure on $C^\infty(S^1, \mathcal{B}_n^+)$ for the $\hat{C}_n^{(1)}$ and $\hat{A}_{2n-1}^{(2)}$ hierarchies given in [11] is

$$\{F_1, F_2\}_1^\wedge(q) = \langle [\beta, \nabla F_1(q)], \nabla F_2(q) \rangle, \qquad (66)$$

$$\{F_1, F_2\}_2^\wedge(q) = \langle [\partial_x + b + q, \nabla F_1(q)], \nabla F_2(q) \rangle, \qquad (67)$$

where

$$\beta = \begin{cases} e_{1,2n}, & \text{for } \hat{C}_n^{(1)}, \\ \frac{1}{2}(e_{1,2n-1} + e_{2,2n}), & \text{for } \hat{A}_{2n-1}^{(2)}. \end{cases} \qquad (68)$$

Using the same proof as in [49], we see that the bi-Hamiltonian structure is invariant under the gauge action of the group $C^\infty(S^1, N_n^+)$, i.e., if F_1, F_2 are invariant under the gauge action, then $\{F_1, F_2\}_i^\wedge$ is also invariant for $i = 1, 2$. Since $C^\infty(S^1, V_n)$ is the orbit space of this gauge action, we can identify functionals F on $C^\infty(S^1, V_n)$ with invariant functionals $\hat{F}$ on $C^\infty(S^1, \mathcal{B}_n^+)$, where

$$\hat{F}(\triangle * u) = F(u).$$

Hence

$$\{F_1, F_2\}_i(u) = \{\hat{F}_1, \hat{F}_2\}_i^\wedge(u)$$

are Poisson structures on $C^\infty(S^1, V_n)$ for $i = 1, 2$.

Given a functional $F : C^\infty(S^1, V_n) \to \mathbb{R}$, let $\nabla F(u)$ be the unique map from $S^1 \to V_n^t$ satisfying

$$dF_u(v) = \langle \nabla F(u), v \rangle = \oint \text{tr}(\nabla F(u) v) \, dx$$

for all $v \in C^\infty(S^1, V_n)$.

Again we use the same proof as in [49,50] to write $\{,\}_i$ in terms of the linear operator P_u:

$$\{F_1, F_2\}_1(u) = \langle [\beta, P_u(\nabla F_1(u))], P_u(\nabla F_2(u)) \rangle,$$
$$\{F_1, F_2\}_2(u) = \langle [\partial_x + b + u, P_u(\nabla F_1(u))], P_u(\nabla F_2(u)) \rangle,$$

where β is given by (68) These give a bi-Hamiltonian structure for the $\hat{C}_n^{(1)}$-KdV flows.

The first bracket is always zero and $\{,\}_2$ is a Poisson structure for the $\hat{A}_{2n-1}^{(2)}$-hierarchy. There is a standard way (cf. [56]) to generate a sequence of compatible invariant Poisson structures $\{,\}_j^\wedge$, $j \geq 1$ on $C^\infty(S^1, \mathcal{B}_n^+)$. It can be checked that the induced structure $\{,\}_{2i+1}$ on $C^\infty(S^1, V_n)$ is always zero for the $\hat{A}_{2n-1}^{(2)}$-KdV hierarchy, but $\{,\}_{2i}$ are non-trivial Poisson structure. So $(\{,\}_2, \{,\}_4)$ gives a bi-Hamiltonian structure for the $\hat{A}_{2n-1}^{(2)}$-KdV flows. Since the formulas are tedious and do not give us useful information, we omit the discussion of $\{,\}_4$ for the $\hat{A}_{2n-1}^{(2)}$-KdV hierarchy.

Since $[\partial_x + b + u, P_u(\nabla F_1(u))]$ is in $C^\infty(S^1, V_n)$ and $\pi_0(P_u(\nabla F_2(u))) = \nabla F_2(u)$, we have

$$\{F_1, F_2\}_2(u) = \langle [\partial_x + b + u, P_u(\nabla F_1(u))], \nabla F_2(u) \rangle.$$

So the Hamiltonian flow for a functional F with respect to $\{,\}_2$ is

$$u_t = [\partial_x + b + u, P_u(\nabla F(u))].$$

The following results can be proved by a similar computation as in [49] for the $\hat{A}_n^{(1)}$-KdV hierarchy:

Theorem 21. *Set*

$$F_{2j-1}(u) = -\oint \mathrm{tr}(P_{2j-1,-1}(u) e_{1,2n}) dx,$$
$$H_{2j-1}(u) = -\frac{1}{2} \oint \mathrm{tr}(Q_{2j-1,-1}(u)(e_{1,2n+1} + e_{2,2n})) dx.$$

Then we have

$$\nabla F_{2j-1}(u) = \pi_0(P_{2j-1,0}(u)), \quad \nabla H_{2j-1}(u) = \pi_0(Q_{2j-1,0}(u)),$$

where π_0 is the projection onto V_n^t defined by (44). Moreover, we also have:

(i) *The Hamiltonian equation for F_{2j-1} (H_{2j-1} resp.) with respect to $\{,\}_2$ is the $(2j-1)$-th $\hat{C}_n^{(1)}$-KdV ($\hat{A}_{2n-1}^{(2)}$-KdV resp.) flow for $j \geq 1$.*

(ii) *The Hamiltonian equation for $F_{2(n+j)-1}$ with respect to $\{,\}_1$ is the $(2j-1)$-th $\hat{C}_n^{(1)}$-KdV flow for $j > n$.*

Remark 3. *The bi-Hamiltonian structure on $C^\infty(S^1, V_1)$ for the $\hat{C}_1^{(1)}$-KdV hierarchy is the standard bi-Hamiltonian structure for the KdV hierarchy (cf. [52]).*

Example 13. Bi-Hamiltonian structure for the $\hat{C}_2^{(1)}$-KdV hierarchy

Let $u = u_1e_{23} + u_2e_{14}$, $\zeta = \zeta_1e_{32} + \zeta_2e_{41}$, $\eta = \eta_1e_{32} + \eta_2e_{41}$, $C = (C_{ij}) = P_u(\zeta)$, and $D = (D_{ij}) = P_u(\eta)$. We use Example 9 to write down the following Hamiltonian structures:

$$\{F_1, F_2\}_1(u) = \langle [e_{14}, C], D \rangle$$
$$= \oint (3\zeta_2''' + 4\zeta_1' - u_1\zeta_2')\eta_2 + 4\zeta_2'\eta_1 + u_1\zeta_2\eta_2' dx,$$
$$= \oint (3\zeta_2''' + 4\zeta_1' - 2u_1\zeta_2' - u_1'\zeta_2)\eta_2 + 4\zeta_2'\eta_1 dx,$$
$$\{F_1, F_2\}_2(u) = \langle [\partial_x + b + u, C], D \rangle$$
$$= \oint (C_{14}' - 2u_2C_{11})\eta_2 + (C_{23}' + 2C_{13} + u_1\zeta_1')\eta_1 dx,$$

where C_{ij}'s are written in terms of ζ_1 and ζ_2 as in Example 9.

Example 14. Conservation laws for the $\hat{C}_n^{(1)}$-KdV hierarchy
Let
$$f_{2j-1}(u) = \text{tr}(P_{2j-1,-1}(u)e_{1,2n})$$
denote the density of F_{2j-1}.

(1) For $n = 2$, we have
$$f_1 = u_1, \quad f_3 = u_2 + \frac{1}{8}u_1^2, \quad f_5 = -\frac{1}{32}u_1^3 + u_1u_2 - \frac{3}{32}u_1u_1''.$$

(2) For general n, the first two densities of conservation laws are
$$f_1 = u_1, \quad f_3 = u_2 + \frac{2n-3}{4n}u_1^2.$$

Example 15. Conservation laws for the $\hat{A}_{2n-1}^{(2)}$-KdV hierarchy
Let
$$h_{2j-1}(u) = \frac{1}{2}\text{tr}(Q_{2j-1,-1}(u)(e_{1,2n+1} + e_{2,2n})).$$

(1) For $n = 2$, we have
$$h_1 = u_1, \quad h_3 = u_2, \quad h_5 = \frac{1}{3}(\frac{2}{3}u_1(u_1)_{xx} - 4u_1u_2 - \frac{4}{9}u_1^3).$$

(2) For general n, the first two densities of conservation laws are
$$h_1 = u_1, \quad h_3 = u_2 + \frac{n-2}{2n-1}u_1^2.$$

Example 16. Hamiltonian flows for F_3 and H_3
A simple computation implies that $\nabla F_3(u) = \frac{1}{4}u_1e_{32} + e_{41}$, where $u = u_1e_{23} + u_2e_{14}$. We use notations and formulas as in Example 9 to compute $P_u(\nabla F_3(u))$ and obtain

$$C_{11} = -\frac{3}{8}u_1', \quad C_{13} = -\frac{3}{8}(u_1)_x^{(3)} + u_2',$$
$$C_{14} = -\frac{3}{8}(u_1)_x^{(4)} + (u_2)_{xx} + \frac{3}{8}u_1(u_1)_{xx} - \frac{3}{4}u_1u_2,$$
$$C_{23} = -\frac{1}{2}(u_1)_{xx} + u_2 + \frac{1}{4}u_1^2.$$

The Hamiltonian flow of F_3 with respect to $\{\,,\,\}_2$ is
$$u_t = [\partial_x + b + u, P_u(\nabla F_3(u))]. \tag{69}$$

We use the formula for $P_u(\nabla F_3(u))$ to compute directly and see that (69) is the following system for u_1, u_2,

$$\begin{cases} (u_1)_t = C'_{23} + 2C_{13} + \frac{1}{4}u_1 u'_1, \\ (u_2)_t = C'_{14} - 2u_2 C_{11}. \end{cases}$$

Substitute C_{ij} into the above equation to see that it is (5).

Similarly, we use the same notations and formulas as in Example 9 to compute $P_u(\nabla H_3(u))$. Here $\nabla H_3(u) = e_{32}$. We see thatbe

$$C_{11} = 0, \quad C_{13} = u'_2, \quad C_{14} = u''_2 - u_1 u_2, \quad C_{23} = u_2.$$

So the Hamiltonian flow for H_3 with respect to $\{\,,\,\}_2$ written in terms of u_1, u_2 is (6).

Remark 4. We use the pullback $\{\,,\,\}_i^\wedge$ of the Poisson structures $\{\,,\,\}_i$ on $C^\infty(S^1, V_n)$ by the Lagrangian curvature map Ψ for $i = 1, 2$, to get a bi-Hamiltonian structure on $\mathcal{M}_{2n}$. In other words, given a functional F_i on $C^\infty(S^1, V_n)$, let

$$\hat{F} = F \circ \Psi : \mathcal{M}_{2n} \to \mathbb{R}$$

be functionals on $\mathcal{M}_{2n}$. Then

$$\{\hat{F}, \hat{G}\}_i^\wedge(\gamma) = \{F, G\}_i(\Psi(\gamma)), \quad i = 1, 2$$

are the pullback bi-Hamiltonian on $\mathcal{M}_{2n}$. As a consequence of Theorem 21, we have the following:
(1) The Lagrangian curve flow (39) and (40) are Hamiltonian flows for the Hamiltonians

$$\hat{F}_{2j-1} := F_{2j-1} \circ \Psi, \quad \hat{H}_{2j-1}(u) := H_{2j-1} \circ \Psi$$

with respect to $\{\,,\,\}_2^\wedge$ respectively.
(2) The Lagrangian curve flows of C-type (A-type resp.) are commuting Hamiltonian flows on $\mathcal{M}_{2n}$.

9. Review and Open Problems

In this section, we give an outline of the construction of $\hat{\mathcal{G}}^{(1)}$-KdV hierarchy (cf. [11,53]), explain the key steps needed in constructing curve flows whose differential invariants satisfy the $\hat{\mathcal{G}}^{(1)}$-KdV, and give some open problems.

Let G be a non-compact, real simple Lie group, $\mathcal{G}$ its Lie algebra, and

$$\hat{\mathcal{G}}^{(1)} = \mathcal{L}(\mathcal{G}) = \{\sum_{i \leq n_0} \xi_i \lambda^i \mid n_0 \text{ an integer}, \xi_i \in \mathcal{G}\}.$$

Let

$$\hat{\mathcal{G}}_+^{(1)} = \{\sum_{i \geq 0} \xi_i \lambda^i \in \mathcal{L}(\mathcal{G})\}, \quad \hat{\mathcal{G}}_-^{(1)} = \{\sum_{i < 0} \xi_i \lambda^i \in \mathcal{L}(\mathcal{G})\}.$$

Then $(\hat{\mathcal{G}}_+^{(1)}, \hat{\mathcal{G}}_-^{(1)})$ is a splitting of $\hat{\mathcal{G}}^{(1)}$.

Let $\{\alpha_1, \ldots, \alpha_n\}$ be a simple root system of $\mathcal{G}$, and $\mathcal{B}_+, \mathcal{B}_-, \mathcal{N}_+$ the Borel subalgebras of $\mathcal{G}$ of non-negative roots, non-positive roots, and positive roots respectively. Let B_+, B_-, N_+ be connected subgroups of G with Lie algebras $\mathcal{B}_+, \mathcal{B}_-, \mathcal{N}_+$ respectively. Let

$$J = \beta \lambda + b, \tag{70}$$

where $b = -\sum_{i=1}^n \alpha_i$ and β is the highest root.

The construction of $\hat{\mathcal{C}}_n^{(1)}$-hierarchy in Section 3 works for $\hat{\mathcal{G}}^{(1)}$ except that the generating function $P(q, \lambda)$ in Proposition 2 should satisfy

$$\begin{cases} [\partial_x + b + q, S(q, \lambda)] = 0, \\ m(S(q, \lambda)) = 0, \end{cases} \tag{71}$$

where m is the minimal polynomial of J defined by (70).

Assume that there is a sequence of increasing positive integers $\{n_j \mid j \geq 1\}$ such that J^{n_j} lies in $\hat{\mathcal{G}}_+^{(1)}$ for all $j \geq 1$. Write

$$S^{n_j}(q, \lambda) = \sum_i S_{n_j, i}(q) \lambda^i.$$

Then the n_j-th flow in the $\hat{\mathcal{G}}^{(1)}$-hierarchy is

$$q_t = [\partial_x + b + q, S_{n_j, 0}(q)] \tag{72}$$

for $q : \mathbb{R}^2 \to \mathcal{B}_+$.

Using the same kind of proofs for the $\hat{\mathcal{C}}_n^{(1)}$-hierarchy, we obtain the following properties of the $\hat{\mathcal{G}}^{(1)}$-hierarchy:

(i) The existence of a Lax pair, $[\partial_x + J + q, \partial_t + (S^{n_j}(q, \lambda))_+] = 0$ for (72).
(ii) The $\hat{\mathcal{G}}^{(1)}$-flows are invariant under the gauge action of $C^\infty(\mathbb{R}, N_+)$ on $C^\infty(\mathbb{R}, \mathcal{B}_+)$.
(iii) If we find a linear subspace V of $\mathcal{G}$ such that $C^\infty(\mathbb{R}, V)$ is a cross-section of the gauge action of $C^\infty(\mathbb{R}, N_+)$ on $C^\infty(\mathbb{R}, \mathcal{B}_+)$. Then we can push down the $\hat{\mathcal{G}}^{(1)}$-flows to the cross-section $C^\infty(\mathbb{R}, V)$ along gauge orbits and obtain a $\hat{\mathcal{G}}^{(1)}$-KdV hierarchy on $C^\infty(\mathbb{R}, V)$. Moreover, there exists a polynomial differentials $\zeta_j(u)$ such that the n_j-th flow in the $\hat{\mathcal{G}}^{(1)}$-KdV hierarchy is

$$u_t = [\partial_x + b + u, S_{n_j, 0}(u) - \zeta_j(u)]. \tag{73}$$

The $\hat{\mathcal{G}}^{(1)}$-KdV hierarchies constructed from two different cross-sections are not the same but are gauge equivalent.

(iv) The bi-Hamiltonian structure $(\{,\}_1^\wedge, \{,\}_2^\wedge)$ on $C^\infty(\mathbb{R}, \mathcal{B}_+)$ is given by (66), (67).
(v) The Poisson structures $\{,\}_1^\wedge$ and $\{,\}_2^\wedge$ are invariant under the gauge group action. So there is an induced bi-Hamiltonian structure on $C^\infty(S^1, V)$ for the $\hat{\mathcal{G}}^{(1)}$-KdV hierarchy, which will be denoted by $(\{,\}_1, \{,\}_2)$.
(vi) $F_{n_j}(q) = -\oint (S_{n_j, -1}(q) \beta) dx$ is the Hamiltonian for the n_j-th flow with respect to $\{,\}_2^\wedge$.

Although properties (i)–(vi) can be proved in a unified way for any $\hat{\mathcal{G}}^{(1)}$, the following results need to be proved case by case depending on $\mathcal{G}$:

(1) Find a linear subspace V such that $C^\infty(\mathbb{R}, V)$ is a cross-section of the gauge action of $C^\infty(\mathbb{R}, N_+)$ on $C^\infty(\mathbb{R}, \mathcal{B}_+)$.
(2) Suppose $\mathcal{G}$ is a subalgebra of $gl(n)$ and $C^\infty(\mathbb{R}, V)$ is a cross-section of the gauge action. We consider the following class of curves in $\mathbb{R}^n$:

$$\mathcal{M} = \{ge_1 \mid g \in C^\infty(\mathbb{R}, G) \text{ satisfying } g^{-1} g_x = b + u,$$
$$\text{for some } u \in C^\infty(\mathbb{R}, V)\}.$$

Find geometric properties of curves in $\mathcal{M}$ that characterize $\gamma \in \mathcal{M}$ (so g is the moving frame and u is the differential invariant of γ under the group G). For example, for the $\hat{\mathcal{C}}_n^{(1)}$ case, it is easy to see that if $\gamma \in \mathcal{M}$, then γ is Lagrangian (see Definition 1). Conversely, if γ is Lagrangian then $g \in \mathcal{M}$.
(3) Identify the tangent space of $\mathcal{M}$ at γ.

(4) Show that
$$\gamma_t = g S_{n_j,0}(u) e_1 \tag{74}$$
is a flow on $\mathcal{M}$, i.e., the right hand side is tangent to $\mathcal{M}$.

(5) Show that if $\gamma(x,t)$ is a solution of (74), then the differential invariants $u(\cdot, t)$ satisfies the $\hat{\mathcal{G}}^{(1)}$-KdV flow (73). This also gives a natural interpretation of the $\hat{\mathcal{G}}^{(1)}$-KdV.

(6) Write down the formula for the induced bi-Hamiltonian structure for the $\hat{\mathcal{G}}^{(1)}$-KdV hierarchy.

(7) We pull back the bi-Hamiltonian structure on $C^\infty(S^1, V)$ to $\mathcal{M}$ via the curvature map $\Psi : \mathcal{M} \to C^\infty(S^1, V)$ defined by $\Psi(\gamma) = u$ the differential invariant of γ. Then soliton properties of $\hat{\mathcal{G}}^{(1)}$-KdV can be also pulled back to the curve flows (74) on $\mathcal{M}$.

(8) Prove an analogue of Theorem 5, i.e., if $C : \mathbb{R} \to \mathcal{G}$ satisfies $[\partial_x + b + u, C] \in C^\infty(\mathbb{R}, V)$, then
  (a) C is determined by Ce_1,
  (b) C is determined by the projection of C onto V^t, where $u \in C^\infty(\mathbb{R}, V)$.

We need this result to give a precise description of the tangent space of $\mathcal{M}$ at γ and to write down the formula for the induced bi-Hamiltonian structure on $C^\infty(\mathbb{R}, V)$ for the $\hat{\mathcal{G}}^{(1)}$-KdV hierarchy.

(9) To construct Darboux transforms, we need to find rational maps $g : \mathbb{R} \to G_\mathbb{C}$ satisfies $\overline{g(\bar{\lambda})} = g(\lambda)$ with minimal number of poles and work out the factorization formula explicitly.

Let σ be an involution of $\mathcal{G}$, and $\mathcal{K}, \mathcal{P}$ the $1, -1$ eigenspaces of σ. The $\hat{\mathcal{G}}^{(2)}$-hierarchy is constructed from the splitting $(\hat{\mathcal{G}}_+^{(2)}, \hat{\mathcal{G}}_-^{(2)})$ of $\hat{\mathcal{G}}^{(2)}$, where

$$\hat{\mathcal{G}}^{(2)} = \{\xi(\lambda) = \sum_i \xi_i \lambda^i \mid \overline{\xi(\bar{\lambda})} = \xi(\lambda), \sigma(\xi(-\lambda)) = \xi(\lambda)\},$$

$$\hat{\mathcal{G}}_+^{(2)} = \hat{\mathcal{G}}^{(2)} \cap \hat{\mathcal{G}}_+^{(1)}, \quad \hat{\mathcal{G}}_-^{(2)} = \hat{\mathcal{G}}^{(2)} \cap \hat{\mathcal{G}}_-^{(1)}.$$

Assume that there is a simple root system of $\mathcal{G}$ so that $\beta \in \mathcal{P}$ and $b \in \mathcal{K}$. Then $C^\infty(\mathbb{R}, \mathcal{K} \cap \mathcal{B}_+)$ is invariant under the $\hat{\mathcal{G}}^{(1)}$-hierarchy. The $\hat{\mathcal{G}}^{(2)}$-hierarchy is the restriction of the $\hat{\mathcal{G}}^{(1)}$-hierarchy to $C^\infty(\mathbb{R}, \mathcal{K} \cap \mathcal{B}_+)$. Most properties of the $\hat{\mathcal{G}}^{(1)}$-hierarchy hold for the $\hat{\mathcal{G}}^{(2)}$-hierarchy except the bi-Hamiltonian structure $\{,\}_1^\wedge$ is zero on $C^\infty(S^1, \mathcal{K} \cap \mathcal{B}_+)$. To obtain the other Poisson structure, we need to review briefly a general method to construct a sequence of compatible Poisson structures from a bi-Hamiltonian structure: Let Ξ_i denote the Poisson operator for $\{,\}_i^\wedge$ on $C^\infty(\mathbb{R}, \mathcal{B}_+)$, i.e., $(\Xi_i)_q : C^\infty(S^1, \mathcal{B}_-) \to C^\infty(S^1, \mathcal{B}_+)$ is defined by

$$\{F_1, F_2\}_i^\wedge(q) = \langle (\Xi_1)_q(\nabla F_1(q)), \nabla F_2(q) \rangle$$

for $i = 1, 2$. It is known (cf. [55,56]) that

$$\{F_1, F_2\}_j^\wedge(q) = \langle (\Xi_j)_q(\nabla F_1(q)), \nabla F_2(q) \rangle$$

is again a Poisson structure and are compatible, where

$$\Xi_j := \Xi_2 (\Xi_1^{-1} \Xi_2)^{j-2}.$$

It can be checked that $\Xi_{2i+1} = 0$ on $C^\infty(S^1, \mathcal{K} \cap \mathcal{B}_+)$, and Ξ_{2i} is a Poisson structure for the $\hat{\mathcal{G}}^{(2)}$-hierarchy for all $i \geq 1$. So $(\{,\}_2^\wedge, \{,\}_4^\wedge)$ is a bi-Hamiltonian structure for the $\hat{\mathcal{G}}^{(2)}$-hierarchy and it induces a bi-Hamiltonian structure $(\{,\}_2, \{,\}_4)$ for the $\hat{\mathcal{G}}^{(2)}$-KdV hierarchy.

Finally we give a list of open problems:

⋄ Find integrable curve flows on $\mathbb{R}^{2n,1}$ whose differential invariants satisfy the $\hat{B}_n^{(1)}$-KdV flows.

⋄ Find integrable curve flows on $\mathbb{R}^{k,2n-k}$ whose differential invariants satisfy the $\hat{D}_n^{(1)}$-KdV flows.

- Find integrable curve flows on $\mathbb{R}^{2n}$ whose differential invariants satisfy the $\hat{D}_n^{(2)}$-KdV flows.
- Find integrable curve flows on $\mathbb{R}^8$ whose differential invariants satisfy the $\hat{D}_4^{(3)}$-KdV flows.
- Find integrable curve flows on $\mathbb{R}^7$ whose differential invariants satisfy the $\hat{G}_2^{(1)}$-KdV flows.
- Calini and Ivey constructed finite gap solutions for the VFE in [57]. It would be interesting to construct finite-gap solutions for central affine curve flows, isotropic curve flows, and Lagrangian curve flows.
- The Gauss-Codazzi equations of submanifolds occurring in soliton theory are often given by the first level flows of the soliton hierarchy, i.e., the commuting flows generated by degree one (in λ) elements in the vacuum sequence. It would be interesting to see whether the flows of the $\hat{\mathcal{G}}^{(1)}$-KdV hierarchy generated by degree one elements in the vacuum sequence also arise as the Gauss-Codazzi equations for some class of submanifolds.

Author Contributions: Both authors are equally responsible for all results in this paper. All authors have read and agreed to the published version of the manuscript.

Funding: This research received no external funding.

Conflicts of Interest: The authors declare no conflict of interest.

References

1. Zabusky, N.J.; Kruskal, M.D. Interaction of solitons in a collisionless plasma and the recurrence of initial states. *Phys. Rev. Lett.* **1965**, *15*, 240–243. [CrossRef]
2. Gardner, C.S.; Greene, J.M.; Kruskal, M.D.; Miura, R.M. Method for solving the Korteweg-de Vries equation. *Phys. Rev. Lett.* **1967**, *19*, 1095–1097. [CrossRef]
3. Lax, P.D. Integrals of nonlinear equations of evolution and solitary waves. *Commun. Pure Appl. Math.* **1968**, *21*, 467–490. [CrossRef]
4. Zakharov, V.E.; Faddeev, L.D. Korteweg-de Vries equation, a completely integrable Hamiltonian system. *Func. Anal. Appl.* **1971**, *5*, 280–287. [CrossRef]
5. Zakharov, V.E.; Shabat, A.B. Exact theory of two-dimensional self-focusing and one-dimensional of waves in nonlinear media. *Sov. Phys. JETP* **1972**, *34*, 62–69.
6. Adler, M. On a trace functional for formal pseudo-differential operators and the symplectic structure of the Korteweg-de Vries Type Equations. *Invent. Math.* **1979**, *50*, 219–248. [CrossRef]
7. Adler, M.; van Moerbeke, P. Completely integrable systems, Euclidean Lie algebras and curves. *Adv. Math.* **1980**, *38*, 267–317. [CrossRef]
8. Kostant, B. The solution to a generalized Toda lattice and representation theory. *Adv. Math.* **1979**, *34*, 195–338. [CrossRef]
9. Symes, W.W. Systems of Toda type, Inverse spectral problems, and representation theory. *Invent. Math.* **1980**, *59*, 13–51. [CrossRef]
10. Kupershmidt, B.A.; Wilson, G. Modifying Lax equations and the second Hamiltonian structure. *Invent. Math.* **1981**, *62*, 403–436. [CrossRef]
11. Drinfeld, V.G.; Sokolov, V.V. Lie algebras and equations of Korteweg-de Vries type. (Russ.) *Curr. Probl. Math.* **1984**, *24*, 81–180.
12. Ablowitz, M.J.; Clarkson, P.A. *Solitons, Nonlinear Evolution Equations and Inverse Scattering*; London Mathematical Society Lecture Note Series, 149; Cambridge University Press: Cambridge, UK, 1991.
13. Beals, R.; Deift, P.; Tomei, C. *Direct and Inverse Scattering on the Line, Mathematical Surveys and Monographs*; American Mathematical Society: Providence, RI, USA, 1988; Volume 28.
14. Dickey, L.A. *Soliton Equations and Hamiltonian Systems*, 2nd ed.; Advanced Series in Mathematical Physics; World Scientific Publishing Co. Inc.: River Edge, NJ, USA, 2003; p. 26.
15. Faddeev, L.D.; Takhtajan, L.A. *Hamiltonian Methods in the Theory of Solitons*; Springer: Berlin, Germany, 1987.
16. Magri, F. *Eight Lectures on Integrable Systems*; Written in Collaboration with P. Casati, G. Falqui and M. Pedroni, Lecture Notes in Phys 495, Integrability of Nonlinear Systems; Springer: Berlin, Germany, 1997; pp. 256–296.
17. Palais, R.S. The symmetries of solitons. *Bull. AMS* **1997**, *34*, 339–403. [CrossRef]
18. Terng, C.L.; Uhlenbeck, K. Poisson actions and scattering theory for integrable systems. *Surv. Differ. Geom.* **1999**, *4*, 315–402. [CrossRef]
19. Darboux, G. *Leçons sur les Systèmes Orthogonaux et les Coordonnées Curvilignes. Principes de Géométrie Analytique*, 2nd ed.; Gauthier-Villars: Paris, France, 1910.
20. Tenenblat, K.; Terng, C.L. Bäcklund's theorem for n-dimensional submanifolds of R^{2n-1}. *Ann. Math.* **1980**, *111*, 477–490. [CrossRef]

21. Terng, C.L. A higher dimensional generalization of the sine-Gordon equation and its soliton theory. *Ann. Math.* **1980**, *111*, 491–510. [CrossRef]
22. Ferus, D.; Pedit, F. Isometric immersions of space forms and soliton theory. *Math. Ann.* **1996**, *305*, 329–342. [CrossRef]
23. Tenenblat, K. Bäcklund's theorem for submanifolds of space forms and a generalized wave equation. *Bull. Soc. Brasil. Math.* **1985**, *16*, 67–92. [CrossRef]
24. Terng, C.L.; Wang, E. Transformations of flat Lagrangian immersions and Egoroff nets. *Asian J. Math.* **2008**, *12*, 99–119. [CrossRef]
25. Donaldson, N.; Terng, C.L. Conformally flat submanifolds in spheres and integrable systems. *Tohoku Math. J.* **2011**, *63*, 277–302. [CrossRef]
26. Burstall, F.E. *Isothermic Surfaces: Conformal Geometry, Clifford Algebras and Integrable Systems*; Integrable Systems, Geometry, and Topology, AMS/IP Stud. Adv. Math.; American Mathematical Society: Providence, RI, USA, 2006; Volume 36, pp. 1–82.
27. Bruck, M.; Du, X.; Park, J.; Terng, C.L. Submanifold geometry of real Grassmannian systems. *Mem. AMS* **2002**, *155*, 102.
28. Donaldson, N.; Terng, C.L. Isothermic submanifolds. *J. Geom. Anal.* **2012**, *22*, 827–844. [CrossRef]
29. Terng, C.L. *Geometries and Symmetries of Soliton Equations and Integrable Elliptic Systems*; Surveys on Geometry and Integrable Systems; Advanced Studies in Pure Mathematics Mathematical Society of Japan: Tokyo, Japan, 2008; Volume 51, pp. 401–488.
30. Hasimoto, R. A soliton on a vortex filament. *J. Fluid Mech.* **1972**, *51*, 477–485. [CrossRef]
31. Terng, C.L. *Dispersive Geometric Curve Flows*; Surveys in Differential Geometry 2014. Regularity and Evolution of Nonlinear Equations, Surv. Differ. Geom.; International Press: Somerville, MA, USA, 2015; Volume 19, pp. 179–229.
32. Langer, J.; Perline, R. Poisson geometry of the filament equation. *J. Nonlinear Sci.* **1991**, *1*, 71–93. [CrossRef]
33. Langer, J.; Perline, R. Local geometric invariants of integrable evolution equations. *J. Math. Phys.* **1994**, *35*, 1732–1737. [CrossRef]
34. Langer, J.; Perline, P. Geometric realizations of Fordy-Kulish non- linear Schrödinger systems. *Pac. J. Math.* **2000**, *195*, 157–178. [CrossRef]
35. Doliwa, A.; Santini, P.M. An elementary geometric characterization of the integrable motions of a curve. *Phys. Lett. A* **1994**, *185*, 373–384. [CrossRef]
36. Ferapontov, E.V. Isoparametric hypersurfaces in spheres, integrable non-diagonalizable systems of hydrodynamic type, and N-wave systems. *Differ. Geom. Appl.* **1995**, *5*, 335–369. [CrossRef]
37. Yasui, Y.; Sasaki, N. Differential geometry of the vortex filament equation. *J. Geom. Phys.* **1998**, *28*, 195–207. [CrossRef]
38. Chou, K.-S.; Qu, C. Integrable equations arising from motions of plane curves. *Phys. D* **2002**, *162*, 9–33. [CrossRef]
39. Chou, K.-S.; Qu, C. Integrable equations arising from motions of plane curves, II. *J. Nonlinear Sci.* **2003**, *13*, 487–517. [CrossRef]
40. Anco, S.C. Hamiltonian flows of curves in $G/SO(N)$ and vector soliton equations of mKdV and sine-Gordon type. *SIGMA Symmetry Integr. Geom. Methods Appl.* **2006**, *2*, 044. [CrossRef]
41. Sanders, J.A.; Wang, J.P. Integrable systems in n-dimensional Riemannian geometry. *Mosc. Math. J.* **2003**, *3*, 1369–1393. [CrossRef]
42. Terng, C.L.; Thorbergsson, G. Completely integrable curve flows on adjoint orbits. *Results Math.* **2001**, *40*, 286–309. [CrossRef]
43. Terng, C.-L.; Uhlenbeck, C. Schrödinger flows on Grassmannians, Integrable systems, geometry, and topology. *AMS/IP Stud. Adv. Math.* **2006**, *36*, 235–256.
44. Terng, C.L. Geometric Airy curve flow on $\mathbb{R}^n$. *Surv. Differ. Geom.* **2018**, *23*, 277–303. [CrossRef]
45. Mari Beffa, G. Poisson geometry of differential invariants of curves in some non- semi-simple homogeneous spaces. *Proc. Am. Math. Soc.* **2006**, *134*, 779–791. [CrossRef]
46. Mari Beffa, G. Geometric Hamiltonian structures on flat semisimple homogeneous manifolds. *Asian J. Math.* **2008**, *12*, 1–33. [CrossRef]
47. Pinkall, U. Hamiltonian flows on the space of star-shaped curves. *Results Math.* **1995**, *27*, 328–332. [CrossRef]
48. Calini, A.; Ivey, T.; Marí Beffa, G. Integrable flows for starlike curves in centroaffine space. *SIGMA Symmetry Integr. Geom. Methods Appl.* **2013**, *9*, 21. [CrossRef]
49. Terng, C.L.; Wu, Z. N-dimension central affine curve flows. *J. Differ. Geom.* **2019**, *111*, 145–189. [CrossRef]
50. Terng, C.L.; Wu, Z. Isotropic geometric curve flows on $\mathbb{R}^{n+1,n}$. to appear in Comm. Anal. Geom.
51. Calini, A.; Ivey, T.; Mari Beffa, G. Remarks on KdV-type flows on star-shaped curves. *Physica D* **2009**, *238*, 788–797. [CrossRef]
52. Terng, C.L.; Wu, Z. Central affine curve flow on the plane. *J. Fixed Point Theory Appl.* **2013**, *14*, 375–396. [CrossRef]
53. Terng, C.L.; Uhlenbeck, K. The $n \times n$ KdV hierarchy. *J. Fixed Point Theory Appl.* **2011**, *10*, 37–61. [CrossRef]
54. Terng, C.L.; Uhlenbeck, K. Bäcklund transformations and loop group actions. *Comm. Pure Appl. Math.* **2000**, *53*, 1–75. [CrossRef]
55. Magri, F. On the geometry of Soliton equations. *Acta Aplic. Math.* **1995**, *41*, 247–270.
56. Terng, C.L. Soliton equations and differential geometry. *J. Differ. Geom.* **1997**, *45*, 407–445. [CrossRef]
57. Calini, A.; Ivey, T. *Finite-Gap Solutions of the Vortex Filament Equation: Isoperiodic Deformations*. *J. Nonlinear Sci.* **2007**, *17*, 527–567. [CrossRef]

Article

Generalized Navier–Stokes Equations and Dynamics of Plane Molecular Media †

Alexei Kushner [1,2,*] and Valentin Lychagin [3]

1 Faculty of Physics, Lomonosov Moscow State University, Leninskie Gory, 119991 Moscow, Russia
2 Institute of Mathematics and Informatics, Moscow Pedagogical State University, 14 Krasnoprudnaya, 107140 Moscow, Russia
3 V.A. Trapeznikov Institute of Control Sciences of Russian Academy of Sciences, 65 Profsoyuznaya, 117997 Moscow, Russia; valentin.lychagin@uit.no
* Correspondence: kushner@physics.msu.ru
† Dedicated to memory of our friend Rem Gasparov.

Abstract: The first analysis of media with internal structure were done by the Cosserat brothers. Birkhoff noted that the classical Navier–Stokes equation does not fully describe the motion of water. In this article, we propose an approach to the dynamics of media formed by chiral, planar and rigid molecules and propose some kind of Navier–Stokes equations for their description. Examples of such media are water, ozone, carbon dioxide and hydrogen cyanide.

Keywords: Navier–Stokes equations; media with inner structures; plane molecules; water; Levi–Civita connections

Citation: Kushner, A.; Lychagin, V. Generalized Navier–Stokes Equations and Dynamics of Plane Molecular Media. *Symmetry* **2021**, *13*, 288. https://doi.org/10.3390/sym13020288

Academic Editor: Rahmat Ellahi
Received: 10 January 2021
Accepted: 4 February 2021
Published: 8 February 2021

Publisher's Note: MDPI stays neutral with regard to jurisdictional claims in published maps and institutional affiliations.

Copyright: © 2021 by the authors. Licensee MDPI, Basel, Switzerland. This article is an open access article distributed under the terms and conditions of the Creative Commons Attribution (CC BY) license (https://creativecommons.org/licenses/by/4.0/).

1. Introduction

It was the Cosserat brothers, [1], who first analyzed media formed by "rigid microelements", and G. Birkhoff [2] who noted that the classical Navier–Stokes equations give us uncomplete descriptions of water flows (see also [3]). In papers [4,5] the authors gave a general approach to dynamics of media having some inner structure and proposed some generalizations of the Euler and Navier–Stokes equations.

In this paper, we consider the dynamics of media formed by chiral, planar and rigid molecules (we call them CPR-molecules) molecules and propose some kind of Navier–Stokes equations for their description. Recall that a molecule is called planar if it is formed by atoms lying in the same plane and it is chiral and rigid if its symmetry group belongs to $SO(3)$. Hence, we consider a molecule as a rigid body on an oriented plane, the mechanical properties of which are specified by the tensor of inertia.

2. The Configuration Space of a CPR-Molecule

We will assume that all CPR-molecules under consideration have the trivial point symmetry group. Then a position of such a CPR-molecule is defined, up to rotations, by an oriented plane in the three-dimensional space, passing through of the center of mass of the molecule, or by the unit vector perpendicular to this plane or by a point on the unit sphere S^2.

Such molecules include, for example, molecules of ortho-water, i.e., molecules of water with different spins of hydrogen atoms [6].

Let $a \in S^2$ be a fixed point and let $T_a S^2$ be the tangent space to the sphere at the point a. The position of a CPR molecule on the oriented plane is uniquely determined by a rotation, and therefore, by a point on the unit circle on the tangent space $T_a S^2$.

Thus, the configuration space of a planar molecule with a fixed center of mass is the circle bundle of the tangent bundle for the unit two-dimensional sphere. For our goal it is more convenient to use the cotangent bundle $T_a^* S^2$ instead of the tangent one. We denote the circle bundle of the cotangent bundle by N and it will be the configuration space of the molecule.

Let us introduce local coordinates on the configuration space. The position of a rigid body in the space is determined by the position of its center of mass and angular parameters (the Euler angles) showing its position relative to the center of mass. Let us choose a Cartesian coordinate system x, y, z in the space $\mathbb{R}^3$ so that its axes coincide with the principal axes of inertia tensor of the molecule. The metric tensor has the form $g = dx^2 + dy^2 + dz^2$, and the Lie algebra $\mathfrak{so}(3)$ can be represented by the triple of vector fields on $\mathbb{R}^3$:

$$X = z\frac{\partial}{\partial y} - y\frac{\partial}{\partial z}, \quad Y = x\frac{\partial}{\partial z} - z\frac{\partial}{\partial z}, \quad Z = y\frac{\partial}{\partial x} - x\frac{\partial}{\partial y}, \qquad (1)$$

corresponding to the rotations around the axes OX, OY, OZ respectively.

In spherical coordinates ϕ, ψ, r in $\mathbb{R}^3$:

$$x = r\cos\psi \sin\phi, \quad y = r\sin\psi \sin\phi, \quad z = r\cos\phi,$$

where

$$\phi = \arccos\left(\frac{z}{r}\right), \quad \psi = \arctan\left(\frac{y}{x}\right), \quad r = \sqrt{x^2 + y^2 + z^2},$$

vector fields (1) will take the following form:

$$R_X = \sin\psi \frac{\partial}{\partial \phi} + \cot\phi \cos\psi \frac{\partial}{\partial \psi}, \quad R_Y = -\cos\psi \frac{\partial}{\partial \phi} + \cot\phi \sin\psi \frac{\partial}{\partial \psi}, \quad R_Z = -\frac{\partial}{\partial \psi}$$

respectively, and the metric tensor takes the form

$$g = r^2 \left(d\phi^2 + \sin^2\phi \, d\psi^2 \right)$$

in spherical coordinates. The metric g generates the invariant tensor field (the inverse metric)

$$g^{-1} = \frac{1}{r^2}\left(\partial_\phi^2 + \frac{1}{\sin^2\phi} \partial_\psi^2 \right).$$

which defines the metric on the cotangent bundle $T_a^*\mathbb{R}^3$. The metric g^{-1} induces the metric

$$g_1^{-1} = \partial_\phi^2 + \frac{1}{\sin^2\phi} \partial_\psi^2$$

on the cotangent bundle $T_a^* S^2$ of a sphere of unit radius $r = 1$.

Let $q_1 = \phi, q_2 = \psi, p_1, p_2$ be the canonical coordinates on the cotangent bundle $T_a^* S^2$, and

$$\Omega = dq_1 \wedge dp_1 + dq_2 \wedge dp_2$$

be the structure differential 2-form that defines the symplectic structure on $T_a^* S^2$.

Then the Hamiltonian, corresponding to the metric g_1^{-1}, has the form

$$H = p_1^2 + \frac{1}{\sin^2 q_1} p_2^2.$$

The Hamiltonians of the vector fields R_X, R_Y, R_Z are

$$H_X = p_1 \sin q_2 + p_2 \cot q_1 \cos q_2, \quad H_Y = -p_1 \cos q_2 + p_2 \cot q_1 \sin q_2, \quad H_Z = -p_2$$

respectively, and therefore, corresponding Hamiltonian vector fields are

$$X_1 = \sin q_2 \frac{\partial}{\partial q_1} + \cot q_1 \cos q_2 \frac{\partial}{\partial q_2} + p_2 \frac{\cos q_2}{\sin^2 q_1} \frac{\partial}{\partial p_1} - (p_1 \cos q_2 - p_2 \cot q_1 \sin q_2) \frac{\partial}{\partial p_2},$$

$$X_2 = -\cos q_2 \frac{\partial}{\partial q_1} + \cot q_1 \sin q_2 \frac{\partial}{\partial q_2} + p_2 \frac{\sin q_2}{\sin^2 q_1} \frac{\partial}{\partial p_1} - (p_1 \sin q_2 + p_2 \cot q_1 \cos q_2) \frac{\partial}{\partial p_2},$$

$$X_3 = -\frac{\partial}{\partial q_2}.$$

Thus, we have the representation of the Lie algebra $\mathfrak{so}(3)$ by Hamiltonian vector fields X_1, X_2, X_3 with the commutation relations:

$$[X_1, X_2] = X_3, \quad [X_1, X_3] = -X_2, \quad [X_2, X_3] = X_1.$$

It is easy to see these fields are tangential to N: $X_1(H) = X_2(H) = X_3(H) = 0$.

Thus the motion of a molecule relative to its center of mass corresponds to the motion of a point on the level surface N. We take q_1, q_2 and

$$q_3 = \arctan\left(\frac{p_2}{p_1 \sin q_1}\right).$$

as local coordinates on the configuration space $N = \{H = 1\}$.

3. Metric and Levi–Civita Connection, Associated with a CPR-Molecule

The restrictions of the vector fields X_1, X_2, X_3 on the level surface N are

$$E_1 = \sin q_2 \frac{\partial}{\partial q_1} + \cot q_1 \cos q_2 \frac{\partial}{\partial q_2} - \frac{\cos q_2}{\sin q_1} \frac{\partial}{\partial q_3},$$

$$E_2 = -\cos q_2 \frac{\partial}{\partial q_1} + \cot q_1 \sin q_2 \frac{\partial}{\partial q_2} - \frac{\sin q_2}{\sin q_1} \frac{\partial}{\partial q_3},$$

$$E_3 = -\frac{\partial}{\partial q_2}$$

respectively.

Any motion of a CPR-molecule around the center of mass occurs along the trajectory of vector fields, which are linear combinations of vector fields E_1, E_2, E_3.

The basis dual to E_1, E_2, E_3 is formed by the differential 1-forms

$$\Omega_1 = \sin q_2 dq_1 - \cos q_2 \sin q_1 dq_3,$$
$$\Omega_2 = -\cos q_2 dq_1 - \sin q_2 \sin q_1 dq_3,$$
$$\Omega_3 = -dq_2 - \cos q_1 dq_3,$$

such that the Maurer–Cartan relations hold:

$$d\Omega_1 = -\Omega_2 \wedge \Omega_3, \quad d\Omega_2 = \Omega_1 \wedge \Omega_3, \quad d\Omega_3 = -\Omega_1 \wedge \Omega_2.$$

The vector fields E_1, E_2, E_3 and the differential 1-forms $\Omega_1, \Omega_2, \Omega_3$ give us the base (over $\mathbb{R}$) in the space of left-invariant vector fields and correspondingly left invariant differential 1-forms on the configuration space. Moreover, any left invariant tensor on N is a linear combination of tensor products of these vector fields and differential 1-forms with constant coefficients.

Let Λ be the inertial tensor of a molecule. It can be considered as a positive self adjoint operator acting on the Lie algebra $\mathfrak{so}(3)$. Let positive numbers $\lambda_1, \lambda_2, \lambda_3$ be eigenvalues of Λ. The inertia tensor defines the metric tensor on the Lie algebra $\mathfrak{so}(3)$:

$$g_\lambda = \frac{1}{2}\left(\lambda_1 \Omega_1^2 + \lambda_2 \Omega_2^2 + \lambda_3 \Omega_3^2\right),$$

where Ω_i^2 are the symmetric squares of the 1-forms. The inertia tensor has the following coordinate representation:

$$\begin{aligned}g_\lambda =& (\lambda_1 \sin^2 q_2 + \lambda_2 \cos^2 q_2) dq_1^2 + \lambda_3 dq_2^2 \\ &+ (\lambda_1 \sin^2 q_1 \cos^2 q_2 + \lambda_2 \sin^2 q_1 \sin^2 q_2 + \lambda_3 \cos^2 q_1) dq_3^2 \\ &+ 2(\lambda_2 - \lambda_1) \sin(q_2) \cos q_2 \sin q_1 dq_1 \cdot dq_3 \\ &+ 2\lambda_3 \cos q_1 dq_2 \cdot dq_3.\end{aligned}$$

Here the dot $\cdot$ means the operation of symmetric multiplication.

Let ∇^λ be the Levi–Civita connection [7] associated with the metric g_λ and ∇_i^λ be the covariant derivative along vector field E_i. Then

$$\nabla_i^\lambda(E_j) = \sum_k \Gamma_{ij}^k E_k,$$

where Γ_{ij}^k are the Christoffel symbols. Direct calculations show that

$$\Gamma_{12}^3 = \frac{\lambda - \lambda_1}{\lambda_3}, \quad \Gamma_{21}^3 = -\frac{\lambda - \lambda_2}{\lambda_3},$$
$$\Gamma_{23}^1 = \frac{\lambda - \lambda_2}{\lambda_1}, \quad \Gamma_{32}^1 = -\frac{\lambda - \lambda_3}{\lambda_1},$$
$$\Gamma_{31}^2 = \frac{\lambda - \lambda_3}{\lambda_2}, \quad \Gamma_{13}^2 = -\frac{\lambda - \lambda_1}{\lambda_2}. \tag{2}$$

where

$$\lambda = \frac{\lambda_1 + \lambda_2 + \lambda_3}{2}.$$

All other Christoffel symbols equal to zero.

4. Metric Associated with the Media

Let $\mathbb{R}^3$ be the 3-dimensional Euclidian space, endowed with the standard metric tensor g. Consider a medium, formed by CPR-molecules filling a region $D \subset \mathbb{R}^3$. The configuration space for this type of media is the $SO(3)$-bundle $\pi : \Phi \longrightarrow D$, where $\Phi = N \times D$.

The group $SO(3)$ acts in the natural way on fibers of the projection π and we will continue to use notation E_1, E_2, E_3 for the induced vertical vector fields on Φ. These fields form the basis in the module of vertical vector fields on Φ, and accordingly differential 1-forms $\Omega_1, \Omega_2, \Omega_3$ define the dual basis in the space of differential forms on N.

The medium is also characterized by a $SO(3)$-connection in the bundle π, (see [4,5]). We call this connection the media connection and denote it by ∇^μ. The media connection allows us to compare molecules at different points of the region D.

The connection ∇^μ depends on the properties of the medium and establishes a relation between the translational motion of the molecule and its motion relative to the center of mass. Such a relation can be caused, for example, by physical inhomogeneity of space or by the presence of effects on the environment. Let us show how it can be defined (see [5]). The connection form ω we will consider as a matrix

$$\omega = \begin{Vmatrix} 0 & -\omega_3 & \omega_2 \\ \omega_3 & 0 & -\omega_1 \\ -\omega_2 & \omega_1 & 0 \end{Vmatrix}$$

where $\omega_1, \omega_2, \omega_3$ are differential 1-forms on D. In other words, connection ∇^μ shows that a molecule is subject to rotation along vector $(\omega_1(X)E_1 + \omega_2(X)E_2 + \omega_3(X)E_3)$ on the angle

$$\varphi = \sqrt{\omega_1(X)^2 + \omega_2(X)^2 + \omega_3(X)^2}$$

when we transport it on the vector X in D.

Let (x_1, x_2, x_3) be the standard Euclidian coordinates on D and $(\partial_1, \partial_2, \partial_3)$ and (d_1, d_2, d_3) be the corresponding frame and coframe respectively. Here $\partial_i = \dfrac{\partial}{\partial x_i}$ and $d_i = dx_i$. In these coordinates we have

$$\omega = \begin{Vmatrix} 0 & -\omega_{31} & \omega_{21} \\ \omega_{31} & 0 & -\omega_{11} \\ -\omega_{21} & \omega_{11} & 0 \end{Vmatrix} d_1 + \begin{Vmatrix} 0 & -\omega_{32} & \omega_{22} \\ \omega_{32} & 0 & -\omega_{12} \\ -\omega_{22} & \omega_{12} & 0 \end{Vmatrix} d_2 + \begin{Vmatrix} 0 & -\omega_{33} & \omega_{23} \\ \omega_{33} & 0 & -\omega_{13} \\ -\omega_{23} & \omega_{13} & 0 \end{Vmatrix} d_3.$$

This connection allows us to split tangent spaces $T_b\Phi$ into the direct sum

$$T_b\Phi = V_b \bigoplus H_b,$$

where V_b is the vertical part with basis $E_{1,b}, E_{2,b}, E_{3,b}$, and the horizontal space H_b is generated by the following vector fields:

$$\partial_1 - \omega_{11}E_1 - \omega_{21}E_2 - \omega_{31}E_3,$$
$$\partial_2 - \omega_{12}E_1 - \omega_{22}E_2 - \omega_{32}E_3,$$
$$\partial_3 - \omega_{13}E_1 - \omega_{23}E_2 - \omega_{33}E_3.$$

The horizontal distribution

$$H : \Phi \ni b \longrightarrow H_b \subset T_b\Phi$$

could be also defined as the kernel of the following system of differential 1-forms on Φ:

$$\theta_1 = \Omega_1 + \omega_{11}d_1 + \omega_{12}d_2 + \omega_{13}d_3,$$
$$\theta_2 = \Omega_2 + \omega_{21}d_1 + \omega_{22}d_2 + \omega_{23}d_3,$$
$$\theta_3 = \Omega_3 + \omega_{31}d_1 + \omega_{32}d_2 + \omega_{33}d_3.$$

Define a metric g^μ on the manifold Φ as a direct sum of the metric g_λ on the vertical space V and the standard metric $g_0 = dx_1^2 + dx_2^2 + dx_3^2$ on the horizontal space H:

$$g^\mu = \frac{1}{2}\sum_{i=1}^{3}\left(\lambda_i\Omega_i^2 + d_i^2\right).$$

Note that the frame $(E_1, E_2, E_3, \partial_1 - \omega(\partial_1), \partial_2 - \omega(\partial_2), \partial_3 - \omega(\partial_3))$ and the coframe $(\Omega_1, \Omega_2, \Omega_3, d_1, d_2, d_3)$ are dual and their elements are pairwise orthogonal with respect to the metric g^μ.

5. Levi–Civita Connection Associated with the Homogeneous Media

A media is said to be homogeneous if components of the connection form ω and the inertia tensor Λ are constants. Below we consider only homogeneous media.

Let ∇ be the Levi–Civita connection on the configuration space Φ associated with the metric g^μ.

For basic vector fields E_i and ∂_j, where $i,j = 1,2,3$, we have the following commutation relations:

$$[\partial_i, \partial_j] = [\partial_i, E_j] = 0, [E_1, E_2] = E_3, [E_1, E_3] = -E_2, [E_2, E_3] = E_1. \quad (3)$$

Therefore, the Levi–Civita connection ∇ on the configuration space Φ associated with the metric g^μ and homogeneous media has the form wherein the non trivial Christoffel symbols are given by Formula (2).

The operator of the covariant differential d_∇ associated with the Levi–Civita connection acts on the basis vectors as follows:

$$d_\nabla(\partial_i) = 0 \quad (i=1,2,3),$$
$$d_\nabla(E_1) = \Gamma_{31}^2 E_2 \otimes \Omega_3 + \Gamma_{21}^3 E_3 \otimes \Omega_2,$$
$$d_\nabla(E_2) = \Gamma_{32}^1 E_1 \otimes \Omega_3 + \Gamma_{12}^3 E_3 \otimes \Omega_1,$$
$$d_\nabla(E_3) = \Gamma_{23}^1 E_1 \otimes \Omega_2 + \Gamma_{13}^2 E_2 \otimes \Omega_1,$$

and on the basic differential 1-forms:

$$d_\nabla(d_i) = 0 \quad (i=1,2,3).$$
$$d_\nabla(\Omega_1) = -\Gamma_{32}^1 \Omega_2 \otimes \Omega_3 - \Gamma_{23}^1 \Omega_3 \otimes \Omega_2;$$
$$d_\nabla(\Omega_2) = -\Gamma_{31}^2 \Omega_1 \otimes \Omega_3 - \Gamma_{13}^2 \Omega_3 \otimes \Omega_1;$$
$$d_\nabla(\Omega_3) = -\Gamma_{21}^3 \Omega_1 \otimes \Omega_2 - \Gamma_{12}^3 \Omega_2 \otimes \Omega_1.$$

6. Thermodynamic State of Media

The motion of the medium will be described by the trajectories of vector fields on the configuration space, which preserve the bundle $\pi : \Phi \longrightarrow D$,

$$U = \sum_{i=1}^{3}(X_i(t,x)\partial_i + Y_i(t,x,q)E_i).$$

The tensor $\Delta = d_\nabla U$ is called the rate of deformation tensor [4]. Following [5,8], this tensor bears an enormous thermodynamic quantity. Using properties of covariant derivative we get:

$$\Delta = \sum_{i,j=1}^{3}\left(\partial_j(X_i)\partial_i \otimes d_j + \partial_j(Y_i)E_i \otimes d_j + E_j(Y_i)E_i \otimes \Omega_j\right) + \sum_{i=1}^{3}Y_i d_\nabla(E_i).$$

The matrix corresponding to the tensor Δ has the block structure:

$$\Delta = \begin{Vmatrix} \Delta_H & 0 \\ \Delta_{HV} & \Delta_V \end{Vmatrix}$$

where

$$\Delta_H = \begin{Vmatrix} \partial_1(X_1) & \partial_2(X_1) & \partial_3(X_1) \\ \partial_1(X_2) & \partial_2(X_2) & \partial_3(X_1) \\ \partial_1(X_3) & \partial_2(X_3) & \partial_3(X_1) \end{Vmatrix}, \quad \Delta_{HV} = \begin{Vmatrix} \partial_1(Y_1) & \partial_2(Y_1) & \partial_3(Y_1) \\ \partial_1(Y_2) & \partial_2(Y_2) & \partial_3(Y_2) \\ \partial_1(Y_3) & \partial_2(Y_3) & \partial_3(Y_3) \end{Vmatrix},$$

$$\Delta_V = \begin{Vmatrix} E_1(Y_1) & E_2(Y_1) + \Gamma_{23}^1 Y_3 & E_3(Y_1) + \Gamma_{32}^1 Y_2 \\ E_1(Y_2) + \Gamma_{13}^2 Y_3 & E_2(Y_2) & E_3(Y_2) + \Gamma_{31}^2 Y_1 \\ E_1(Y_3) + \Gamma_{12}^3 Y_2 & E_2(Y_3) + \Gamma_{21}^3 Y_1 & E_3(Y_3) \end{Vmatrix}.$$

The metric tensor g^μ defines the canonical isomorphism between vector fields and differential 1-forms on Φ: a vector field X on Φ is associated with the differential 1-form $X^\flat$ on Φ and vice versa: with any differential 1-form ω on Φ we can associate the vector field $\omega^\flat$. We have

$$E_i^\flat = \lambda_i \Omega_i, \quad \Omega_i^\flat = \frac{1}{\lambda_i} E_i, \quad \partial_i^\flat = d_i, \quad d_i^\flat = \partial_i \quad i = 1, 2, 3.$$

For fields of endomorphisms we put $(X \otimes \omega)^\flat = \omega^\flat \otimes X^\flat$. Then we have:

$$\Delta^\flat = \sum_{i,j=1}^{3} \left(\partial_j(X_i) \partial_j \otimes d_i + \lambda_i \partial_j(Y_i) \partial_j \otimes \Omega_i + \frac{\lambda_i}{\lambda_j} E_j(Y_i) E_j \otimes \Omega_i \right) + \sum_{i=1}^{3} Y_i d_\nabla^\flat(E_i),$$

where

$$d_\nabla^\flat(E_1) = \frac{\lambda_2}{\lambda_3} \Gamma_{31}^2 E_3 \otimes \Omega_2 + \frac{\lambda_3}{\lambda_2} \Gamma_{21}^3 E_2 \otimes \Omega_3,$$

$$d_\nabla^\flat(E_2) = \frac{\lambda_1}{\lambda_3} \Gamma_{32}^1 E_3 \otimes \Omega_1 + \frac{\lambda_3}{\lambda_1} \Gamma_{12}^3 E_1 \otimes \Omega_3,$$

$$d_\nabla^\flat(E_3) = \frac{\lambda_1}{\lambda_2} \Gamma_{23}^1 E_2 \otimes \Omega_1 + \frac{\lambda_2}{\lambda_1} \Gamma_{13}^2 E_1 \otimes \Omega_2.$$

Let σ be a stress tensor which can be considered as a field of endomorphisms on the tangent bundle. Let $\sigma^\flat$ be field of endomorphisms on the tangent bundle $T\Phi$ dual to σ. The following differential 1-form

$$\psi = ds - \frac{1}{T}(d\epsilon - \text{Tr}(\sigma^\flat d\Delta) - \zeta d\rho)$$

defines the contact structure on the thermodynamic phase space of medium

$$\Psi = \mathbb{R}^5 \times \text{End}(T^*\Phi) \times \text{End}(T\Phi)$$

with coordinates $s, T, \epsilon, \zeta, \rho, \sigma, \Delta$. Here ρ, s, ϵ are the densities of the media, entropy and inner energy respectively, T and ζ are temperature and chemical potential respectively (see [4,9]). Since $\dim \text{End}(T^*\Phi) = \dim \text{End}(T\Phi) = 9$ we get $\dim \Psi = 23$. Legendrian manifolds L we call thermodynamic states of the media, in given case $\dim L = 11$.

Consider only those thermodynamic states for which T, ρ, Δ can be selected as coordinates.

Let $h = \epsilon - Ts$ be the density of Helmholtz free energy. Then we have the following description of the Legendrian manifold:

$$s = h_T, \quad \sigma = h_\Delta, \quad \zeta = h_\rho.$$

In this case when the media is Newtonian and satisfies the Hooke law, the Helmholtz free energy is a quadratic function of Δ and has the form [4]:

$$h = \frac{1}{2} \left(a_1 \text{Tr}(\Delta^2) + a_2 \text{Tr}(\Delta \Delta^\flat) + a_3 (\text{Tr}\Delta)^2 + a_4 (\text{Tr}(\Delta\Pi))^2 + a_5 \text{Tr}(\Delta^\flat \Delta \Pi) + a_6 \text{Tr}(\Delta \Delta^\flat \Pi) \right)$$
$$+ b_1 \text{Tr}(\Delta) + b_2 \text{Tr}(\Delta\Pi) + c,$$

where Π is the projector to the vertical component and $a_1, \ldots, a_6, b_1, b_2, c$ are some functions of ρ, T.

In this case the stress tensor has the form

$$\sigma = a_1 \Delta^\flat + a_2 \Delta + (a_3 \text{Tr}(\Delta) + b_1) + (a_4 \text{Tr}(\Delta\Pi) + b_2)\Pi + a_5 \Delta\Pi + a_6 \Pi\Delta.$$

7. Divergence of Operator Fields

In order to write the momentum conservation law, we need a notation of the divergence of the endomorphism field on Φ (see [4]). The covariant differential of an endomorphism field $A \in T\Phi \otimes T^*\Phi$ is the tensor field $d_\nabla A \in T\Phi \otimes T^*\Phi \otimes T^*\Phi$. Taking the contraction, the first and third indices of this tensor, we get the differential 1-form which is called the divergence of the operator field A:

$$\mathrm{div}\, A = c_{1,3}(d_\nabla A).$$

For decomposable fields $A = X \otimes \omega$, where X is a vector field and ω is a differential 1-form, the divergence operator can be calculated by the following formula:

$$\mathrm{div}(X \otimes \omega) = (\mathrm{div}\, X)\omega + \nabla_X(\omega). \tag{4}$$

Note that
$$\mathrm{div}(fX \otimes \omega) = f\,\mathrm{div}(X \otimes \omega) + X(f)\omega.$$

The following formula gives an explicit form of the divergence operator. If the operator has the form

$$A = \sum_{i,j=1}^{3} \left(a_{ij} \partial_i \otimes d_j + b_{ij} E_i \otimes \Omega_j \right),$$

then

$$\mathrm{div}\, A = \sum_{i,j=1}^{3} \partial_i(a_{ij}) d_j + \sum_{\sigma \in S_3} \left(E_{\sigma(2)}\left(b_{\sigma(2)\sigma(1)} \right) - \Gamma^{\sigma(3)}_{\sigma(2)\sigma(1)} b_{\sigma(2)\sigma(3)} \right) \Omega_{\sigma(1)}. \tag{5}$$

Here a_{ij}, b_{ij} are functions on Φ.

For endomorphisms that are linear combinations of tensors $\partial_i \otimes \Omega_j$ and $E_i \otimes d_j$, the divergence is zero.

8. Conservation Laws

8.1. The Momentum Conservation Law

Let
$$\frac{d}{dt} = \frac{\partial}{\partial t} + \nabla_U$$

be a material derivative; then [4] the momentum conservation law, or Navier–Stocks equation, takes the form

$$\rho \frac{dU}{dt} = (\mathrm{div}\,\sigma)^\flat + F,$$

or, equivalently,

$$\rho \left(\frac{\partial U}{\partial t} + \nabla_U(U) \right) = (\mathrm{div}\,\sigma)^\flat + F. \tag{6}$$

Here F is a density of exterior volume forces.

Let us calculate the covariant derivative $\nabla_U(U)$. We have

$$\nabla_U(U) = \sum_{j=1}^{3} \left(X_j \nabla_{\partial_j}(U) + Y_j \nabla_{E_j}(U) \right)$$

and
$$\nabla_{\partial_i}(\partial_j) = \nabla_{\partial_i}(E_j) = 0,$$
$$\nabla_{E_i}(\partial_j) = \nabla_{E_i}(E_i) = 0,$$
$$\nabla_{E_1}(E_2) = \Gamma_{12}^3 E_3, \quad \nabla_{E_1}(E_3) = \Gamma_{13}^2 E_2,$$
$$\nabla_{E_2}(E_1) = \Gamma_{21}^3 E_3, \quad \nabla_{E_2}(E_3) = \Gamma_{23}^1 E_1,$$
$$\nabla_{E_3}(E_1) = \Gamma_{31}^2 E_2, \quad \nabla_{E_3}(E_2) = \Gamma_{32}^1 E_1.$$

Therefore,
$$\nabla_{\partial_j}(U) = \sum_{i=1}^{3} \partial_j(X_i)\partial_i + \partial_j(Y_i)E_i, \quad j = 1,2,3;$$
$$\nabla_{E_1}(U) = E_1(Y_1)E_1 + (E_1(Y_2) + \Gamma_{13}^2 Y_3)E_2 + (E_1(Y_3) + \Gamma_{12}^3 Y_2)E_3$$
$$\nabla_{E_2}(U) = (E_2(Y_1) + \Gamma_{23}^1 Y_3)E_1 + E_2(Y_2)E_2 + (E_2(Y_3) + \Gamma_{21}^3 Y_1)E_3,$$
$$\nabla_{E_3}(U) = (E_3(Y_1) + \Gamma_{32}^1 Y_2)E_1 + (E_3(Y_2) + \Gamma_{31}^2 Y_1)E_2 + E_3(Y_3)E_3,$$

and
$$\nabla_U(U) = \sum_{i,j=1}^{3} \left(X_j \partial_j(X_i)\partial_i + (X_j \partial_j(Y_i) + Y_j E_j(Y_i))E_i \right)$$
$$+ (\Gamma_{23}^1 + \Gamma_{32}^1)Y_2 Y_3 E_1 + (\Gamma_{13}^2 + \Gamma_{31}^2)Y_1 Y_3 E_2 + (\Gamma_{12}^3 + \Gamma_{21}^3)Y_1 Y_2 E_3$$

Moreover, we have
$$\nabla_{\partial_i}(d_j) = \nabla_{\partial_i}(\Omega_j) = \nabla_{E_i}(d_j) = \nabla_{E_i}(\Omega_i) = 0 \quad i,j = 1,2,3;$$
$$\nabla_{E_1}(\Omega_2) = -\Gamma_{13}^2 \Omega_3, \quad \nabla_{E_1}(\Omega_3) = -\Gamma_{12}^3 \Omega_2,$$
$$\nabla_{E_2}(\Omega_1) = -\Gamma_{23}^1 \Omega_3, \quad \nabla_{E_2}(\Omega_3) = -\Gamma_{21}^3 \Omega_1,$$
$$\nabla_{E_3}(\Omega_1) = -\Gamma_{32}^1 \Omega_2, \quad \nabla_{E_3}(\Omega_2) = -\Gamma_{31}^2 \Omega_1.$$

The momentum conservation law takes the form:
$$\begin{cases} \rho\left(\partial_t(X_i) + \sum_{j=1}^{3} X_j \partial_j(X_i)\right) = ((\text{div}\sigma)^\flat + F)_{d_i} \quad i = 1,2,3; \\ \\ \rho\left(\partial_t(Y_1) + \sum_{j=1}^{3} (X_j \partial_j(Y_1) + Y_j E_j(Y_1)) + (\Gamma_{23}^1 + \Gamma_{32}^1)Y_2 Y_3\right) = ((\text{div}\sigma)^\flat + F)_{\Omega_1}; \\ \\ \rho\left(\partial_t(Y_2) + \sum_{j=1}^{3} (X_j \partial_j(Y_2) + Y_j E_j(Y_2)) + (\Gamma_{13}^2 + \Gamma_{31}^2)Y_1 Y_3\right) = ((\text{div}\sigma)^\flat + F)_{\Omega_2}; \\ \\ \rho\left(\partial_t(Y_3) + \sum_{j=1}^{3} (X_j \partial_j(Y_3) + Y_j E_j(Y_3)) + (\Gamma_{12}^3 + \Gamma_{21}^3)Y_1 Y_2\right) = ((\text{div}\sigma)^\flat + F)_{\Omega_3}; \end{cases} \quad (7)$$

where $((\text{div}\sigma)^\flat + F)_\omega$ is the coefficient of the right-hand side of (6) at the differential 1-form ω. The divergence div can be found by Formula (5). We do not give explicit formulas due to their cumbersomeness.

Equation (7) is the Navier–Stokes equation for the CPR-molecular medium.

8.2. The Mass Conservation Law

The mass conservation law has the form

$$\frac{\partial \rho}{\partial t} + U(\rho) + \rho \operatorname{div} U = 0,$$

where

$$\operatorname{div} U = \operatorname{Tr}(d_\nabla U) = \operatorname{Tr} \Delta = \sum_{i=1}^{3} \left(\frac{\partial X_i}{\partial x_i} + E_i(Y_i) \right).$$

The coordinate representation of this equation is as follows:

$$\frac{\partial \rho}{\partial t} + \sum_{i=1}^{3} \left(X_i \frac{\partial \rho}{\partial x_i} + Y_i E_i(\rho) \right) + \rho \sum_{i=1}^{3} \left(\frac{\partial X_i}{\partial x_i} + E_i(Y_i) \right) = 0. \qquad (8)$$

8.3. The Energy Conservation Law

We suppose that there are no internal energy sources in the media. Then the conservation law of energy has the form (see [5])

$$\frac{\partial \epsilon}{\partial t} + \epsilon \operatorname{div}(U) - \operatorname{div}(\chi \operatorname{grad}(T)) + \operatorname{Tr}(\sigma^\flat \Delta) = 0. \qquad (9)$$

Here $\chi \in \operatorname{End} T\Phi$ is the thermal conductivity of the medium.

Equations (7)–(9), and the equation of thermodynamic states of the media

$$s = h_T, \quad \sigma = h_\Delta, \quad \xi = h_\rho$$

describe the motion and thermodynamics of the CPR-molecular medium.

Author Contributions: Conceptualization, V.L.; Formal analysis, A.K.; Investigation, V.L. and A.K.; Writing—original draft, A.K. Both authors have read and agreed to the published version of the manuscript.

Funding: This work was partially supported by the Russian Foundation for Basic Research (project 18-29-10013).

Conflicts of Interest: The authors declare no conflict of interest.

References

1. Cosserat, E.; Cosserat, F. *Théorie des Corps Déformables*; A. Hermann et Fils: Paris, France, 1909.
2. Birkhoff, G. *Hydrodynamics: A study in Logic, Fact, and Similitude*, 2nd ed.; Princeton University Press: Princeton, NJ, USA, 1960; 184p.
3. Altenbach, H.; Maugin, G.A.; Verichev, N. (Eds.) *Mechanics of Generalized Continua*; Springer: Berlin/Heidelberg, Germany, 2011.
4. Duyunova, A.; Lychagin, V.; Tychkov, S. Continuum mechanics of media with inner structures. *Differ. Geom. Appl.* **2021**, *74*, 101703. [CrossRef]
5. Lychagin, V. Euler equations for Cosserat media. *Glob. Stoch. Anal.* **2020**, *7*, 197–208.
6. Kilaj, A.; Gao, H.; Rösch, D.; Küpper, J.; Willitsch, S. Observation of different reactivities of para and ortho-water towards trapped diazenylium ions. *Nat. Commun.* **2018**, *9*, 2096. [CrossRef] [PubMed]
7. Chern, S.S.; Chen W.H.; Lam, K.S. *Lectures on Differential Geometry*; Series on University Mathematics, 1; World Scientific Publishing Co., Inc.: River Edge, NJ, USA, 1999.
8. Lychagin, V. Contact Geometry, Measurement, and Thermodynamics. In *Nonlinear PDEs, Their Geometry and Applications*; Kycia, R., Schneider, E., Ulan, M., Eds.; Birkhäuser: Cham, Switzerland, 2019; pp. 3–52.
9. Gibbs, J.W. A Method of Geometrical Representation of the Thermodynamic Properties of Substances by Means of Surfaces. *Trans. Connect. Acad.* **1873**, *1*, 382–404.

Article

On Monotonic Pattern in Periodic Boundary Solutions of Cylindrical and Spherical Kortweg–De Vries–Burgers Equations

Alexey Samokhin

Trapeznikov Institute of Control Scienties, Russian Academy of Sciences, 65 Profsoyuznaya Street, 117997 Moscow, Russia; a.samohin@mstuca.aero or samohinalexey@gmail.com

Abstract: We studied, for the Kortweg–de Vries–Burgers equations on cylindrical and spherical waves, the development of a regular profile starting from an equilibrium under a periodic perturbation at the boundary. The regular profile at the vicinity of perturbation looks like a periodical chain of shock fronts with decreasing amplitudes. Further on, shock fronts become decaying smooth quasi-periodic oscillations. After the oscillations cease, the wave develops as a monotonic convex wave, terminated by a head shock of a constant height and equal velocity. This velocity depends on integral characteristics of a boundary condition and on spatial dimensions. In this paper the explicit asymptotic formulas for the monotonic part, the head shock and a median of the oscillating part are found.

Keywords: Korteweg–de Vries–Burgers equation; cylindrical and spherical waves; saw-tooth solutions; periodic boundary conditions; head shock wave

MSC: 35Q53; 35B36

1. Introduction

The well known Korteweg–de Vries (KdV)–Burgers equation for flat waves is of the form
$$u_t = -2uu_x + \varepsilon^2 u_{xx} + \delta u_{xxx}. \tag{1}$$

Its cylindrical and spherical analogues are
$$u_t + \frac{1}{2t}u = -2uu_x + \varepsilon^2 u_{xx} + \delta u_{xxx}, \tag{2}$$

and
$$u_t + \frac{1}{t}u = -2uu_x + \varepsilon^2 u_{xx} + \delta u_{xxx}. \tag{3}$$

respectively, see [1,2].

The behavior of solutions of the Korteweg–de Vries (KdV) and KdV–Burgers equations was intensively studied for about fifty years. However, these equations remain subjects of various recent studies, mostly in the case of flat waves in one spatial dimension [3–7]. However, cylindrical and spherical waves have a variety of applications (e.g., waves generated by a downhole vibrator), and are studied much less.

We consider the initial value boundary problem (IVBP) for the KdV–Burgers equation on a finite interval:
$$u(x,0) = f(x),\ u(a,t) = l(t),\ u(b,t) = L(t),\ u_x(b,t) = R(t),\ x \in [a,b]. \tag{4}$$

In the case $\delta = 0$ (that is, for the Burgers equation), it becomes

$$u(x,0) = f(x), \ u(a,t) = l(t), \ u(b,t) = R(t), \ x \in [a,b]. \tag{5}$$

The case of the boundary conditions $u(a,t) = A\sin(\omega t)$, $u(b,t) = 0$ and the related asymptotics are of a special interest here. For numerical modeling we use $x \in [0,b]$ instead of $\mathbb{R}^+$ for appropriately large b.

For the flat wave Burgers equation ($\delta = 0$) the resulting asymptotic profile looks like a periodical chain of shock fronts with a decreasing amplitude (weak breaks or sawtooth waves). If dispersion is non-zero, each wavefront ends with high-frequency micro-oscillations. Further from the oscillator, shock fronts become decaying smooth quasi-periodic oscillations. After the oscillations cease, the wave develops as a constant height and velocity shock. It almost coincides with a traveling wave solution (TWS) of the Burgers equation [8,9].

A traveling wave solution is the solution of the form $u = u(x + Vt)$. Such a solution travels with a constant velocity V along the x-axis, unchanged in its form. The well-known examples are solitons for KdV, shock waves for the Burgers equation. For the existence of TWS for all values of the parameter V it is necessary that an equation has Galillean symmetry.

In the case $\delta = 0$, the Burgers equation has traveling wave solutions, vanishing at $x \to +\infty$. They are given by the formula [10]

$$u_B(x,t) = \frac{V}{2}\left[1 - \tanh\left(\frac{V}{2\varepsilon^2}(x - Vt + s)\right)\right]; \tag{6}$$

it is used below.

Our aim is to obtain a similar description of a long-time asymptote for cylindrical and spherical waves with periodic boundary conditions. We demonstrate that, in the case of the above IVBP, the perturbation of the equilibrium state for Equations (2) and (3) ultimately takes a form similar to this shock.

This paper is organized as follows. In Section 2, we demonstrate graphs of our numerical experiments for cylindrical/spherical Burgers/KdV–Burgers equations for different combinations to show their the common patterns. In particular we demonstrate that, after the oscillation cease, a solution becomes a monotonic convex line terminated by a head shock.

In Section 3, we find symmetries to Equations (2) and (3). No Galilean symmetry is found, so no real TWS exists. Then equations are brought to a conservation law form, which is later used to obtain rough estimates for the median parameters of the solution. This rough estimate becomes exact for constant boundary conditions, and in Section 4, a very close asymptote for said solution is found in self-similar or homothetic form $u = u(x/t)$.

Yet, at the head shock this asymptotic is unsatisfactory. This head shock moves in unchanged form and with numerically equal velocity and amplitude—exactly as the Burgers traveling wave solution does. In Section 5, using a simple combination of a self-similar approximation and the Burgers traveling wave solution, we obtain the compact closed form approximation. It coincides with a solution in its monotonic part; and this approximation correctly represents the median of the solution in its oscillating part. The quality of the approximation is verified numerically. Connection between the velocity of the solution's head shock and the median value at the start is obtained.

In the section "Conclusions" we formulate main result and discuss the remaining open questions.

2. Typical Examples

Here we demonstrate typical graphs for cylindrical and spherical Burgers waves (see Figures 1 and 2) and for cylindrical and spherical KdV–Burgers (Figures 3 and 4).

We obtained these graphs using the *Maple PDETools package*. The mode of operation used was the default Euler method, which is a centered implicit scheme.

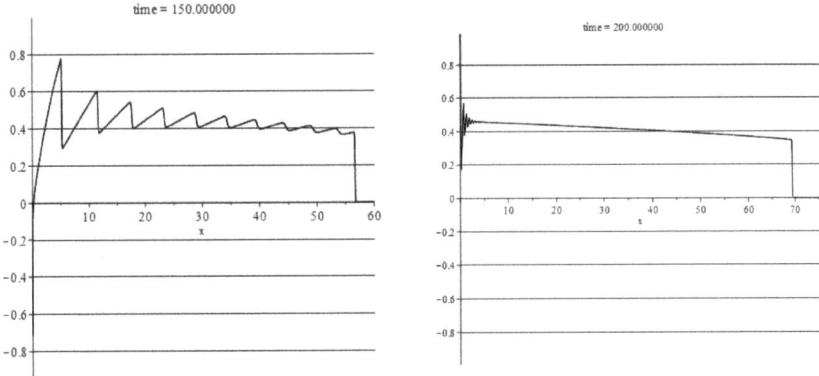

Figure 1. Cylindrical Burgers. $\varepsilon = 0.1$, **Left**: $u_0 = \sin t$, $t = 150$. **Right**: $u_0 = \sin 10t$, $t = 200$.

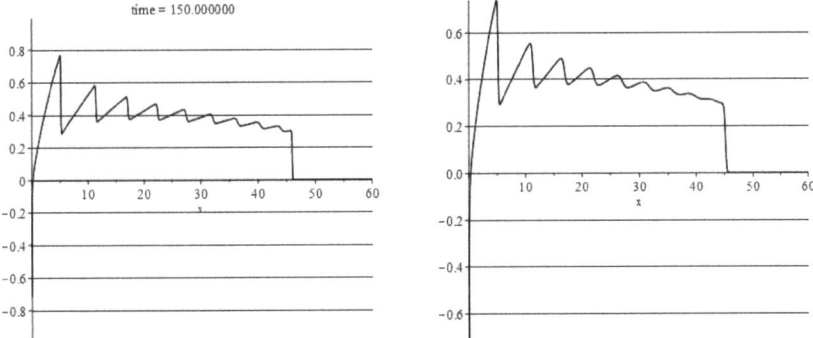

Figure 2. Spherical Burgers, $u_0 = \sin t$. **Left**: $\varepsilon = 0.1$, $t = 150$. **Right**: $\varepsilon^2 = 0.3$, $t = 150$.

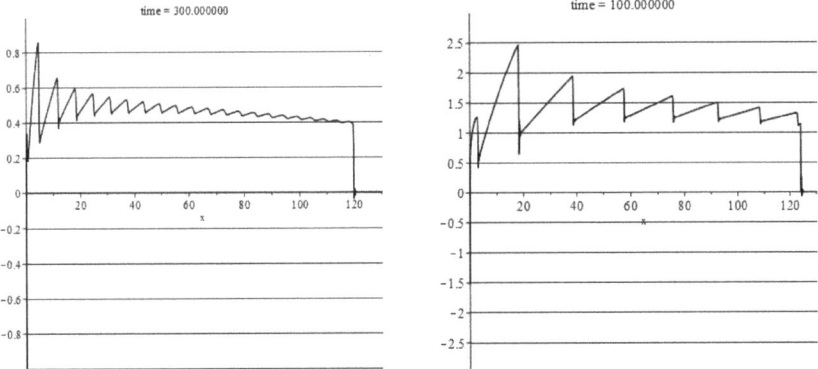

Figure 3. Cylindrical KdV–Burgars. **Left**: $u_0 = \sin t$, $t = 300$, $\varepsilon = 0.1$, $\delta = 0.001$. **Right**: $u_0 = 3\sin t$, $t = 100$, $\varepsilon = 0.1$, $\delta = 0.001$.

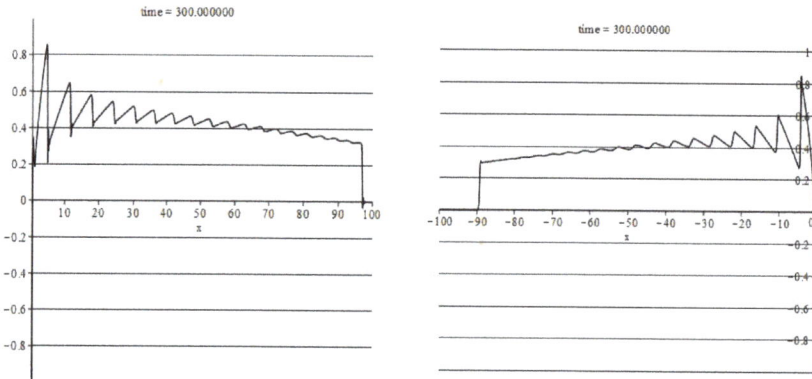

Figure 4. Spherical KdV–Burgers, $u_0 = \sin t$. **Left**: $t = 300$, $\varepsilon = 0.1, \delta = 0.001$. **Right**: $u \leftrightarrow -u$, $t = 300$, $\varepsilon^2 = 0.02$, $\delta = 0.001$, $\varepsilon^2 = 0.2$

The solution usually starts with a periodical chain of shock fronts with decreasing amplitudes (sawtooth waves). This weak breaks/sawtooth profile is inherent to periodic waves in dissipative media. Sawtooth waves, their decay, amplitudes, width, etc., were intensively studied in 1970 (see [1,2]) and later. One can also see a common pattern, previously not described, emerging on these figures. After the decay of initial oscillations, graphs become monotonic declining convex lines, terminated by a shock. Recall that for flat waves this monotonic part almost coincides with a constant height traveling wave solution of Burgers equation [7]. The new feature of convex declining lines is caused by the space divergence. We obtain an analytical description of this pattern below.

3. Symmetries and Conservation Laws

3.1. Symmetries

Since cylindrical and spherical equations explicitly depend on time, their stock of symmetries is scarce. For the algorithm of symmetry calculations, see [11]. We found that the algebras of classical symmetries are generated by the following vector fields:

$$X = \frac{\partial}{\partial x}, \quad Y = x\frac{\partial}{\partial x} + 2t\frac{\partial}{\partial t} - u\frac{\partial}{\partial u}, \quad Z = \sqrt{t}\frac{\partial}{\partial x} + \frac{1}{4\sqrt{t}}\frac{\partial}{\partial u}, \quad W = \ln(t)\frac{\partial}{\partial x} + \frac{1}{2t}\frac{\partial}{\partial u}.$$

This list does not contain the Galilean symmetry, so no real traveling wave solution exists.

In particular, symmetry algebra for:

- Cylindrical Burgers is generated by X, Y, Z;
- Cylindrical KdV–Burgers is generated by X, Z;
- Spherical Burgers is generated by X, Y, W;
- Spherical KdV–Burgers is generated by X, W.

3.2. Conservation Laws

First rewrite Equations (1)–(3) into an appropriate conservation law form

$$[t^n \cdot u]_t = [t^n \cdot (-u^2 + \varepsilon^2 u_x + \delta u_{xx})]_x, \tag{7}$$

where $n = 0, 1/2, 1$ for flat, cylindrical and spherical cases, respectively.

Hence, for solutions of the above equations we have

$$\oint_{\partial D} t^n \cdot [u\,dx + (\varepsilon^2 u_x - u^2 + \delta u_{xx})\,dt] = 0, \tag{8}$$

where $\mathcal{D}$ is a rectangle
$$\{0 \leq x \leq L,\ 0 \leq t \leq T\}.$$

While bearing in mind the initial value/boundary conditions $u(x,0) = u(+\infty, t) = 0$, for $L = +\infty$ the integrals read

$$\int_{+\infty}^{0} T^n u(x,T)\,dx + \int_{T}^{0} t^n (\varepsilon^2 u_x(0,t) - u^2(0,t) + \delta u_{xx}(0,t))\,dt = 0.$$

Thus

$$\int_{0}^{+\infty} u(x,T)\,dx = \frac{1}{T^n} \int_{0}^{T} t^n (-\varepsilon^2 u_x(0,t) + u^2(0,t) - \delta u_{xx}(0,t))\,dt. \tag{9}$$

Subsequently

$$\frac{1}{T} \int_{0}^{+\infty} u(x,T)\,dx = \frac{1}{T} \int_{0}^{T} \frac{1}{T^n} t^n (-\varepsilon^2 u_x(0,t) + u^2(0,t) - \delta u_{xx}(0,t))\,dt. \tag{10}$$

The right-hand side of Equation (10) can be computed in some simple cases or estimated. For instance, assume that $\varepsilon^2 u_x(0,t) + \delta u_{xx}(0,t)$ is negligible compared to $u^2(0,t)$. Then

$$\frac{1}{T} \int_{0}^{+\infty} u(x,T)\,dx \approx \frac{1}{T} \int_{0}^{T} \frac{1}{T^n} t^n (u^2(0,t))\,dt = \frac{1}{T} \int_{0}^{T} \frac{1}{T^n} t^n (A \sin^2(\omega t))\,dt. \tag{11}$$

It follows that

$n = 0 \Rightarrow \lim_{T \to \infty} \frac{1}{T} \int_{0}^{T} A^2 \sin^2(\omega t)\,dt == \frac{A^2}{2};$

$n = \frac{1}{2} \Rightarrow \lim_{T \to \infty} \frac{1}{T} \int_{0}^{T} \frac{1}{T^{\frac{1}{2}}} t^{\frac{1}{2}} (A \sin^2(\omega t))\,dt = \frac{A^2}{3};$

$n = 1 \Rightarrow \lim_{T \to \infty} \frac{1}{T} \int_{0}^{T} \frac{1}{T} t (A^2 \sin^2(\omega t))\,dt = \frac{A^2}{4}.$

Another example of exact estimation of right-hand side of Equation (10) is the case of constant boundary conditions.

Consider boundary condition $u(0,t) = M$. The graphs of solution are shown in Figure 5, left (compare their rates of decay caused solely by the spacial dimensions.)

For the resulting compression wave $u_x(0,t) = 0$, the right-hand side of Equation (10) equals

$$\frac{1}{T} \int_{0}^{T} \frac{M^2}{T^n} t^n\,dt = \frac{M^2}{n+1} \tag{12}$$

As the Figures 1–4 show, for a periodic boundary condition, after the decay of initial oscillations, graphs become monotonic convex lines. These convex lines break at $x = V \cdot T$ and at the height V. These monotonic lines are similar to the graphs of constant-boundary solutions; see Figure 5.

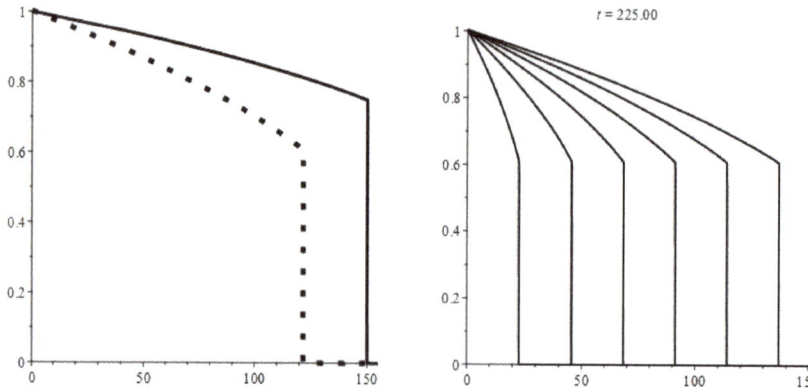

Figure 5. Constant boundary solutions to the Burgers equation, $\varepsilon = 0.1, t = 200$. **Left**: solid line—cylindrical; dotted line—spherical. **Right**: A trace of movement to the right of the spherical solution at moments $t = 37.5 \cdot k, k = 1\ldots 6$.

4. Self-Similar Approximations To Solutions

By observing the solution's graphs, one can clearly see (e.g., on Figure 5, right) that the monotonic part and its head shock develops as a homothetic transformation of the initial configuration (by t as a homothety parameter). Hence, we seek solutions in the self-similar form, $u(x,t) = y(\frac{x}{t})$. By substituting it into Equations (1)–(3), we get the equation:

$$-y'\frac{x}{t^2} + \frac{ny}{t} = \frac{2yy'}{t} + \frac{\varepsilon^2 y''}{t^2} + \frac{\delta y'''}{t^3}, \qquad (13)$$

or

$$-\xi y' + ny = 2yy' + \frac{\varepsilon^2 y''}{t} + \frac{\delta y'''}{t^2}, \qquad (14)$$

for $y = y(\xi)$ and $n = 0, 1/2, 1$. For sufficiently large t we may omit last two terms. It follows that appropriate solutions of these truncated ordinary differential equations are given by

$$u_1(x,t) = C_1, \ C_1 \in \mathbb{R}, \ n = 0, \text{ for flat waves equation;}$$

$$u_2(x,t) = -\frac{2 + \sqrt{C_2\xi + 4}}{C_2}, \ C_2 \in \mathbb{R}, \ n = \frac{1}{2}, \text{ for cylindrical and}$$

$$u_3(x,t) = \exp\left(\text{LambertW}\left(-\frac{\xi}{2}e^{-\frac{C_3}{2}}\right) + \frac{C_3}{2}\right), \ C_3 \in \mathbb{R}, \ n = 1 \text{ for spherical equation.}$$

(The *Lambert W* function, also called the omega function or product logarithm, is a multivalued function, namely, the branches of the inverse relation of the function $f(w) = we^w$, where w is any complex number.

For each integer k there is one branch, denoted by $W_k(z)$, which is a complex-valued function of one complex argument. W_0 is known as the principal branch. When dealing with real numbers the $W_0 = \text{LambertW}$ function satisfies $\text{LambertW}(x) \cdot e^{\text{LambertW}(x)} = x$. The *Lambert W* function, introduced in 1758, has numerous applications in solving equations, mathematical physics, statistics, etc.; for more detail, see [12].)

Let V be the velocity of the signal propagation in the medium. Since at the head shock we have $x = Vt$ and $u = V$, we obtain the condition for finding C_i. It is $y(V) = V$. It follows then that

$$C_1 = V, \ C_2 = -\frac{3}{V}, \ C_3 = \ln(V) + \frac{1}{2}.$$

For flat waves, it corresponds to a traveling wave solution of the classical Burgers equation.

For the cylindrical waves, the monotonic part is given by

$$u_2 = \frac{1}{3}\left(2V + V\sqrt{4 - \frac{3x}{Vt}}\right);$$

and for spherical waves

$$u_3 = V\sqrt{e}\exp\left(\text{LambertW}\left(-\frac{x}{2Vt\sqrt{e}}\right)\right).$$

Note that

$$u_2|_{x=0} = \frac{4V}{3} \text{ and } u_3|_{x=0} = V\sqrt{e} \approx 1.65V. \quad (15)$$

These formulas show that the velocity is proportional to the value of a constant boundary solution at $x = 0$.

The corresponding graphs visually coincide with the graphs obtained by numerical modeling; for instance, see a comparison to the solution (at $t = 100$) for the problem

$$u_t = 0.01 u_{xx} - 2uu_x - u/t, \ u(0,t) = 1, u(75,t) = 0, u(x,0) = 0 \quad (16)$$

in Figure 6, left.

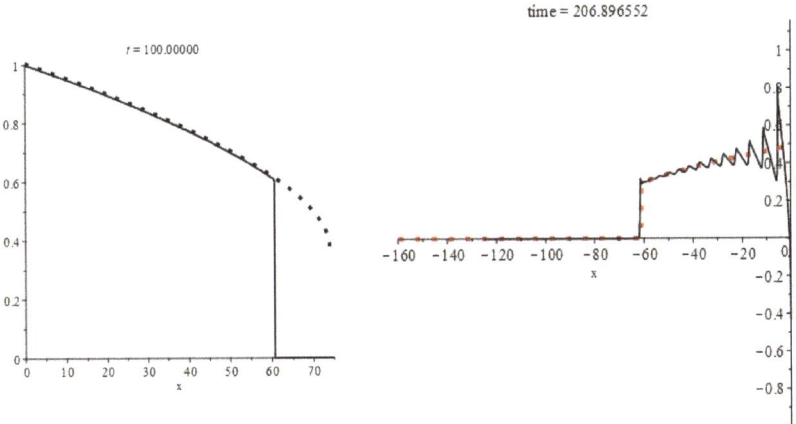

Figure 6. Left: solid line—solution to Equation (16); dotted line—its u_3 approximation. **Right**: solid line—solution to spherical KdV, $x \to -x$, $\varepsilon^2 = 0.02$, $\delta = 0.002$; dotted line—its $\tilde{u}_3$ approximation; both at $t = 200$.

5. Median Approximation

Yet, the monotonic part of the periodic boundary solution ends with a breaking, which travels with a constant velocity and amplitude, very much like the head of the Burgers' traveling wave solution (Equation (6)). A rather natural idea is to truncate a self-similar solution, multiplying it by a (normalized) formula for the Burgers TWS. Namely, put

- For the cylindrical waves take

$$\tilde{u}_2 = \frac{1}{2}[1 - \tanh(\frac{V}{\varepsilon^2}(x - Vt))] \cdot \frac{1}{3}\left(2V + V\sqrt{4 - \frac{3x}{Vt}}\right); \quad (17)$$

- For spherical waves,

$$\tilde{u}_3 = \frac{1}{2}[1 - \tanh(\frac{V}{\varepsilon^2}(x - Vt))] \cdot V\sqrt{e}\exp\left(\text{LambertW}\left(-\frac{x}{2Vt\sqrt{e}}\right)\right). \quad (18)$$

This construction produces an approximation of astonishing accuracy (see Figure 6, right and Figure 7); these graphs correspond to the spherical KdV–Burgers problem (it comes from Equation) (3) after the change $x \to -x$.

$$u_t = 0.02u_{xx} + 2uu_x - u/t - 0.002u_{xxx}, \quad u(0,t) = \sin t, u(10,t) = 0, u(x,0) = 0. \quad (19)$$

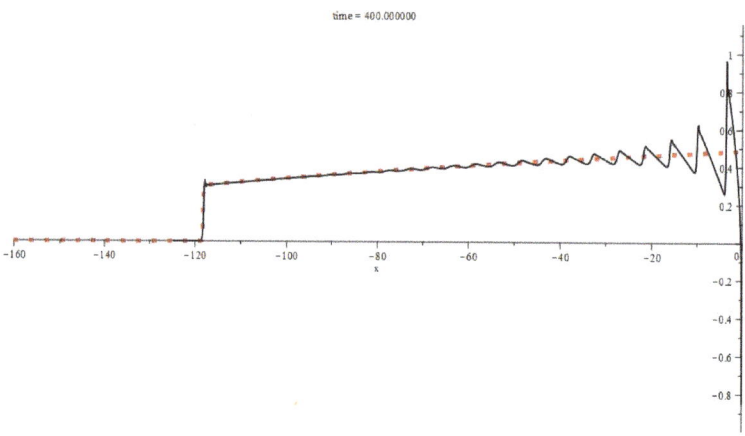

Figure 7. Solid line—solution to spherical KdV, $x \to -x$, $\varepsilon^2 = 0.02$, $\delta = 0.002$, dotted line—its $\tilde{u}_3$ approximation; both at $t = 400$.

Moreover, it is evident that the graphs of $\tilde{u}_2, \tilde{u}_3$ neatly represent the median lines of the approximated solutions over their whole ranges. By median we mean

$$M(x) = (2\pi n/\omega)^{-1} \int_0^{2\pi n/\omega} u(x,t)\, dt, \quad n \in \mathbb{N}, \, n \gg 1 \, (u(0,t) = \sin \omega t).$$

Let us assess the quality of $\tilde{u}_2, \tilde{u}_3$ approximations numerically.
Evaluate the trapezoid area under $\tilde{u}_2, \tilde{u}_3$ graphs:

- For the cylindrical equation

$$\int_0^{Vt} \left[\frac{1 - \tanh(\frac{V}{\varepsilon^2}(x - Vt))}{2}\right]\frac{1}{3}\left(2V + V\sqrt{4 - \frac{3x}{Vt}}\right) dx = \frac{32}{27}V^2 t;$$

- For the spherical equation

$$\int_0^{Vt} \left[\frac{1 - \tanh(\frac{V}{\varepsilon^2}(x - Vt))}{2}\right] V\sqrt{e}\exp\left(\text{LambertW}\left(\frac{-x}{2Vt\sqrt{e}}\right)\right) dx = \frac{V^2 t \cdot e}{2}.$$

Hence, the mean value of the left-hand side of (10) can be estimated as follows.

$$\frac{1}{T}\int_0^{+\infty} u(x,T)\, dx = \frac{1}{T}\int_0^{VT} u(x,T)\, dx \approx \begin{cases} \frac{32}{27}V^2 & \text{in cylindrical case;} \\ \frac{V^2 \cdot e}{2} & \text{in spherical case,} \end{cases} \quad (20)$$

This mean value can be also evaluated numerically. In the case illustrated by Figure 1 the direct numerical evaluation of the integral differs from the estimation (20) by 1%. It confirms the quality of the approximation.

For constant-boundary waves, it follows from Equation (12) that

$$\frac{M^2}{n+1} = \begin{cases} \frac{32}{27}V^2 & \text{in cylindrical case;} \\ \frac{V^2 \cdot e}{2} & \text{in spherical case;} \end{cases} \qquad (21)$$

see Equation (12); of course this result coincides with Equation (15). Hence, the mean value M of an arbitrary solution at the start of oscillations (or in a vicinity of the oscillator) is linearly linked to the velocity of the head shock.

However, to find this mean value for an arbitrary border condition is a tricky task, because the integrands u_x and u_{xx} of the right-hand side of Equation (10) have numerous breaks. Still, one may get an (admittedly rough) estimation for M using Equations (11) and (21). It follows that

$$\frac{M^2}{n+1} \approx \frac{A^2}{k}, \quad k = 2, 3, 4 \qquad (22)$$

for flat, cylindrical and spherical cases. In all these cases it results in $M \approx A\frac{\sqrt{2}}{2} \approx 0.71A$.

Numerical experiments also show (e.g., see Figure 3) that for the $u|_{x=0} = A\sin(t)$ boundary condition such a value is $M \approx A \cdot a$, where $a \approx 0.467$ is the mean value for $1 \cdot \sin(t)$ condition. That is, M depends on A almost linearly.

Note that this value may be obtained via the velocity V of the head shock, which, in turn, can be measured with great accuracy by the distance passed by the head shock after a sufficiently long time.

6. Conclusions

In this paper, we studied the pattern formation in periodic boundary solutions of spherical and cylindrical KdV–Burgers equations. Such a solution usually starts with a periodical chain of shock fronts with a decreasing amplitude. When oscillations decay and cease, a solution proceeds as a monotonic convex line that ends with a head shock. This last pattern was not described previously and it is the main subject of the paper.

We obtained simple explicit formulas describing the monotonic part of the solution and its head break. These approximate formulas have great accuracy. Moreover, their graphs neatly represent the median lines of the approximated solutions on their entire ranges. (By median line we mean the level around which the periodical oscillations occur).

To obtain these approximations we used self-similar solutions of the dissipationless and dispersionless KdV–Burgers equation and a traveling wave solution of the flat Burgers equation. Formulas depend on only one parameter: either on the velocity of the signal propagation or on the median value of the solution in the vicinity of the periodic boundary.

Some open questions remain. Our approximations are very good for the one-parameter class of constant boundary solutions. The existence of a one-parameter family of solutions points to the existence of a suitable symmetry, but the classical symmetry analysis was, so far, unhelpful. Conservation laws allows us to assess the value of the approximation's parameter using the boundary condition, but the resulting estimation is rough.

Funding: This work was partially supported by the Russian Basic Research Foundation grant 18-29-10013.

Conflicts of Interest: The author declares no conflict of interest.

Abbreviations

The following abbreviations are used in this manuscript:

KdV   Korteweg–de Vries
IVBP   Initial value | boundary problem
TWS   Traveling wave solution

References

1. Leibovich, S.; Seebass, R. (Eds.) *Nonlinear Waves*; Cornell University Press: London, UK, 1974; Chapter 4.
2. Sachdev P.L.; Seebass, R. Propagation of spherical and cylindrical N-waves. *J. Fluid. Mech.* **1973**, *58*, 1973. [CrossRef]
3. Grava, T. Whitham modulation equations and application to small dispersion asymptotics and long time asymptotics of nonlinear dispersive equations. *arXiv* **2016**, arXiv:1701.00069v1.
4. Chugainova A.P.; Shargatov, V.A. Traveling waves and undercompressive shocks in solutions of the generalized Korteweg–de Vries–Burgers equation with a time-dependent dissipation coefficient distribution. *Eur. Phys. J. Plus* **2020**, *135*, 635. [CrossRef]
5. Bendaasa, S.; Alaa, N. Periodic Wave Shock Solutions of Burgers Equations, A New Approach. *Int. J. Nonlinear Anal. Appl.* **2020**, *10*, 119–129.
6. Chugainova, A.P.; Shargatov, V.A. Stability of the breaks structure described by the generalized Kortweg-de Vries-Burgers equation. *Comp. Math. Math. Phys.* **2016**, *56*, 259–274 [CrossRef]
7. Samokhin, A. Periodic boundary conditions for KdV-Burgers equation on an interval. *J. Geom. Physics* **2017**, *113*, 250–256. [CrossRef]
8. Samokhin, A. Nonlinear waves in layered media: solutions of the KdV—Burgers equation. *J. Geom. Physics* **2018**, *130*, 33–39. [CrossRef]
9. Samokhin, A. On nonlinear superposition of the KdV-Burgers shock waves and the behavior of solitons in a layered medium. *J. Diff. Geom. Its Appl.* **2017**, *54*, 91–99. [CrossRef]
10. Whitham, G.B. *Linear and Nonlinear Waves*; John Wiley & Sons: Hoboken, NJ, USA, 1999; ISBN 9781118032954. [CrossRef]
11. Krasil'shchik I.S.; Vinogradov, A.M. (Eds.) *Symmetries and Conservation Laws for Differential Equations of Mathematical Physics*; American Mathematical Society: Providence RI, USA, 1999; ISSN 0065-9282.
12. Bronstein, M.; Corless, R.M.; Davenport, J.H.; Jeffrey, D.J. Algebraic properties of the Lambert W function from a result of Rosenlicht and of Liouville. In *Integral Transforms and Special Functions*; Taylor & Francis: Abingdon, UK; Volume 19, pp. 709–712. [CrossRef]

 symmetry

Article

Contact Symmetries of a Model in Optimal Investment Theory

Daniel J. Arrigo * and Joseph A. Van de Grift

Department of Mathematics, University of Central Arkansas, Conway, AR 72035, USA; jvandegrift@cub.uca.edu
* Correspondence: darrigo@uca.edu

Abstract: It is generally known that Lie symmetries of differential equations can lead to a reduction of the governing equation(s), lead to exact solutions of these equations and, in the best case scenario, lead to a linearization of the original equation. In this paper, we consider a model from optimal investment theory where we show the governing equation possesses an extensive contact symmetry and, through this, we show it is linearizable. Several exact solutions are provided including a solution to a particular terminal value problem.

Keywords: contact symmetry; optimal investment theory; linearization; exact solutions

Citation: Arrigo, D.J.; Van de Grift, J.A. Contact Symmetries of a Model in Optimal Investment Theory. *Symmetry* **2021**, *13*, 217. https://doi.org/10.3390/sym13020217

Academic Editor: José Carlos R. Alcantud

Received: 31 December 2020
Accepted: 26 January 2021
Published: 28 January 2021

Publisher's Note: MDPI stays neutral with regard to jurisdictional claims in published maps and institutional affiliations.

Copyright: © 2021 by the authors. Licensee MDPI, Basel, Switzerland. This article is an open access article distributed under the terms and conditions of the Creative Commons Attribution (CC BY) license (https://creativecommons.org/licenses/by/4.0/).

1. Introduction

Nonlinear partial differential equations (NLPDEs) play an integral part in describing the world around us. They can be found in the fields of nonlinear diffusion, wave propagation, Mathematical Biology, ray optics, solid mechanics, and financial mathematics to name just a few (see, for example, refs. [1–5] and the references within). However, obtaining exact solutions of these equations is usually a difficult task and techniques for obtaining solutions is a current area of research. One popular technique are symmetry methods probably due to the fact that the method is rather algorithmic and thus computer algebra systems such as *Maple* and *Mathematica* can be used. Symmetry methods have been extensively used in a number of fields, and we refer the reader to the books by Arrigo [6], Bluman et al. [7,8], Bordag [9], Cantwell [10], Cherniha et al. [11], and Olver [12] .

In this paper, we are interested in a model from optimal investment theory. Consider an investment portfolio consisting of $n + 1$ assets. Let the first asset be a bond and the next n assets be stocks, all of which are traded continuously. In the simplest case where $n = 1$, the value of the portfolio, $u(t, x)$, for time t and investment amount x, one model presented by see Yong [13] is the NPDE

$$u_t + rxu_x - \frac{(b-r)u_x^2}{2\sigma u_{xx}} = 0, \qquad (1)$$

or

$$(u_t + rxu_x)u_{xx} - \theta u_x^2 = 0. \qquad (2)$$

where $\theta = \frac{b-r}{2\sigma}$ and the variables r, b, σ represent the interest rate, appreciation rate, and volatility, respectively, and are assumed constant with $\sigma > 0$ and $b - r > 0$.

A classical symmetry analysis was performed by Yang and Xu [14] who were able to show that (2) admitted the symmetry generator

$$\Gamma = T\frac{\partial}{\partial t} + X\frac{\partial}{\partial x} + U\frac{\partial}{\partial u}, \qquad (3)$$

where

$$\begin{aligned} T &= c_1, \\ X &= c_2 x + c_3 e^{rt}, \\ U &= c_4 u + c_5, \end{aligned} \qquad (4)$$

where c_i, $i = 1...5$ are arbitrary constants. Through their analysis, they were able to exploit several of these symmetries to obtain a number of reductions and, in some cases, construct exact solutions. It is natural to ask whether (2) admits symmetries that are more general than Lie point symmetries. In this paper, we consider contact symmetries of (2), and we will show that, in fact, (2) admits a rather large contact symmetry which leads to its linearization. In addition to recovering known solutions, we will obtain new exact solutions. We also solve a particular terminal value problem.

2. Contact Symmetries

In this section, we construct contact symmetries of (2). If we denote this original NLPDE by Δ so

$$\Delta = (u_t + rxu_x)u_{xx} - \theta u_x^2 = 0, \tag{5}$$

then contact symmetries are given by

$$\Gamma^{(2)} \Delta|_{\Delta=0} = 0, \tag{6}$$

where the infinitesimal generator Γ is

$$\Gamma = U \frac{\partial}{\partial u}, \tag{7}$$

where $U = U(t, x, u, u_t, u_x)$; the first and second extensions of the generator are

$$\begin{aligned}
\Gamma^{(1)} &= \Gamma + D_t U \frac{\partial}{\partial u_t} + D_x U \frac{\partial}{\partial u_x}, \\
\Gamma^{(2)} &= \Gamma^{(1)} + D_t^2 U \frac{\partial}{\partial u_{tt}} + D_t D_x U \frac{\partial}{\partial u_{tx}} + D_x^2 U \frac{\partial}{\partial u_{xx}},
\end{aligned} \tag{8}$$

where the operators D_t and D_x are

$$\begin{aligned}
D_t &= \frac{\partial}{\partial t} + u_t \frac{\partial}{\partial u} + u_{tt} \frac{\partial}{\partial u_t} + u_{tx} \frac{\partial}{\partial u_x} + u_{ttt} \frac{\partial}{\partial u_{tt}} + u_{ttx} \frac{\partial}{\partial u_{tx}} + u_{txx} \frac{\partial}{\partial u_{xx}} \cdots \\
D_x &= \frac{\partial}{\partial x} + u_x \frac{\partial}{\partial u} + u_{tx} \frac{\partial}{\partial u_t} + u_{xx} \frac{\partial}{\partial u_x} + u_{ttx} \frac{\partial}{\partial u_{tt}} + u_{txx} \frac{\partial}{\partial u_{tx}} + u_{xxx} \frac{\partial}{\partial u_{xx}} \cdots
\end{aligned} \tag{9}$$

This leads to the set of determining equations

$$\begin{aligned}
U_{u_t u_t} &= 0, \\
\theta u_x^2 U_{u_t u_x} + (rxu_x + u_t) U_{xu_t} + u_x (rxu_x + u_t) U_{uu_t} &= 0, \\
(u_t + rxu_x)^2 U_{xx} + 2u_x (u_t + rxu_x)^2 U_{xu} + u_x^2 (u_t + rxu_x)^2 U_{uu} + \\
2\theta u_x^2 (u_t + rxu_x) U_{xu_x} + 2\theta u_x^3 (u_t + 2rxu_x) U_{uu_x} + \\
\theta^2 u_x^4 U_{u_x u_x} - \theta u_x (2u_t + rxu_x) U_x - r\theta u_x^3 U_{u_x} + \theta u_x^2 U_t &= 0.
\end{aligned} \tag{10}$$

Although somewhat a laborious calculation, we find the solution of (10) to be

$$\begin{aligned}
U &= F(t,u_x) + c_1 u_t + c_2 x u_x + c_3(x u_x - u) \\
&+ c_4\left(-\frac{x u_x - u}{2\theta}\ln u_x + \frac{t}{2\theta}((\theta-r)x u_x + (\theta+r)u)\right) \\
&+ c_5\left(t u_t - \frac{1}{4\theta}((\theta-r)x u_x + (\theta+r)u)\ln u_x + \frac{(\theta+r)^2}{4\theta}t(x u_x - u)\right) \\
&+ c_6\left(t^2 u_t + \frac{x u_x - u}{4\theta}\ln^2 u_x - \frac{t}{2\theta}((\theta-r)x u_x + (\theta+r)u)\ln u_x\right. \\
&\left.+ \left(\frac{(r+\theta)^2 t^2 - 2\theta t}{4\theta}\right)(x u_x - u)\right)
\end{aligned} \quad (11)$$

where $c_i, i = 1...6$ are arbitrary constants and the function $F(t, u_x)$ satisfies

$$\theta u_x^2 F_{u_x u_x} - r u_x F_{u_x} + F_t = 0. \quad (12)$$

Equation (12) is linear and possesses an infinite number of solutions, which means that there are an infinite number of symmetries to (2). Furthermore, since there is a particular function F in (11) that satisfies a linear PDE, this suggests that (2) can be transformed to a linear PDE (Bluman and Kumei [15]). As both t and u_x are independent variables in (12), this indicates that maybe we should use these as new independent variables.

3. A Linearization

Since the symmetry obtained in the last section contains an arbitrary function that satisfies a linear PDE, this suggests that the original PDE is linearizable. Introducing the new variables

$$t = T, \quad x = U_X, \quad u = X U_X - U, \quad (13)$$

where $U = U(T, X)$, derivatives transform as

$$u_t = -U_T, \quad u_x = X, \quad u_{xx} = \frac{1}{U_{XX}}, \quad (14)$$

and (2) becomes

$$\theta X^2 U_{XX} - r X U_X + U_T = 0, \quad (15)$$

which is exactly (12). Interestingly enough, (15) looks remarkably similar to the Black–Scholes–Merton equation [16,17], which is known to be mappable to the linear heat equation ([16,18]), so it should come as no surprise that the same is true for (15). However, it would make the contact transformation (13) more complicated and, thus, we will not pursue this line any further.

In the next section, we obtain simple exact solutions of (2) in addition to exploiting Lie symmetries of the linearized Equation (15) to obtain additional solutions.

4. Exact Solutions

We have shown that the nonlinear PDE

$$(u_t + r x u_x) u_{xx} - \theta u_x^2 = 0 \quad (16)$$

can be transformed to the linear PDE

$$\theta X^2 U_{XX} - r X U_X + U_T = 0 \quad (17)$$

via the transformation

$$t = T, \quad x = U_X, \quad u = X U_X - U. \quad (18)$$

We are now in a position to obtain a number of exact solutions to (16). For example, (17) admits separable solutions of the form

$$U = F(T)G(X) \tag{19}$$

where F and G satisfies

$$F' - kF = 0, \tag{20a}$$
$$\theta X^2 G'' - rXG' + kG = 0 \tag{20b}$$

where k is a separation constant. Equation (20a) is easily solved giving

$$F = F_0 e^{kT} \tag{21}$$

for some arbitrary constant F_0. Equation (20b) possesses solutions of the form X^m, where m is a solution of

$$\theta m^2 - (r+\theta)m + k = 0. \tag{22}$$

For example, if $k = sm$ (where s is some constant), then the solution of (22) is

$$m_1 = 0, \quad m_2 = \frac{r + \theta - s}{\theta}, \tag{23}$$

leading to the exact solution

$$U = c_1 + c_2 X^m e^{mT}, \quad m = \frac{r + \theta - s}{\theta} \tag{24}$$

Passing (24) through the transformation (18) (resetting the constants c_1 and c_2) gives

$$u = c_1 + c_2 e^{\alpha(st - \ln x)}, \quad \alpha = \frac{r+\theta-1}{1-r}, \tag{25}$$

which recovers the exact solution presented by Yang and Xu [14] by choosing $c_1 = 0, c_2 = 1$ and $s = 1$.

As a second example, if $k = r$, then, from (22), we obtain $m = 1, r/\theta \, (\neq 1)$ and we obtain the solution to (20b) as

$$G = g_1 X + g_2 X^{r/\theta} \tag{26}$$

and the general solution to (17)

$$U = \left(g_1 X + g_2 X^{r/\theta}\right) e^{rT}. \tag{27}$$

Passing (27) through the transformation (18) leads to the exact solution

$$u = a(x + be^{rt})^{\frac{r}{r-\theta}} e^{-\frac{r\theta}{r-\theta}t} \tag{28}$$

of (16), which we believe to be new. In the cases where $r = \theta$, the solution of (20b) is

$$G = g_1 X + g_2 X \ln X \tag{29}$$

and the general solution to (17)

$$U = (c_1 X + c_2 X \ln X) e^{rT}. \tag{30}$$

Passing (30) through the transformation (18) leads to the exact solution

$$u = c_1 e^{rt + c_2 x e^{-rt}}, \tag{31}$$

which was given in [14] in the case of $r = \theta$.

Of course, other choices of m that satisfy (22) would lead to exact solutions of (16) which we will not pursue here.

We also note that symmetries of (17) can be used to generate new solutions of (17), which could lead to new solutions of (16). It is well known (Broadbridge and Arrigo [19]) that, if (17) possesses symmetries with the generator

$$\Gamma = \Sigma \frac{\partial}{\partial T} + \Phi \frac{\partial}{\partial X} + (\Xi U + Q(T, X)) \frac{\partial}{\partial U}, \tag{32}$$

where Σ, Φ, and Ξ have some particular forms and Q satisfies the original PDE (17), then, if one has one seed solution, say $U = U_0(T, X)$, then additional solutions can be obtained from

$$Q = \Sigma \frac{\partial U_0}{\partial T} + \Phi \frac{\partial U_0}{\partial X} - \Xi U_0 \tag{33}$$

For example, (17) admits the symmetry generator (32) where Σ, Φ, and Ξ are given by

$$\Sigma = c_1 + 2c_2 T + c_3 T^2$$
$$\Phi = ((c_2 + c_3 T) \ln X + c_4 + c_5 T) X$$
$$\Xi = \left(\frac{c_3}{4\theta} \ln^2 X + \frac{(c_2 + c_3 T)(r + \theta) + c_5}{2\theta} \ln X \right. \tag{34}$$
$$\left. + \frac{(r + \theta)^2 (2c_2 + c_3 T) T + 2c_5(r + \theta) T}{4\theta} - \frac{c_3 T}{2} + c_6 \right) U$$

where $c_i, i = 1...6$ are arbitrary constants.

One particularly simple solution of (17) is

$$U = 1. \tag{35}$$

From (33), we obtain the solution

$$Q = -\left(\frac{c_3}{4\theta} \ln^2 X + \frac{(c_2 + c_3 T)(r + \theta) + c_5}{2\theta} \ln X \right. \tag{36}$$
$$\left. + \frac{(r + \theta)^2 (2c_2 + c_3 T) T + 2c_5(r + \theta) T}{4\theta} - \frac{c_3 T}{2} + c_6 \right).$$

Passing (36) through the transformation (18) leads to a solution that is parametric in its nature (which we do not list here); however, setting $c_3 = 0$, we obtain (omitting translational constants)

$$u = \frac{(r + \theta)(c_2(r + \theta) + c_5)}{2\theta} t - \frac{c_2(r + \theta) + c_5}{\theta} \ln x \tag{37}$$

which we believe is new. Of course, other seed solutions could lead to an abundance of exact solutions to (17) which, in turn, would lead to exact solutions to (16).

5. A Particular Terminal Value Problem

A particular problem of interest is one that is given in Koleva and Vulkow [20], which is to solve (2) subject to the terminal condition

$$u(x, t^*) = 1 - e^{-\mu x}. \tag{38}$$

Here, we introduce a slight variation of (18)

$$t = T + t^*, \quad x = \frac{U_X}{\mu}, \quad u = X U_X - U + 1. \tag{39}$$

Under this transformation, the PDE (2) still transforms to (17); however, the terminal condition (38) turns into the initial condition

$$U(X,0) = X - X \ln X. \tag{40}$$

At this point, we exploit the symmetries obtained in the previous section. With these symmetries, we associate an invariant surface condition

$$\Sigma U_T + \Phi U_X = \Xi U, \tag{41}$$

where Σ, Φ and Ξ are given in (34). From (40) and (17), we obtain the initial conditions

$$U_X(X,0) = -\ln X, \quad U_T(X,0) = \theta X - rX \ln X. \tag{42}$$

Requiring that (40) and (42) satisfy (41) on the boundary $T = 0$ gives

$$c_3 = 0, \quad c_4 = (\theta - r)c_1 - c_2, \quad c_5 = (\theta - r)c_2, \quad c_6 = \theta c_1, \tag{43}$$

where c_1 and c_2 are arbitrary. Here, we choose $c_1 = 1$, $c_2 = 0$, leading to the invariant surface condition

$$U_T + (\theta - r)XU_X = \theta U. \tag{44}$$

This is easily solved giving

$$U = e^{\theta T} F(\ln X + (r - \theta)T). \tag{45}$$

Imposing the initial condition (40) on the solution (45) gives

$$F(\ln X) = X - X \ln X, \tag{46}$$

and, if we let $\lambda = \ln X$, we obtain

$$F(\lambda) = (1 - \lambda)e^{\lambda} \tag{47}$$

and, from (45), we obtain

$$U = X(1 - \ln X + (\theta - r)T)e^{rT} \tag{48}$$

and one can indeed verify that (48) does satisfy (17) and (42). As the final step, we pass (48) through the transformation (39), leading to

$$u = 1 - e^{-\theta(t^* - t) - \mu x e^{r(t^* - t)}}, \tag{49}$$

which satisfies the original PDE (2) and the terminal condition (38).

6. Conclusions

It is well known that classical Lie symmetries can lead to a reduction of a given PDE and sometimes lead to exact solutions of the equation. The best case scenario, albeit rare, indicates that the original equation is linearizable. In this paper, we constructed the contact symmetries of a model from an optimal investment theory, which led to a linearization of the given PDE. Several exact solutions were obtained. The symmetries of the linearized equation were also considered where an additional exact solution was obtained in addition to solving a particular terminal value problem.

Author Contributions: The authors contributed equally to this work. All authors have read and agreed to the published version of the manuscript.

Funding: This research received no external funding.

Acknowledgments: We thank the reviewers for their valuable comments and suggestions. The second author gratefully acknowledges the support from the University of Central Arkansas in terms of a graduate assistantship.

Conflicts of Interest: The authors declare no conflict of interest.

References

1. Arrigo, D.J. *Analytical Techniques for Solving Nonlinear Partial Differential Equations*; Morgan-Claypool: San Rafael, CA, USA, 2019.
2. Debnath, L. *Nonlinear Partial Differential Equations for Scientists and Engineers*, 3rd ed.; Springer: New York, USA 2012.
3. Logan, J.D. *An Introduction to Nonlinear Partial Differential Equations*, 2nd ed.; John Wiley & Sons, Inc.: Hoboken, NJ, USA, 2008.
4. Murray, J.D. *Mathematical Biology 1: An Introduction*; Springer: Berlin/Heidelberg, Germany, 2002.
5. Murray, J.D. *Mathematical Biology II: Spatial Models and Biomedical Applications*; Springer: Berlin/Heidelberg, Germany, 2003.
6. Arrigo, D.J. *Symmetry Analysis of Differential Equations: An Introduction*; John Wiley & Sons, Inc.: Hoboken, NJ, USA, 2015.
7. Bluman, G.; Kumei, S. *Symmetries and Differential Equations*; Springer: New York, NY, USA, 1989.
8. Bluman, G.; Anco, S.C. *Symmetry and Integration Methods for Differential Equations*; Springer: Berlin/Heidelberg, Germany, 2002.
9. Bordag, L.A. *Geometrical Properties of Differential Equations. Applications of the Lie Group Analysis in Financial Mathematics*; World Scientific: Singapore, 2015.
10. Cantwell, B.J. *Introduction to Symmetry Analysis*; Cambridge Texts in Applied Mathematics: Cambridge, UK, 2002.
11. Cherniha, R.; Mykola, S.; Pliukhin, O. *Nonlinear Reaction-Diffusion-Convection Equations: Lie and Conditional Symmetry, Exact Solutions and Their Applications*; CRC Press: Boca Raton, FL, USA, 2018.
12. Olver, P. *Applications of Lie Groups to Differential Equations*; Springer: New York, NY, USA, 1986.
13. Yong, J. Introduction to mathematical finance. In *Mathematical Finance-Theory and Applications*; Young, J., Cont, R., Eds.; Higher Education Press: Bejing, China, 2000.
14. Yang, S.; Xu, T. Lie symmetry analysis for a parabolic Monge-Ampere equation in the optimal investment theory. *J. Comput. Appl. Math.* **2019**, *346* 483–489. [CrossRef]
15. Bluman, G.; Kumei, S. When nonlinear differential equations are equivalent to linear differential equations. *SIAM J. Appl. Math.* **1982**, *42*, 1157–1173.
16. Merton, R.C. Theory of Rational Option Pricing. *Bell J. Econ. Manag. Sci.* **1973**, *4*, 141–183. [CrossRef]
17. Black, F.; Scholes, M. The pricing of Options and Corporate Liabilities. *J. Political Econ.* **1973**, *81*, 637–654. [CrossRef]
18. Gazizov, R.K.; Ibragimov, N.H. Lie Symmetry Analysis of Differential Equations in Finance. *Nonlinear Dyn.* **1998**, *17*, 387–407. [CrossRef]
19. Broadbridge, P.; Arrigo, D.J. All solutions of standard symmetric linear partial differential equations have a classical Lie symmetry. *J. Math. Anal. Appl.* **1999**, *234*, 109–122. [CrossRef]
20. Koleva, M.N.; Vulkov, L.G. A numerical study of a parabolic Monge-Ampere equation in mathematical finance. In *Numerical Methods and Applications. Lecture Notes in Computer Science*; Dimov, I., Dimova, S., Kolkovska, N., Eds.; Springer: Berlin/Heidelberg, Germany, 2010; Volume 6046.

Article

Quotients of Euler Equations on Space Curves

Anna Duyunova *,†, Valentin Lychagin † and Sergey Tychkov †

V. A. Trapeznikov Institute of Control Sciences of RAS, Profsoyuznaya Street, 65, 117997 Moscow, Russia; valentin.lychagin@uit.no (V.L.); sergey.lab06@ya.ru (S.T.)
* Correspondence: anna.duyunova@yahoo.com
† These authors contributed equally to this work.

Abstract: Quotients of partial differential equations are discussed. The quotient equation for the Euler system describing a one-dimensional gas flow on a space curve is found. An example of using the quotient to solve the Euler system is given. Using virial expansion of the Planck potential, we reduce the quotient equation to a series of systems of ordinary differential equations (ODEs). Possible solutions of the ODE system are discussed.

Keywords: Euler equation; quotient equation

MSC: 76N99; 35B06

Citation: Duyunova, A.; Lychagin, V.; Tychkov, S. Quotients of Euler Equations on Space Curves. *Symmetry* **2021**, *13*, 186. https://doi.org/10.3390/sym13020186

Academic Editor: Calogero Vetro
Received: 31 December 2020
Accepted: 21 January 2021
Published: 25 January 2021

Publisher's Note: MDPI stays neutral with regard to jurisdictional claims in published maps and institutional affiliations.

Copyright: © 2021 by the authors. Licensee MDPI, Basel, Switzerland. This article is an open access article distributed under the terms and conditions of the Creative Commons Attribution (CC BY) license (https://creativecommons.org/licenses/by/4.0/).

1. Introduction

In this paper, we continue the study of the Euler equation describing gas flows on space curves in a constant gravity field. Symmetry algebras and differential invariant fields, as well as their dependence on thermodynamic state equations and the form of a space curve, were considered in [1]. Here, we find a quotient PDE for the Euler equation and show its role in solving the original equation.

Recall that the system of PDEs describing such flows is the following:

$$\begin{cases} \rho(u_t + uu_a) = -p_a - \rho g h', \\ \rho_t + (\rho u)_a = 0, \\ \rho\theta(s_t + us_a) - k\theta_{aa} = 0, \end{cases} \quad (1)$$

where $u(t,a)$ is the flow velocity, $p(t,a)$, $\rho(t,a)$, $s(t,a)$, and $\theta(t,a)$ are the pressure, density, specific entropy, and temperature of the fluid, respectively, k is the constant thermal conductivity, g is the gravitational acceleration, and $h(a)$ is the z-component of a naturally parametrized space curve.

System (1) is incomplete, i.e., it has two more unknown functions than equations. In the present paper, we put aside the question of classification of possible thermodynamic relations, since it was described in detail in [1]. We assume that these relations are given either in the forms $p = P(\rho, \theta)$ and $s = S(\rho, \theta)$, or in terms of the Planck potential [2]. In particular, we consider the ideal gas equation.

This paper is organized as follows. In Section 2, the notion of PDE quotients is discussed. In Section 3, we recall the symmetry algebra and differential invariants for the Euler system. In Section 4, we find the quotient for the Euler equation and discuss possible symmetries and solutions.

All calculations for this paper were performed with the DifferentialGeometry package in Maple. The corresponding Maple files can be found on the webpage http://d-omega.org/appendices/.

2. PDE Quotients

2.1. Algebraic Structures in PDE Geometry

Let $\pi\colon E(\pi) \to M$ be a smooth bundle over a manifold M and let $\pi_k\colon J^k(\pi) \to M$, $k = 0, 1, \ldots$, be the k-jet bundles of sections of the bundle π. To simplify the notations, we use $\mathbf{J}^k$ instead of $J^k(\pi)$.

Depending on $\dim \pi$, the jet geometry [3] is defined by the following pseudogroups.

1. If $\dim \pi = 1$, it is defined by the pseudogroup $\mathrm{Cont}(\pi)$ of the local contact transformations of the manifold $\mathbf{J}^1$.
2. For $\dim \pi \geq 2$, the jet geometry is defined by the pseudogroup $\mathrm{Point}(\pi)$ of the local point transformations, i.e., local diffeomorphisms of the manifold $\mathbf{J}^0$.

It is also known that the prolongations of these pseudogroups to the jet bundles exhaust all Lie transformations, i.e., local diffeomorphisms of jet spaces that preserve the Cartan distributions (see, for example, [3]).

Moreover, bundles $\pi_{k,k-1}\colon \mathbf{J}^k \to \mathbf{J}^{k-1}$ ($k \geq 2$ when $\dim \pi \geq 2$, and $k \geq 3$ when $\dim \pi = 1$) have affine structures, which are invariant with respect to the Lie transformations, and prolongations of the pseudogroups $\mathrm{Cont}(\pi)$ or $\mathrm{Point}(\pi)$ are given by rational functions of u_σ^i in the standard jet coordinates (x, u_σ^i).

The last statement means that, in the case of $\dim \pi \geq 2$, the fibers $\mathbf{J}_\theta^{k,0}$ of the projections $\pi_{k,0}\colon \mathbf{J}^k \to \mathbf{J}^0$ at a point $\theta \in \mathbf{J}^0$ are algebraic manifolds, and the stationary subgroup $\mathrm{Point}_\theta(\pi) \subset \mathrm{Point}(\pi)$ gives us birational isomorphisms of the manifold.

In the case of $\dim \pi = 1$, the fibers $\mathbf{J}_\theta^{k,1}$ of the projections $\pi_{k,1}\colon \mathbf{J}^k \to \mathbf{J}^1$ at a point $\theta \in \mathbf{J}^1$ are *algebraic manifolds* too, and the stationary subgroup $\mathrm{Cont}_\theta(\pi) \subset \mathrm{Cont}(\pi)$ gives us birational isomorphisms of the manifold.

Following this picture, we say that a differential equation $\mathcal{E}_k \subset \mathbf{J}^k$ is algebraic if fibers $\mathcal{E}_{k,\theta}$ of the projections $\pi_{k,0}\colon \mathcal{E}_k \to \mathbf{J}^0$ are algebraic manifolds when $\dim \pi \geq 2$, or $\pi_{k,1}\colon \mathcal{E}_k \to \mathbf{J}^1$ when $\dim \pi = 1$.

All differential equations here are assumed to be formally integrable; then, the prolongations $\mathcal{E}_k^{(l)} = \mathcal{E}_{k+l} \subset \mathbf{J}^{k+l}$ of an algebraic equation $\mathcal{E}_k \subset \mathbf{J}^k$ are algebraic, too.

By a symmetry algebra of an algebraic differential equation, we mean the Lie algebra $\mathrm{Sym}(\mathcal{E}_k)$ of point vector fields if $\dim \pi \geq 2$ or contact vector fields if $\dim \pi = 1$ that act transitively on $\mathbf{J}^0$ in the case of $\dim \pi \geq 2$ or $\mathbf{J}^1$ in the case of $\dim \pi = 1$. Moreover, the stationary sub-algebra $\mathrm{Sym}_\theta(\mathcal{E}_k)$ where $\theta \in \mathbf{J}^0$ or $\theta \in \mathbf{J}^1$ produces actions of algebraic Lie algebras on algebraic manifolds $\mathcal{E}_{l,\theta}$ for all $l \geq k$.

2.2. The Rosenlicht Theorem

Let B be an algebraic manifold, i.e., an irreducible variety without singularities over a field of characteristic zero, let G be an algebraic group, and let $G \times B \to B$ be an algebraic action.

Denote by $\mathcal{F}(B)$ the field of rational functions on the manifold B, and, by $\mathcal{F}(B)^G \subset \mathcal{F}(B)$, denote the field of rational G-invariants on B.

We say that an orbit $Gb \subset B$ is regular (as well as a point b itself) if there are $m = \mathrm{codim}\, Gb$ G-invariants $x_1, \ldots, x_m$ such that their differentials are linearly independent at the points of the orbit.

Let $B_0 = B \setminus \mathrm{Sing}$ be the set of all regular points and let $Q(B) = B_0/G$ be the set of all regular orbits.

The Rosenlicht theorem [4] states that B_0 is open and dense in B.

Moreover, if the above invariants $x_1, \ldots, x_m$ are considered as local coordinates on the quotient $Q(B)$ at the point $Gb \in Q(B)$, then on the intersections of the coordinate charts, the coordinates are connected by rational functions. In other words, $Q(B)$ is an algebraic manifold of the dimension $m = \mathrm{codim}\, Gb$, and the rational map $\varkappa\colon B_0 \to Q(B)$ of algebraic manifolds gives us the field isomorphism $\mathcal{F}(B)^G = \pi^*(\mathcal{F}(Q(B)))$.

To apply this theorem to algebraic differential equations, we should reformulate it for the case of Lie algebras.

Let B be an algebraic manifold and let $\mathfrak{g}$ be a Lie sub-algebra of the Lie algebra of the vector fields on B.

We say that $\mathfrak{g}$ is an algebraic Lie algebra if there is an algebraic action of an algebraic group G on B such that $\mathfrak{g}$ coincides with the image of Lie algebra $\text{Lie}(G)$ under this action.

By an algebraic closure $\tilde{\mathfrak{g}}$ of a Lie algebra $\mathfrak{g}$, we mean the intersection of all algebraic Lie algebras that contain $\mathfrak{g}$.

Example 1. *Let $B = \mathbb{R}$; then, Lie algebra*
$$\mathfrak{g} = \mathfrak{sl}_2 = \langle \partial_x, x\partial_x, x^2 \partial_x \rangle$$
is algebraic because it corresponds to the projective action of the algebraic group $\mathbf{SL}_2(\mathbb{R})$.

Example 2. *Let $B = S^1 \times S^1$ be a torus and $\mathfrak{g} = \langle \partial_\phi + \lambda \partial_\psi \rangle$, where ϕ and ψ are the angles, $\lambda \in \mathbb{R}$. Then, $\mathfrak{g}$ is algebraic if and only if $\lambda \in \mathbb{Q}$. Otherwise, $\tilde{\mathfrak{g}} = \langle \partial_\phi, \partial_\psi \rangle$. A similar situation occurs in the case of $B = \mathbb{R}^2$ and*
$$\mathfrak{g} = \langle x\partial_x + \lambda y \partial_y \rangle,$$
where $\tilde{\mathfrak{g}} = \mathfrak{g}$ if $\lambda \in \mathbb{Q}$, and $\tilde{\mathfrak{g}} = \langle x\partial_x, y\partial_y \rangle$ otherwise.

The Rosenlicht theorem is also true for algebraic Lie algebras or for their algebraic closure in the case of general Lie algebras.

Let us be given a Lie algebra $\mathfrak{g}$ of vector fields on an algebraic manifold B and let $\tilde{\mathfrak{g}} \supset \mathfrak{g}$ be its algebraic closure. Then, the field $\mathcal{F}(B)^{\mathfrak{g}}$ of rational $\mathfrak{g}$-invariants has a transcendence degree equal to the codimension of regular $\tilde{\mathfrak{g}}$-orbits, and it is also equal to the dimension of the quotient algebraic manifold $Q(B)$.

2.3. *Quotients of Algebraic Differential Equations*

Let $\mathfrak{g}$ be an algebraic symmetry Lie algebra of an algebraic formally integrable differential equation $\mathcal{E}_k$, and let $\mathcal{E}_l$ be the $(l-k)$-th prolongations of $\mathcal{E}_k$. Then, all equations $\mathcal{E}_l \subset J^l$ are algebraic, and we have the tower of algebraic bundles:

$$\mathcal{E}_k \longleftarrow \mathcal{E}_{k+1} \longleftarrow \cdots \longleftarrow \mathcal{E}_l \longleftarrow \mathcal{E}_{l+1} \longleftarrow \cdots .$$

Let $\mathcal{E}_l^0 \subset \mathcal{E}_l$ be the set of strongly regular points and let $Q_l(\mathcal{E})$ be the set of all strongly regular $\mathfrak{g}$-orbits, where, by a strongly regular point (and orbit), we mean such points of $\mathcal{E}_l$ that are regular with respect to $\mathfrak{g}$-action and whose projections on $\mathcal{E}_{l-1}$ are regular, too.

Then, as we have seen, $Q_l(\mathcal{E})$ are algebraic manifolds, and the projections $\varkappa_l \colon \mathcal{E}_l^0 \to Q_l(\mathcal{E})$ are rational maps such that the fields $\mathcal{F}(Q_l(\mathcal{E}))$ (the field of rational functions on $Q_l(\mathcal{E})$), and $\mathcal{F}(\mathcal{E}_l^0)^{\mathfrak{g}}$ (the field of rational functions on $\mathcal{E}_l^0$), which are $\mathfrak{g}$-invariants (rational differential invariants), coincide: $\varkappa_l^*(\mathcal{F}(Q_l(\mathcal{E}))) = \mathcal{F}(\mathcal{E}_l^0)^{\mathfrak{g}}$.

The $\mathfrak{g}$-action preserves the Cartan distributions $C(\mathcal{E}_l)$ on the equations, and therefore, projections $\varkappa_l$ define distributions $C(Q_l)$ on the quotients $Q_l(\mathcal{E})$.

Finally, we get the tower of algebraic bundles of the quotients

$$Q_k(\mathcal{E}) \overset{\pi_{k+1,k}}{\longleftarrow} Q_{k+1}(\mathcal{E}) \longleftarrow \cdots \longleftarrow Q_l(\mathcal{E}) \overset{\pi_{l+1,l}}{\longleftarrow} Q_{l+1}(\mathcal{E}) \longleftarrow \cdots$$

such that the projection of the distribution $C(Q_l)$ belongs to $C(Q_{l-1}(\mathcal{E}))$.

2.4. *Tresse Derivatives*

Let $\omega \in \Omega^1(J^k)$ be a differential 1-form on a k-jet manifold. Then, the class

$$\omega^h = \pi_{k+1,k}^*(\omega) \mod \text{Ann}\, C_{k+1},$$

is called a horizontal part of ω.

In the standard jet coordinates (x, u_σ^j), the horizontal part has the following representation:

$$\omega = \sum_i a_i dx_i + \sum_{\substack{|\sigma| \le k \\ j \le m}} a_\sigma^j du_\sigma^j \implies \omega^h = \sum_i a_i dx_i + \sum_{\substack{|\sigma| \le k \\ j \le m, i \le n}} a_\sigma^j u_{\sigma+1_i}^j dx_i,$$

where $n = \dim M$ and $m = \dim \pi$.

As a particular case of this construction, we get the total differential $f \in C^\infty(\mathbf{J}^k) \implies \widehat{d}f = (df)^h$, or, in the standard jet coordinates,

$$\widehat{d}f = \sum_{i \le n} \frac{df}{dx_i} dx_i,$$

where

$$\frac{d}{dx_i} = \frac{\partial}{\partial x_i} + \sum_{j,\sigma} u_{\sigma+1_i}^j \frac{\partial}{\partial u_\sigma^j}$$

are the total derivations.

It is important to observe that the operation of taking a horizontal part, as well as the total differential, is invariant with respect to the point and contact transformations.

We say that functions $f_1, \ldots, f_n \in C^\infty(\mathbf{J}^k)$ are in general position on a domain D if

$$\widehat{d}f_1 \wedge \cdots \wedge \widehat{d}f_n \ne 0$$

on this domain.

Let f be a smooth function on this domain; then, we get decomposition in D:

$$\widehat{d}f = \sum_{i \le n} F_i \widehat{d}f_i,$$

where F_i are smooth functions on the domain $\pi_{k+1,k}^{-1}(D) \subset \mathbf{J}^{k+1}$.

We call them Tresse derivatives [5] and denote them by

$$\frac{df}{df_i}.$$

As we have seen, the operation of taking a horizontal part, as well as the total differential, is invariant with respect to the point and contact transformations.

Therefore, we have the following.

Proposition 1. *Let $f_1, \ldots, f_n$ be $\mathfrak{g}$-invariants of order $\le k$ that are in general position. Then, for any $\mathfrak{g}$-invariant f of order $\le k$, the Tresse derivatives $\frac{df}{df_i}$ are $\mathfrak{g}$-invariants of order $\le k+1$.*

2.5. The Lie–Tresse Theorem

Theorem 1. *[6] Let $\mathcal{E}_k \subset \mathbf{J}^k$ be a formally integrable algebraic differential equation and let $\mathfrak{g}$ be an algebraic symmetry Lie algebra. Then, there are rational differential $\mathfrak{g}$-invariants $a_1, \ldots, a_n, b^1, \ldots, b^N$ of order $\le l$ such that the field of all rational differential $\mathfrak{g}$-invariants is generated by rational functions of these invariants and their Tresse derivatives $\frac{d^{|\alpha|} b^j}{da^\alpha}$.*

We call invariants $a_1, \ldots, a_n, b^1, \ldots, b^N$ Lie–Tresse coordinates.

It is noteworthy that, in contrast to algebraic invariants, for which we have the algebraic operations only, in the case of differential invariants, we have additional operations, i.e., Tresse derivatives, that allow us to get really new invariants.

Syzygies, in the case of differential invariants, provide us with new differential equations that we call quotient equations.

From the geometrical point of view, the above theorem states that there is a level l and a domain $D \subset Q(\mathcal{E})$ where the invariants a_i and b^j can be considered as local coordinates, and the preimage of D in the tower

$$Q_l(\mathcal{E}) \xleftarrow{\pi_{l+1,l}} Q_{l+1}(\mathcal{E}) \longleftarrow \cdots \longleftarrow Q_r(\mathcal{E}) \xleftarrow{\pi_{r+1,r}} Q_{r+1}(\mathcal{E}) \longleftarrow \cdots \qquad (2)$$

is just an infinitely prolonged differential equation given by the syzygy.

For this reason, we call the quotient tower (2) an algebraic diffiety.

2.6. Relations between Differential Equations and Their Quotients

1. Let $u = f(x)$ be a solution of differential equation $\mathcal{E}$ and let $a_i(f)$ and $b^j(f)$ be values of the invariants a_i and b^j on the section f. Then, locally, $b^j(f) = B^j(a(f))$, and therefore, $b^j = B^j(a)$ is the solution of the quotient equation.
2. The above construction is local. In general, the correspondence between solutions is valid on the level of generalized solutions, i.e., on the level of integral manifolds of the Cartan distributions. In addition, the correspondence will lead us to integral manifolds with singularities.
3. Now let $b^j = B^j(a)$ be a solution of the quotient equation. Then, considering equations $b^j - B^j(a) = 0$ as a differential constraint for the equation $\mathcal{E}$, we get a finite-type equation $\mathcal{E} \cap \{b^j - B^j(a) = 0\}$ with a solution that is a $\mathfrak{g}$-orbit of a solution of $\mathcal{E}$.
4. Symmetries of the quotient equation are Bäcklund-type transformations of the original equation $\mathcal{E}$.

Example 3. *The Lie algebra of the projective transformations of the line $M = \mathbb{R}$, $\mathfrak{g} = \mathfrak{sl}_2 = \langle \partial_x, x\partial_x, x^2\partial_x \rangle$ has the following generators in rational differential invariants for the $\mathfrak{sl}_2$-action on functions:*

$$\left\langle a = u_0,\; b = \frac{u_3}{u_1^3} - \frac{3u_2^2}{2u_1^4},\; \frac{db}{da} = \frac{u_4}{u_1^4} - 6\frac{u_2 u_3}{u_1^5} + 6\frac{u_2^3}{u_1^6}, \ldots \right\rangle.$$

Let

$$F\left(u_0,\; \frac{u_3}{u_1^3} - \frac{3u_2^2}{2u_1^4},\; \frac{u_4}{u_1^4} - 6\frac{u_2 u_3}{u_1^5} + 6\frac{u_2^3}{u_1^6}\right) = 0$$

be a fourth-order $\mathfrak{sl}_2$-invariant equation.

Then, the quotient equation has the first order:

$$F\left(a, b, \frac{db}{da}\right) = 0.$$

Example 4. *The Lie algebra $\mathfrak{g} = \langle \partial_x, \partial_y \rangle$ of translations of the plane has the following Lie–Tresse coordinates for the $\mathfrak{g}$-action on functions:*

$$a_1 = u_{1,0},\; a_2 = u_{0,1},\; b = u_{0,0},\; c = u_{1,1}.$$

Then,

$$b_{1,0} = \delta^{-1}(u_{1,0}u_{0,2} - u_{0,1}u_{1,1}), \quad b_{0,1} = \delta^{-1}(u_{0,1}u_{2,0} - u_{1,0}u_{1,1}),$$
$$c_{1,0} = \delta^{-1}(u_{0,2}u_{2,1} - u_{1,1}u_{1,2}), \quad c_{0,1} = \delta^{-1}(u_{2,0}u_{1,2} - u_{1,1}u_{2,1}),$$

where $\delta = u_{2,0}u_{0,2} - u_{1,1}^2$ is a Hessian determinant, and the syzygy is

$$c^2(b_{1,0}^2 b_{0,2} - 2b_{1,0}b_{0,1}b_{1,1} + b_{0,1}^2 b_{2,0}) + c(b_{1,0}b_{0,1} - a_1 b_{0,1}b_{2,0} - a_2 b_{1,0}b_{0,2} +$$
$$b_{1,1}(a_1 b_{1,0} + a_2 b_{0,1}) - a_1 a_2 b_{1,1} - b_{1,0}b_{0,1}(a_1 c_{1,0} + a_2 c_{0,1}) = 0.$$

Thus, the quotient of an equation $u_{1,1} = C(u_{1,0}, u_{0,1})$ is the last equation for $b(u_{1,0}, u_{0,1})$, where letter C stands for c.

3. Euler Equations on a Curve

In this section, we briefly recall the necessary results obtained in [1].

Consideration of flows of an inviscid medium on a space curve $M = \{x = f(a), y = g(a), z = \lambda a\}$ in a field of constant gravitational force leads to the system

$$\begin{cases} \rho(u_t + uu_a) = -p_a - \rho g \lambda, \\ \rho_t + (\rho u)_a = 0, \\ \rho \theta(s_t + u s_a) - k \theta_{aa} = 0, \end{cases} \quad (3)$$

where p and s are expressed in terms of Planck potential [2] $\Phi(\rho, \theta)$:

$$p(\rho, \theta) = -R\rho^2 \theta \Phi_\rho, \quad s(\rho, \theta) = R(\Phi + \theta \Phi_\theta),$$

where R is the universal gas constant.

To describe this Lie algebra, we consider a Lie algebra $\mathfrak{g}$ of point symmetries of the PDE system (3).

Let $\vartheta : \mathfrak{g} \to \mathfrak{h}$ be the following Lie algebra's homomorphism

$$\vartheta : X \mapsto X(\rho)\partial_\rho + X(s)\partial_s + X(p)\partial_p + X(\theta)\partial_\theta,$$

where $\mathfrak{h}$ is a Lie algebra generated by vector fields that act on the thermodynamic values p, ρ, s, and θ.

It was demonstrated [1] that if $h(a) = \lambda a$, the Lie algebra $\mathfrak{g}$ of point symmetries of the system (1) is generated by the vector fields

$$X_1 = \partial_t, \quad X_2 = \partial_p, \quad X_3 = \partial_s,$$
$$X_4 = \theta \partial_\theta, \quad X_5 = p \partial_p + \rho \partial_\rho - s \partial_s,$$
$$X_6 = \partial_a, \quad X_7 = t \partial_a + \partial_u,$$
$$X_8 = t \partial_t + 2a \partial_a + u \partial_u - 2\rho \partial_\rho - s \partial_s,$$
$$X_9 = \left(\frac{t^2}{2} + \frac{a}{\lambda g}\right) \partial_a + \left(t + \frac{u}{\lambda g}\right) \partial_u - \frac{2\rho}{\lambda g} \partial_\rho.$$

The pure thermodynamic part $\mathfrak{h}_t$ of the symmetry algebra is generated by the vector fields

$$Y_1 = \partial_p, \quad Y_2 = \partial_s, \quad Y_3 = \theta \partial_\theta,$$
$$Y_4 = p \partial_p, \quad Y_5 = \rho \partial_\rho, \quad Y_6 = s \partial_s.$$

Thus, the Euler system has a Lie algebra of point symmetries $\vartheta^{-1}(\mathfrak{h}_t)$.

It has been shown in [1] that, for $h(a) = \text{const}$, $h(a) = \lambda a$, and $h(a) = \lambda a^2$, the basis differential invariants are

$$J_1 = \rho, \quad J_2 = \theta, \quad J_3 = u_a, \quad J_4 = \rho_a, \quad J_5 = \theta_a, \quad J_6 = \theta_t + u\theta_a,$$

and the basis invariant derivatives are

$$\frac{d}{dt} + u\frac{d}{da}, \quad \frac{d}{da}.$$

4. Quotient Equation

Choosing J_1 and J_2 as Lie–Tresse coordinates (x, y) and

$$K(x,y) = J_3, \quad L(x,y) = J_4, \quad M(x,y) = J_5, \quad N(x,y) = J_6$$

as unknown functions, respectively, we get the quotient equation for (3):

$$\begin{cases} Rxy(xK(\Phi_x + y\Phi_{xy}) - N(2\Phi_y + y\Phi_{yy})) + LM_x + MM_y = 0, \\ xKM_x - NM_y + LN_x + M(N_y - K) = 0, \\ xMK_y - xKL_x + xLK_x + 2KL + NL_y = 0, \\ RxyL(\Phi_{xxx}L^2 + 2\Phi_{xxy}ML + \Phi_{xyy}M^2) + \\ RL(xyLL_x + xyML_y + 2xLM + 3yL^2)\Phi_{xx} + \\ RL(xyLM_x + M(xyM_y + 2xM + 3yL))\Phi_{xy} + \\ LR(2yLL_x + 2yML_y + xLM_x + M(xM_y + 3L))\Phi_x + \\ xK^2L_x - KNL_y - (xKM + LN)K_y - 3LK^2 = 0 \end{cases} \qquad (4)$$

Direct computations show that the system (4) has no symmetries if the function Φ is arbitrary. Nevertheless, it is possible to find symmetries for some classes of Φ. Some of these cases are listed below.

Proposition 2. *If the system (4) admits a symmetry of the form*

$$\alpha_1 x \partial_x + (\alpha_2 y + \alpha_3)\partial_y - \alpha_2 K \partial_K + \frac{1}{2}(3\alpha_1 - 2\alpha_2 - \alpha_4)L\partial_L + \frac{1}{2}(\alpha_1 - \alpha_4)M\partial_M,$$

then the function Φ is of the form

$$\Phi(x,y) = C_5 \int \frac{(\alpha_2 y + \alpha_3)^{-\frac{\alpha_4}{\alpha_2}}}{y^2} dy + \frac{C_4 x^{\frac{\alpha_4}{\alpha_1}-1}}{y} + \frac{C_3 y + C_2}{y} + \frac{C_1}{xy},$$

where $C_1, \ldots C_5$ are constants.

Proposition 3. *If the system (4) admits a symmetry of the form*

$$x\frac{\partial}{\partial x} + \alpha_2 y \frac{\partial}{\partial y} - \alpha_2 K \frac{\partial}{\partial K} + \frac{1}{2}(3 - 2\alpha_2 - \alpha_4)L\frac{\partial}{\partial L} + \frac{1}{2}(1 - \alpha_4)M\frac{\partial}{\partial M},$$

then the function Φ is the following

$$\Phi(x,y) = C_5 + \frac{C_4}{y} + C_3 y^{\frac{\alpha_4}{\alpha_2}-1} + \frac{C_2}{xy} + \frac{C_1 x^{\alpha_4 - 1}}{y},$$

where $C_1, \ldots C_5$ are constants.

Particular solutions of (4) for some special classes of the function Φ can be found. For example, consider the Planck potential for the ideal gas model:

$$\Phi(x,y) = \frac{n}{2} \ln y - \ln x, \qquad (5)$$

where n is the number of freedom degrees of a gas particle.

Then, for simplicity, let $N = K = 0$, then these are some of the solutions for L and M:

1. $L = 0, M = f(x)$.
2. $L = \frac{c_1 x}{y}, f(M)x^{\frac{M}{c_1}} = y$.
3. $L = \frac{c_3 x}{(\ln x - c_2)^{c_1}} \left(\frac{-c_5 \ln y + c_4}{c_1} y^{-c_1} \right)^{c_1}, M = \frac{c_3 y(-c_5 \ln y + c_4)}{(\ln x - c_2)^{c_1}(-\ln x + c_2)c_5} \left(\frac{-c_5 \ln y + c_4}{c_1} y^{-1/c_1} \right)^{c_1}$.

Here, $c_1, \ldots, c_5$ are constants and $f(x)$ is an arbitrary function.

Let us illustrate how we can solve the original Euler PDE system using its quotient. To this end, we consider the system (3) for ideal gas together with the solution (for example, $N = K = L = 0, M = x$), which is equivalent to a finite-type system:

$$\begin{cases} \theta_{aa} = 0, & \rho_t = 0, \quad \rho_a = 0, \quad R\rho\theta_a + \rho(g\lambda + u_t) = 0, \\ \theta_a = \rho, & u_a = 0, \quad \theta_t + u\theta_a = 0. \end{cases}$$

Solving the latter, we get

$$\rho = \rho_0, \quad u = u_0 - (\lambda g - R\rho_0)t, \quad \theta = \frac{(R\rho_0 + g\lambda)\rho_0 t^2}{2} + \rho_0(a - u_0 t) + \theta_0,$$

where ρ_0, u_0, θ_0 are arbitrary constants.

Virial Expansion

Another approach we can take is to exploit the fact that it is often possible to consider the Planck potential Φ in the form of virial expansion:

$$\Phi(x, y) = \frac{n}{2} \ln y - \ln x - \sum_{i=1}^{\infty} \frac{x^i}{i} A_i(y).$$

Then, we can find solutions of the system (4) in the form of power series of x:

$$K(x, y) = x^{d_K} \sum_{k=0}^{\infty} K_k(y) x^k, \qquad L(x, y) = x^{d_L} \sum_{k=0}^{\infty} L_k(y) x^k,$$

$$M(x, y) = x^{d_M} \sum_{k=0}^{\infty} M_k(y) x^k, \qquad N(x, y) = x^{d_N} \sum_{k=0}^{\infty} N_k(y) x^k,$$

where d_K, d_L, d_M, and d_N are the integer constants that should be chosen such that (4) can be expanded as a power series of x. It can be shown that $d_K = 1, d_L = 2, d_M = 1$, and $d_N = 1$. Hence, the zeroth-order term of this expansion is a system of ordinary differential equations:

$$\begin{cases} (K_0 M_0 + N_0 L_0) K_0' + (R y L_0 M_0 + K_0 N_0) L_0' + 2 R L_0 M_0 M_0' + \\ L_0 (R y L_0^2 + 2 R L_0 M_0 + K_0^2) = 0, \\ M_0 K_0' + N_0 L_0' + K_0 L_0 = 0, \\ M_0 N_0' - N_0 M_0' + N_0 L_0 = 0, \\ M_0 M_0' - R y K_0 + M_0 L_0 - \dfrac{R n}{2} N_0 = 0. \end{cases} \quad (6)$$

The first-order term of the expansion is a system of linear ordinary differential equations:

$$\begin{cases} (M_0 M_1)' + M_0 L_1 + 2M_1 L_0 + Ry A_1'(2N_0 - yK_0) - \\ Ry K_0 A_1 - \dfrac{R}{2}(nN_1 + 2yK_1) + Ry^2 N_0 A_1'' = 0, \\ M_0 N_1' - N_1 M_0' + M_1 N_0' - N_0 M_1' + M_1 K_0 + N_0 L_1 + 2N_1 L_0 = 0, \\ M_0 K_1' + M_1 K_0' + N_0 L_1' + N_1 L_0' + 2L_0 K_1 = 0, \\ (N_0 L_0 + K_0 M_0) K_1' + (Ry M_0 L_0 + K_0 N_0) L_1' + R L_0 M_0 M_1' + \\ (Ry L_0 L_0' + R L_0 M_0' + K_0 K_0' + 3 R L_0^2) M_1 + \\ (L_0 K_0)' N_1 + (M_0 K_0' + N_0 L_0' + K_0 L_0) K_1 + \\ (4R L_0 M_0 + 4Ry L_0^2 + N_0 K_0' + R M_0 M_0' + Ry L_0 M_0) L_1 + \\ 4Ry L_0^2 (L_0 A_1 + A_1' M_0) + Ry A_1'' L_0 M_0^2 + \\ Ry A_1' M_0 M_0' L_0 + 2Ry L_0 L_0' M_0 A_1 + 4 R L_0^2 M_0 A_1 + \\ 2R A_1' M_0^2 L_0 + R L_0 M_0 M_0' A_1 = 0. \end{cases} \quad (7)$$

The solutions of (6) must be substituted into (7); thus, we obtain more simple differential equations for the functions K_1, L_1, M_1, and N_1. Repeating this process, we can obtain any number of terms in the expansions of the functions K, L, M, and N.

5. Conclusions

In this paper, we gave a brief recollection of the notion of quotient equations. Using previous results regarding invariants of the Euler system in a space, we found its quotient. We found that the quotient has an infinitesimal symmetry for special cases of the thermodynamical state of a medium. We proposed a method for solving the quotient by means of virial expansion of the Planck potential and by reducing it to series of systems of ordinary differential equations.

Author Contributions: Conceptualization, V.L.; writing—original draft preparation, A.D., V.L., and S.T.; writing—review and editing, A.D., V.L., and S.T.; supervision, V.L. All authors have read and agreed to the published version of the manuscript.

Funding: This research was funded by the Russian Foundation for Basic Research grant number 18-29-10013.

Institutional Review Board Statement: Not applicable.

Informed Consent Statement: Not applicable.

Data Availability Statement: Data sharing not applicable.

Conflicts of Interest: The authors declare no conflict of interest.

References

1. Duyunova, A.; Lychagin, V.; Tychkov, S. Symmetries and differential invariants for inviscid flows on a curve. *Lobachevskii J. Math.* **2020**, *41*, 2435–2447.
2. Lychagin, V.; Roop, M. Critical Phenomena in Filtration Processes of Real Gases. *Lobachevskii J. Math.* **2020**, *41*, 382–399. [CrossRef]
3. Krasilchchik, I.; Vinogradov, A.M.; Lychagin, V.V. *Geometry of Jet Spaces and Nonlinear Partial Differential Equations*; Gordon and Breach Science Publishers: New York, NY, USA, 1986.
4. Rosenlicht, M. Some basic theorems on algebraic groups. *Am. J. Math.* **1956**, *78*, 401–443. [CrossRef]
5. Tresse, A.R. Sur les invariants différentiels des groupes continus de transformations. *Acta Math.* **1894**, *18*, 1. [CrossRef]
6. Kruglikov, B.; Lychagin, V. Global Lie–Tresse theorem. *Sel. Math.* **2016**, *22*, 1357–1411. [CrossRef]

Article

Iterated Darboux Transformation for Isothermic Surfaces in Terms of Clifford Numbers

Jan L. Cieśliński * and Zbigniew Hasiewicz

Wydział Fizyki, Uniwersytet w Białymstoku, ul. Ciołkowskiego 1L, 15-245 Białystok, Poland; z.hasiewicz@uwb.edu.pl
* Correspondence: j.cieslinski@uwb.edu.pl

Abstract: Isothermic surfaces are defined as immersions with the curvture lines admitting conformal parameterization. We present and discuss the reconstruction of the iterated Darboux transformation using Clifford numbers instead of matrices. In particulalr, we derive a symmetric formula for the two-fold Darboux transformation, explicitly showing Bianchi's permutability theorem. In algebraic calculations an important role is played by the main anti-automorphism (reversion) of the Clifford algebra $\mathcal{C}(4,1)$ and the spinorial norm in the corresponding Spin group.

Keywords: integrable systems; Darboux-Bäcklund transformation; isothermic immersions; Spin groups; Clifford algebras

1. Introduction

Isothermic surfaces have a very long history. They have been first introduced by Lamé in studies on stationary heat flows (described by the Laplace equation), in the broader context of triply ortogonal systems of coordinates [1]. Then, the main progress towards the theory of isothermic surfaces was done by Bertrand [2], who was first to notice that "in any triply isothermic (in physical sense) orthogonal system in $\mathbb{E}^3$ any coordinate surface admits conformal curvature parameterization" [3]. Transformations of isothemic surfaces, studied by Darboux and Bianchi [4,5], strongly suggested that the related system of nonlinear partial differential equations (see (2) below) is integrable in the sense of the soliton theory [6] and, indeed, such modern formulation of this problem was found [7], which started new developments in this field [8–12]. It is worthwhile to mention that isothermic immersions are invariant with respect to conformal transformations of the ambient space and can be naturally described in terms of conformal geometry (then Darboux transformations correspond to Ribaucour congruences [13]). Studies on isothermic surfaces are still active, see, e.g., [14–19]. In this paper we develop an approach based on using Clifford algebras and Spin groups [20,21] (different from the approach of [12,22]). We re-derive the construction of "multisoliton" surfaces by iterated Darboux transformation. In particular, we present detailed computation of the two-fold Darboux transform.

2. Isothermic Surfaces in $\mathbb{R}^3$

Isothermic surfaces (or, more precisely, isothermic immersions) are characterized as surfaces immersed in E^3 with curvature lines admitting conformal parameterization. It means that there exist coordinates (u,v) in which the isothermic immersion has the following fundamental forms:

$$\begin{aligned} I &= e^{2\theta}(du^2 + dv^2), \\ II &= e^{2\theta}(k_1 du^2 + k_2 dv^2), \end{aligned} \tag{1}$$

where ϑ, k_1, k_2 are functions of u, v, which have to satisfy the following system of nonlinear partial differential equations known as Gauss–Mainardi–Codazzi equations:

$$\vartheta_{,uu} + \vartheta_{,vv} + k_1 k_2 e^{2\vartheta} = 0,$$

$$k_{2,u} + (k_2 - k_1)\vartheta_{,u} = 0, \qquad (2)$$

$$k_{1,v} + (k_1 - k_2)\vartheta_{,v} = 0,$$

where comma denotes partial derivtive. Geometrically, k_1 and k_2 are principal curvatures, and their product $k_1 k_2$ yields the Gaussian curvature. The above nonlinear system can be obtained (see [7]) as compatibility conditions for the following linear problem (or Lax pair):

$$\Psi_{,u} = \tfrac{1}{2} \mathbf{e}_1 (-\vartheta_{,v} \mathbf{e}_2 + k_1 e^{\vartheta} \mathbf{e}_3 + \lambda \sinh \vartheta \mathbf{e}_4 + \lambda \cosh \vartheta \mathbf{e}_5) \Psi,$$
$$\Psi_{,v} = \tfrac{1}{2} \mathbf{e}_2 (-\vartheta_{,u} \mathbf{e}_1 + k_2 e^{\vartheta} \mathbf{e}_3 + \lambda \cosh \vartheta \mathbf{e}_4 + \lambda \sinh \vartheta \mathbf{e}_5) \Psi, \qquad (3)$$

where $\mathbf{e}_1, \ldots, \mathbf{e}_5$ are 4×4 complex matrices (for their exact form see [7] or [23]) that satisfy the relations

$$\mathbf{e}_1^2 = \mathbf{e}_2^2 = \mathbf{e}_3^2 = \mathbf{e}_4^2 = -\mathbf{e}_5^2 = 1, \qquad \mathbf{e}_j \mathbf{e}_k = -\mathbf{e}_k \mathbf{e}_j \quad (j \neq k). \qquad (4)$$

We point out that using a suitable extension of the famous Sym formula (see, e.g., [24]) we can reconstruct the radius vector of the isothermic surface implicitly determined by the fundamental forms (1), for more details see Section 4 and Theorem 1.

3. Clifford Algebras

The matrices $\mathbf{e}_1, \ldots, \mathbf{e}_5$ satisfying (4) can be interpreted as elements of a Clifford algebra, see below. Their exact matrix form is not needed. From technical point of view, it is even easier to use Clifford numbers instead of particular matrix representations [21].

We recall the definition of a Clifford algebra generated by vectors of a Euclidean or pseudo-Euclidean inner product space. Let V be a vector space endowed with a non-degenerate (but not necessarily positive definite) quadratic form Q, see, e.g., [25,26]. If Q is positive definite (the Euclidean case), then $Q(v)$ is the square of the length of v. The associated bilinear form (scalar product) will be denoted by brackets. In particular, $Q(v) \equiv \langle v | v \rangle$. The Clifford algebra $\mathcal{C}(V, Q)$ is generated by products ("Clifford products") of vectors (elements of V). The Clifford product is defined by the following relation:

$$vw + wv = 2\langle v | w \rangle \mathbf{1} \qquad (5)$$

where $\mathbf{1}$ denotes the unit of the Clifford algebra.

Let $\{\mathbf{e}_1, \ldots, \mathbf{e}_n\}$ be an orthonormal basis of V, i.e., $\langle \mathbf{e}_j | \mathbf{e}_k \rangle = 0$ for $j \neq k$, and $\langle \mathbf{e}_k | \mathbf{e}_k \rangle = \pm 1$. The dimension of the Clifford algebra is 2^n. Its standard basis consists of

$$\mathbf{1}, \quad \mathbf{e}_k, \quad \mathbf{e}_{jk} \ (j < k), \quad \mathbf{e}_i \mathbf{e}_j \mathbf{e}_k \ (i < j < k), \ \ldots$$

If the signature of Q is (m, p) (i.e., among $\mathbf{e}_1, \ldots, \mathbf{e}_n$ there are m vectors such that $\mathbf{e}_j^2 = 1$ and p vectors such that $\mathbf{e}_j^2 = -1$, and $m + p = n$), then we denote $\mathcal{C}(V, Q) \equiv \mathcal{C}_{m,p}$.

The Clifford group (or Lipschitz group) $\Gamma(V, Q)$ is the multiplicative group (with respect to the Clifford product) generated by the non-isotropic vectors (we recall that $w \in V$ is isotropic (or null) vector if $\langle w | w \rangle = 0$). The spinor norm of an element $X \in \Gamma(V, Q)$ is defined as

$$N(X) := \beta(X) X, \qquad (6)$$

where β is a reversion, i.e., $\beta(v_1 v_2 \ldots v_M) = v_M v_{M-1} \ldots v_2 v_1$. One can easily see that the spinor norm of a vector is its scalar square, $N(v) = \langle v | v \rangle$, and for any element of $\Gamma(V, Q)$ we have

$$N(v_1 v_2 \ldots v_M) = \langle v_1 | v_1 \rangle \langle v_2 | v_2 \rangle \ldots \langle v_M | v_M \rangle. \tag{7}$$

The group $\text{Pin}(V, Q)$ is a subgroup of $\Gamma(V, Q)$ consisting of elements X such that $N^2(X) = \pm 1$ (i.e., $\Gamma(V, Q)$ contains products of unit vectors), and the group $\text{Spin}(V, Q)$ (a subgroup of $\text{Pin}(V, Q)$) consists of products of even number of unit vectors [26].

4. Spin-Valued Lax Pairs

Our approach to the construction of Darboux transformation consists of two steps. First, we characterize the structure of the Lax pair. Second, we are looking for a transformation preserving the structure [27].

The structure of the Lax pair is characterized by the dependence on λ (e.g., divisor of poles) [6,28], reduction group (loop group) [29] and, possibly, by other invariants of Darboux transformations, like linear and multilinear constraints on coefficients of the Laurent expansion around poles [30].

In this section we present the characterization of the Lax pair (3), see [20,31]. First of all, we consider Clifford numbers instead of matrices. Then, we notice that both matrices of this Lax pair are Clifford bi-vectors linear in λ such that

$$\Psi_{,\mu} = U_\mu \Psi, \qquad U_\mu = \frac{1}{2} \mathbf{e}_\mu (\lambda \mathbf{a}_\mu + \mathbf{b}_\mu) \qquad (\mu = 1, 2) \tag{8}$$

where $\Psi = \Psi(u, v)$, $\Psi_{,1} \equiv \Psi_{,u}$, $\Psi_{,2} \equiv \Psi_{,v}$, $\mathbf{a}_\mu = \mathbf{a}_\mu(u, v) \in W$, $\mathbf{b}_\mu = \mathbf{b}_\mu(u, v) \in V$, V and W are real vector spaces, V is spanned by $\mathbf{e}_1, \mathbf{e}_2$ and $\mathbf{e}_3$, and W is spanned by $\mathbf{e}_4$ and $\mathbf{e}_5$. We assume relations (4), which means that form now on the quadratic form Q, defined on $V \oplus W$, is assumed to have signature $(4, 1)$.

The compatibility conditions for the linear system (8) imply that $\mathbf{a}_1$ and $\mathbf{a}_2$ form an orthogonal basis in W. We can confine ourselves to the particular case (3) without loss of the generality. Indeed, both linear problems are equivalent up to a re-parameterization of independent variables and a discrete transformation in the space W.

The form (8) of the spectral problem can be described in terms of some group constraints ("reduction group", compare [27,29]). First, U_μ are linear combinations of Clifford bi-vectors. In other words, U_μ take values in the Lie algebra of the group $\text{Spin}(V \oplus W, Q)$. In principle, Ψ could be a spinor, but here and in the sequel we assume that it is an element of the Clifford algebra. Without loss of the generality we can confine ourselves to solutions $\Psi \in \Gamma(V, Q)$. The next observation is $\beta(U_\mu) = -U_\mu$.

Lemma 1. *If $\beta(U_\mu) = -U_\mu$ (for $\mu = 1, 2$), and Ψ satisfies $\Psi_{,\mu} = U_\mu \Psi$, then*

$$N(\Psi) \equiv \Psi \beta(\Psi) = \text{const}, \tag{9}$$

Proof. It is sufficient to differentiate $N(\Psi)$:

$$(N(\Psi))_{,\mu} = \Psi_{,\mu} \beta(\Psi) + \Psi \beta(\Psi_{,\mu}) = (U_\mu + \beta(U_\mu)) N(\Psi) = 0, \tag{10}$$

where one has to remember that $N(\Psi)$ is a scalar, so it commutes with any elements. □

Therefore, $\Psi \in \text{Spin}(V \oplus W, Q)$ (for any u, v) provided that Ψ is Spin-valued at some initial point (u_0, v_0). In an analogous way one can show the following loop group conditions:

$$\Psi(-\lambda) = \mathbf{e}_4 \mathbf{e}_5 \Psi(\lambda) \mathbf{e}_4 \mathbf{e}_5, \qquad \overline{\Psi(\lambda)} = \Psi(\bar{\lambda}) \tag{11}$$

(where the bar denotes complex conjugate and, by definition, $\bar{\mathbf{e}}_j = \mathbf{e}_j$). They follow from

$$\mathbf{e}_4 \mathbf{e}_5 U_\mu(\lambda) = U_\mu(-\lambda) \mathbf{e}_4 \mathbf{e}_5, \qquad \overline{U_\mu(\lambda)} = U_\mu(\bar{\lambda}). \tag{12}$$

The properties $\beta(U_\mu) = -U_\mu$ and (12) hold for the Lax pair (8), which can be easily verified using commutation relations (4).

The Sym-Tafel formula $F = 2\Psi^{-1}\Psi_{,\lambda}$ [24], evaluated in $\lambda = 0$, yields a surface immersed into the 6-dimensional space spanned by bi-vectors of the form $\mathbf{e}_k\mathbf{e}_\alpha$ ($k = 1, 2, 3$; $\alpha = 4, 5$). Projecting this surface on especially chosen 3-dimensional subspaces we obtain the original isothermic surface as a linear combination of $\mathbf{e}_k(\mathbf{e}_4 + \mathbf{e}_5)$ ($k = 1, 2, 3$) and its dual (or Christoffel transform) as a combination of $\mathbf{e}_k(\mathbf{e}_4 - \mathbf{e}_5)$ ($k = 1, 2, 3$) [23].

Here we present some details of calculations involving Clifford numbers instead of matrices. They are closely related but not identical to the approach of our earlier papers [20,23,32].

We use the projection $P : \mathcal{C}(V \oplus W) \to \mathcal{C}(V)$ defined as a homomorphism of Clifford algebras such that

$$P(\mathbf{e}_4) = P(\mathbf{e}_5) = 1. \tag{13}$$

Note that

$$P\left(\sum_{j=1}^{3}\sum_{\alpha=4}^{5} c_{j\alpha} \mathbf{e}_j \mathbf{e}_\alpha\right) = \sum_{j=1}^{3}(c_{j4} + c_{j5})\mathbf{e}_j \tag{14}$$

This projection yields an original isothermic surface (the dual surface is a result of another homomorphism P', defined by $P'(\mathbf{e}_4) = -P'(\mathbf{e}_5) = 1$).

Theorem 1. *We assume that Ψ satisfies the linear system (3) and*

$$F := 2\Psi^{-1}\Psi_{,\lambda}\Big|_{\lambda=0}. \tag{15}$$

Then $\mathbf{r} := P(F)$ yields the original isothermic immersion (up to a Euclidean motion), provided that we identify $\mathrm{span}\{\mathbf{e}_1, \mathbf{e}_2, \mathbf{e}_3\}$ with the space $\mathbb{E}^3$.

Proof. The crucial property of the Sym-Tafel formula (15) is a compact form of its derivative (compare [24]):

$$F_{,\mu} = 2\Psi_0^{-1}U_{\mu,\lambda}(0)\Psi_0 = \Psi_0^{-1}\mathbf{e}_\mu\mathbf{a}_\mu\Psi_0, \tag{16}$$

where $\Psi_0 := \Psi(0)$ (i.e., Ψ evaluated at $\lambda = 0$), and we use the slightly more general form (8) of the Lax pair (3). Then

$$\mathbf{r}_{,\mu} = e^\theta \Psi_0^{-1} \mathbf{e}_\mu \Psi_0. \tag{17}$$

Therefore

$$E_\mu = \Psi_0^{-1}\mathbf{e}_\mu\Psi_0 \tag{18}$$

(for $\mu = 1, 2$) form an orthonormal basis in the tangent space and, therefore, the corresponding metric is given by the first equation of (1). Obviously, E_3 (defined by (18) for $\mu = 3$) is orthogonal to E_1 and E_2; hence, it can be identified with the normal vector. In order to derive the second fundamental form we differentiate (17). Taking into account $\Psi_{0,\mu} = \frac{1}{2}\mathbf{e}_\mu \mathbf{b}_\mu \Psi_0$, we obtain

$$\begin{aligned}
\mathbf{r}_{,11} &= e^\theta(\theta_{,1} E_1 + \langle \mathbf{b}_1 | \mathbf{e}_2\rangle E_2 + \langle \mathbf{b}_1 | \mathbf{e}_3\rangle E_3) = e^\theta(\theta_{,1} E_1 - \theta_{,2} E_2 + k_1 e^\theta E_3), \\
\mathbf{r}_{,12} &= e^\theta(\theta_{,2} E_1 - \langle \mathbf{b}_2 | \mathbf{e}_1\rangle E_2) = e^\theta(\theta_{,2} E_1 + \theta_{,1} E_2), \\
\mathbf{r}_{,21} &= e^\theta(-\langle \mathbf{b}_1 | \mathbf{e}_2\rangle E_1 + \theta_{,1} E_2) = e^\theta(\theta_{,2} E_1 + \theta_{,1} E_2), \\
\mathbf{r}_{,22} &= e^\theta(\langle \mathbf{b}_2 | \mathbf{e}_1\rangle E_1 + \theta_{,2} E_2 + \langle \mathbf{b}_2 | \mathbf{e}_3\rangle E_3) = e^\theta(-\theta_{,1} E_1 + \theta_{,2} E_2 + k_2 e^\theta E_3).
\end{aligned} \tag{19}$$

Therefore, coefficients of the second fundamental form (given by $\langle \mathbf{r}_{,ij}| E_3\rangle$) yield the second formula of (1). The proof is completed by applying the Bonnet theorem. □

5. The Darboux-Bäcklund Transformation in the Case of Spin Groups

The Darboux transformation is a gauge-like transformation using the Darboux matrix D (we will keep using the name "matrix" even for D given in terms of Clifford numbers without referring to any matrix representation):

$$\tilde{\Psi} = D\Psi, \qquad \tilde{\Psi}_{,\mu} = \tilde{U}_\mu \tilde{\Psi}, \qquad \tilde{U}_\mu = D_{,\mu} D^{-1} + D U_\mu D^{-1}, \qquad (20)$$

provided that $\tilde{U}_\mu$ has the same dependence on dependent variables as U_μ, see, e.g., [6,33]. In Section 4 we have shown that the form (3) can be derived by imposing a set of constraints on a general linear problem (U_μ are Clifford bi-vectors, linear in λ and belong to the appropriate loop algebra). Then the Darboux transformation has to preserve this structure, which means, in particular, that D should belong to the same group as Ψ.

Different methods of constructing the Darboux matrix need different form of λ-dependence of D (these forms are equivalent up to a λ-dependent scalar factor [30]). In particular, one can assume D as polynomial in λ (*eigenvalues*, corresponding to solitons, are zeros of det D) [34], sum of simple fractions (*eigenvalues*: poles of D and D^{-1}) [6,29], or a "realization" ($D = N + F(\lambda - A)^{-1}G$) [35,36].

Our motivation for dealing with the case of Spin groups came from yet another approach [31]. Multiplying (20) by $D^2(\lambda)$ we get

$$D_{,\mu} D + D U_\mu D = \tilde{U}_\mu D^2. \qquad (21)$$

It is a crucial point that the right-hand side vanishes for λ_+ and λ_- such that $D^2(\lambda_\pm) = 0$. Then, we obtain a solution of the remaining equation: $D(\lambda_\pm) = \rho_\pm \Psi(\lambda_\pm) d_\pm \Psi(\lambda_\pm)^{-1}$, where $d_\pm = $ const, $(d_\pm)^2 = 0$ and $\rho_\pm$ are two scalar functions. Finally, $D(\lambda)$ is given as a linear combination of $D(\lambda_+)$ and $D(\lambda_-)$ with coefficients linear in λ [31], which yields one-soliton Darboux matrix. This approach was extended on the multi-soliton case for 2×2 matrix problems [37].

Generalization of this approach on Spin-valued linear problems is quite natural. Instead of multiplying both sides of (20) by D^2 we multiply them by $D\beta(D)$:

$$D_{,\mu} \beta(D) + D U_\mu \beta(D) = \tilde{U}_\mu D \beta(D). \qquad (22)$$

Note that $\beta(D) = D$ if D is a Clifford vector (which has been usually assumed in earlier papers, like [31]), and in this case Equation (22) assumes the form (21).

Lemma 2. *If an isotropic Clifford vector D' satisfies Equation (22), and G is any Clifford number G (not necessarily constant), then $D = GD'$ satisfies Equation (22) as well.*

Proof. Lemma can be shown by straightforward calculation. First, we have

$$D\beta(D) = GD'\beta(D')\beta(G) = G(D')^2 \beta(G) = 0, \qquad (23)$$

so the right-hand side of (22) vanishes. Then

$$D_{,\mu}\beta(D) + DU_\mu\beta(D) = G_{,\mu} D'\beta(D')\beta(G) + G(D'_{,\mu}\beta(D') + D'U_\mu\beta(D'))\beta(G) = 0, \qquad (24)$$

which ends the proof. □

In this paper we confine ourselves to iterations of the simplest Darboux transformations (defined by D linear in λ). Then we can use the results of [31], where the case of the Clifford vector (here denoted by D') was considered, and the following form of the Darboux transformation was derived:

$$D'(\lambda) = \frac{\lambda - \lambda_-}{\lambda_+ - \lambda_-} D'(\lambda_+) + \frac{\lambda - \lambda_+}{\lambda_- - \lambda_+} D'(\lambda_-), \qquad (25)$$

and $D'(\lambda_\pm)$ can be expressed as

$$D'(\lambda_\pm) = \rho_\pm \Psi(\lambda_\pm) d_\pm \Psi(\lambda_\pm)^{-1} \tag{26}$$

where $\rho_\pm$ are (arbitrary) scalar functions and $d_\pm$ are constant elements such that $d_\pm^2 = 0$. Reductions (12) impose constraints on $\lambda_\pm, \rho_\pm$, and $d_\pm$ (see [31]):

$$\lambda_+ = i\kappa, \quad \lambda_- = -i\kappa \quad (\kappa \in \mathbb{R}), \quad \rho_+ = \rho_- \equiv \rho \in \mathbb{R}, \quad d_\pm = \kappa(p_0 \pm in_0). \tag{27}$$

Moreover we denote (compare [21])

$$p + in := \Psi(i\kappa)(p_0 + in_0)\Psi^{-1}(i\kappa) \tag{28}$$

We assume

$$p_0^2 = n_0^2 = 1$$

(in the Clifford algebra $p^2 = \langle p \mid p \rangle$ etc.). Therefore computing the Clifford square of both sides of (28) and taking into account that p, p_0 anticommute with n, n_0 we get $p^2 = n^2$.

It is convenient to introduce unit vectors $\hat{p}$ and $\hat{n}$

$$\hat{p} := \frac{p}{\sqrt{\langle p \mid p \rangle}}, \quad \hat{n} := \frac{n}{\sqrt{\langle n \mid n \rangle}}, \tag{29}$$

such that $\hat{p}^2 = \hat{n}^2 = 1$. Then the Darboux matrix assumes the form $D'(\lambda) = \lambda \hat{n} + \kappa \hat{p}$. In order to get a Spin-valued D we can take, for instance, $G = \mathbf{e}_4$, obtaining

$$D(\lambda) := \mathbf{e}_4(\lambda \hat{n} + \kappa \hat{p}) \tag{30}$$

Note that

$$D(\lambda)\boldsymbol{\beta}(D(\lambda)) = \lambda^2 + \kappa^2 \tag{31}$$

Remark 1. *It is important to remember that the obtained Darboux matrix D depends on the function Ψ (an exact solution of the linear problem (3)) and constant parameters: κ, p_0, n_0. The notation $D = D_{[\Psi,\kappa,p_0,n_0]}$ would be very awkward, so in the sequel we omit the dependence on p_0 and n_0, writing $D = D_{[\Psi,\kappa]}$.*

Theorem 2. *The transformation $\tilde{\Psi}(\lambda) = \mathbf{e}_4(\lambda \hat{n} + \kappa \hat{p})\Psi(\lambda)$, where $\hat{n}$ and $\hat{p}$ are given by (29) and*

$$\begin{aligned} p &:= \frac{1}{2}\left(\Psi(i\kappa)(p_0 + in_0)\Psi^{-1}(i\kappa) + \Psi(-i\kappa)(p_0 - in_0)\Psi^{-1}(-i\kappa)\right) \\ n &:= \frac{1}{2i}\left(\Psi(i\kappa)(p_0 + in_0)\Psi^{-1}(i\kappa) - \Psi(-i\kappa)(p_0 - in_0)\Psi^{-1}(-i\kappa)\right), \end{aligned} \tag{32}$$

transforms the linear problem (3) into the linear problem of the same form with θ, k_1 and k_2 replaced by

$$\begin{aligned} \tilde{\theta} &= \theta - 2\gamma, \\ \tilde{k}_1 &= e^{2\gamma}\left(k_1 - 2\kappa \langle p \mid \mathbf{e}_3 \rangle e^\theta \sinh(\theta - \gamma)\right), \\ \tilde{k}_2 &= e^{2\gamma}\left(k_2 - 2\kappa \langle p \mid \mathbf{e}_3 \rangle e^\theta \cosh(\theta - \gamma)\right), \end{aligned} \tag{33}$$

where γ is a function parameterizing $\hat{n}$, namely: $\hat{n} = \cosh \gamma \mathbf{e}_4 + \sinh \gamma \mathbf{e}_5$.

We omit the proof, which consists in splitting the equation $D_{,\mu} + DU_\mu = \tilde{U}_\mu D$ into a system of equations by equating coefficients by powers of λ and basis elements of the Clifford algebra.

Theorem 3. *The Darboux transformation for soliton submanifolds (15) reads*

$$\tilde{F} = F + \frac{2}{\kappa}\hat{p}^{-1}\hat{n}, \qquad \tilde{\mathbf{r}} = \mathbf{r} + \frac{2e^{\gamma}}{\kappa}\hat{p}. \tag{34}$$

Proof. Directly applying the Sym formula we get

$$\tilde{F} = 2(D\Psi)^{-1}(D\Psi)_{,\lambda}|_{\lambda=0} = 2\Psi^{-1}\Psi_{,\lambda}|_{\lambda=0} + 2\Psi_0^{-1}(D^{-1}D_{,\lambda})|_{\lambda=0}\Psi_0. \tag{35}$$

Substituting $D_{,\lambda} = \hat{n}$ and $D^{-1}(0) = \kappa^{-1}\hat{p}^{-1}$, we get the first formula of (34). To obtain the second formula we take into account that $\hat{p}^2 = 1$ and $P(\hat{n}) = e^{\gamma}$. □

In the context of soliton surfaces the Darboux transformation is often called the Darboux-Bäcklund transformation [21,31] or the Darboux–Bianchi transformation [23].

6. Iterated Darboux Transformation

The Darboux transformation can be iterated in a natural way. Using the notation introduced in Remark 1 we have the following sequence of solutions to the considered linear problem:

$$\begin{aligned}\Psi^{[1]}(\lambda) &= D_{[\Psi^{[0]},\kappa_1]}(\lambda)\Psi^{[0]}(\lambda), \\ \Psi^{[2]}(\lambda) &= D_{[\Psi^{[1]},\kappa_2]}(\lambda)\Psi^{[1]}(\lambda), \\ &\cdots\cdots\cdots\cdots\cdots, \\ \Psi^{[K]}(\lambda) &= D_{[\Psi^{[K-1]},\kappa_K]}(\lambda)\Psi^{[K-1]}(\lambda).\end{aligned} \tag{36}$$

The last equation can be rewritten in the following, more explicit, way:

$$\Psi^{[K]}(\lambda) = D_{[\Psi^{[K-1]},\kappa_K]}(\lambda)D_{[\Psi^{[K-2]},\kappa_{K-1}]}(\lambda)\ldots D_{[\Psi^{[1]},\kappa_2]}(\lambda)D_{[\Psi^{[0]},\kappa_1]}(\lambda)\Psi^{[0]}(\lambda), \tag{37}$$

where we have to remember that $\Psi^{[1]}, \Psi^{[2]}, \ldots, \Psi^{[K-1]}$ can (and should) be expressed by $\Psi^{[0]}$ and constants $\kappa_1, \ldots, \kappa_{K-1}$. Thus we can use a more compact notation:

$$\Psi^{[K]}(\lambda) = D^{[K]}_{[\Psi^{[0]},\kappa_1,\kappa_2,\ldots,\kappa_{K-1}]}(\lambda), \tag{38}$$

but the explicit expression for $D^{[K]}$ is extremely complicated. The above notation can be shortened into the following, more compact, form:

$$\begin{aligned}\Psi^{[1]}(\lambda) &= D_{[0]1}(\lambda)\Psi^{[0]}(\lambda), \\ \Psi^{[2]}(\lambda) &= D_{[1]2}(\lambda)\Psi^{[1]}(\lambda), \\ &\cdots\cdots\cdots\cdots, \\ \Psi^{[K]}(\lambda) &= D_{[K-1]K}(\lambda)\Psi^{[K-1]}(\lambda).\end{aligned} \tag{39}$$

The index [0] may be often omitted. We have, for example:

$$\begin{aligned}\Psi^{[0]}(\lambda) &\equiv \Psi(\lambda), \\ \Psi^{[1]}(\lambda) &= D_{[0]1}(\lambda)\Psi^{[0]}(\lambda) = D(\lambda)\Psi(\lambda) = D^{[1]}(\lambda)\Psi^{[0]}(\lambda), \\ \Psi^{[2]}(\lambda) &= D_{[1]2}(\lambda)\Psi^{[1]}(\lambda) = D_{[1]2}(\lambda)D_{[0]1}(\lambda)\Psi^{[0]}(\lambda) = D^{[2]}(\lambda)\Psi^{[0]}(\lambda), \\ \Psi^{[3]}(\lambda) &= D_{[2]3}(\lambda)D_{[1]2}(\lambda)D_{[0]1}(\lambda)\Psi^{[0]}(\lambda) = D^{[3]}(\lambda)\Psi^{[0]}(\lambda)\end{aligned} \tag{40}$$

where

$$D(\lambda) = D_{[0]1}(\lambda) = \mathbf{e}_4(\lambda\hat{n}_1 + \kappa_1\hat{p}_1) \tag{41}$$

and $\hat{p}_j$, $\hat{n}_j$ are defined by

$$p_j + in_j := \Psi(i\kappa_j)(p_{0j} + in_{0j})\Psi^{-1}(i\kappa_j) \qquad (j \in \mathbb{N}). \tag{42}$$

Theorem 4. *Two-fold Darboux transformation of the function $\Psi(\lambda)$ is given by $\Psi^{[2]}(\lambda) = D^{[2]}(\lambda)\Psi(\lambda)$, and $D^{[2]}$ can be expressed by $\kappa_1, \kappa_2, \hat{p}_1, \hat{p}_2, \hat{n}_1$ and $\hat{n}_2$ in the form explicitly symmetric with respect to exchange of indices:*

$$D^{[2]}(\lambda) = \frac{C(\lambda) - (\kappa_1^2 - \kappa_2^2)D_{[0]1}(\lambda) \wedge D_{[0]2}(\lambda)}{M}, \tag{43}$$

where $D_{[0]j}(\lambda) = \mathbf{e}_4(\lambda\hat{n}_j + \kappa_j\hat{p}_j)$ $(j = 1, 2)$ and

$$M^2 := 4\kappa_1^2\kappa_2^2(\cos^2\varphi + \cos^2\psi) - 4\kappa_1\kappa_2(\kappa_1^2 + \kappa_2^2)\cos\varphi\cos\psi + (\kappa_1^2 - \kappa_2^2)^2,$$
$$C(\lambda) := \kappa_1\kappa_2\cos\varphi(2\lambda^2 + \kappa_1^2 + \kappa_2^2) - \cos\psi(2\kappa_1^2\kappa_2^2 + \lambda^2(\kappa_1^2 + \kappa_2^2)). \tag{44}$$

Proof. We are going to express in a symmetric form $D^{[2]}(\lambda) = D_{[1]2}(\lambda)D_{[0]1}(\lambda)$. Note that

$$\Psi^{[1]}(i\kappa_2) = D_{[0]1}(i\kappa_2)\Psi(i\kappa_2) = \mathbf{e}_4(i\kappa_2\hat{n}_1 + \kappa_1\hat{p}_1)\Psi(i\kappa_2), \tag{45}$$

and

$$(D_{[0]1}(\lambda))^{-1} = \frac{\lambda\hat{n}_1 + \kappa_1\hat{p}_1}{\lambda^2 + \kappa_1^2}\mathbf{e}_4 \tag{46}$$

Then

$$D_{[1]2}(\lambda) = \mathbf{e}_4(\lambda\hat{n} + \kappa_2\hat{p}), \tag{47}$$

where we still use notation (29), but (within this proof) n and p are associated with the matrix $D_{[1]2}$, i.e.,

$$p + in = \Psi^{[1]}(i\kappa_2)(p_{02} + in_{02})(\Psi^{[1]}(i\kappa_2))^{-1}. \tag{48}$$

Therefore, substituting (45),

$$p + in = \mathbf{e}_4(\kappa_1\hat{p}_1 + i\kappa_2\hat{n}_1)\Psi(i\kappa_2)(p_{02} + in_{02})\Psi(i\kappa_2)^{-1}\left(\frac{\kappa_1\hat{p}_1 + i\kappa_2\hat{n}_1}{\kappa_1^2 - \kappa_2^2}\right)\mathbf{e}_4, \tag{49}$$

which can be rewritten as (compare (29))

$$p + in = \frac{\sqrt{p_2^2}}{\kappa_1^2 - \kappa_2^2}\mathbf{e}_4(\kappa_1\hat{p}_1 + i\kappa_2\hat{n}_1)(\hat{p}_2 + i\hat{n}_2)(\kappa_1\hat{p}_1 + i\kappa_2\hat{n}_1)\mathbf{e}_4, \tag{50}$$

or

$$p + in = \sqrt{p_2^2}\mathbf{e}_4(\kappa_1\hat{p}_1 + i\kappa_2\hat{n}_1)(\hat{p}_2 + i\hat{n}_2)(\kappa_1\hat{p}_1 + i\kappa_2\hat{n}_1)^{-1}\mathbf{e}_4^{-1}, \tag{51}$$

which is a similarity transformation and can be interpreted as an orthogonal transformation in the (complexified) Clifford algebra. Note that for any Clifford vectors v, w we have

$$vwv^{-1} = (2\langle v \mid w\rangle - wv)v^{-1} = -w + 2\frac{\langle v \mid w\rangle}{\langle v \mid v\rangle}v. \tag{52}$$

Therefore

$$p + in = \sqrt{p_2^2}\mathbf{e}_4\left(\frac{2(\kappa_1\langle\hat{p}_1 \mid \hat{p}_2\rangle - \kappa_2\langle\hat{n}_1 \mid \hat{n}_2\rangle)}{\kappa_1^2 - \kappa_2^2}(\kappa_1\hat{p}_1 + i\kappa_2\hat{n}_1) - (\hat{p}_2 + i\hat{n}_2)\right)\mathbf{e}_4. \tag{53}$$

$$p = \frac{\sqrt{p_2^2}}{\kappa_1^2 - \kappa_2^2}\Big((2\kappa_1^2\langle\hat{p}_1\mid\hat{p}_2\rangle - 2\kappa_1\kappa_2\langle\hat{n}_1\mid\hat{n}_2\rangle)\hat{p}_1 - (\kappa_1^2 - \kappa_2^2)\hat{p}_2\Big), \tag{54}$$

$$n = \frac{\sqrt{p_2^2}}{\kappa_1^2 - \kappa_2^2}\mathbf{e}_4\Big((2\kappa_1\kappa_2\langle\hat{p}_1\mid\hat{p}_2\rangle - 2\kappa_2^2\langle\hat{n}_1\mid\hat{n}_2\rangle)\hat{n}_1 - (\kappa_1^2 - \kappa_2^2)\hat{n}_2\Big). \tag{55}$$

Now we can easily compute p^2, n^2 and then $\hat{p}$, $\hat{n}$. Let us denote

$$\cos\varphi := \langle\hat{p}_1\mid\hat{p}_2\rangle, \qquad \cos\psi := \langle\hat{n}_1\mid\hat{n}_2\rangle. \tag{56}$$

Then
$$p^2 = n^2 = \frac{p_2^2 M^2}{(\kappa_1^2 - \kappa_2^2)^2} \tag{57}$$

where M^2 is computed in the straightforward way (taking into account $\langle\hat{p}_j\mid\hat{n}_k\rangle = 0$):

$$M^2 := 4\kappa_1^2\kappa_2^2(\cos^2\varphi + \cos^2\psi) - 4\kappa_1\kappa_2(\kappa_1^2 + \kappa_2^2)\cos\varphi\cos\psi + (\kappa_1^2 - \kappa_2^2)^2. \tag{58}$$

Therefore
$$\hat{p} = \frac{\kappa_1^2 - \kappa_2^2}{M\sqrt{p_2^2}}p, \quad \hat{n} = \frac{\kappa_1^2 - \kappa_2^2}{M\sqrt{p_2^2}}n, \tag{59}$$

$$\begin{aligned}M\hat{p} &= 2(\kappa_1\cos\varphi - \kappa_2\cos\psi)\kappa_1\hat{p}_1 - (\kappa_1^2 - \kappa_2^2)\hat{p}_2, \\ M\hat{n} &= 2(\kappa_1\cos\varphi - \kappa_2\cos\psi)\kappa_2\hat{n}_1 - (\kappa_1^2 - \kappa_2^2)\hat{n}_2.\end{aligned} \tag{60}$$

Thus

$$D_{[1]2}(\lambda) = \frac{2\kappa_2\mathbf{e}_4(\kappa_1\cos\varphi - \kappa_2\cos\psi)(\lambda\hat{n}_1 + \kappa_1\hat{p}_1)}{M} - \frac{\mathbf{e}_4(\kappa_1^2 - \kappa_2^2)(\lambda\hat{n}_2 + \kappa_2\hat{p}_2)}{M}, \tag{61}$$

i.e.,
$$D_{[1]2}(\lambda) = \frac{2\kappa_2(\kappa_1\cos\varphi - \kappa_2\cos\psi)}{M}D_{[0]1} - \frac{(\kappa_1^2 - \kappa_2^2)}{M}D_{[0]2}. \tag{62}$$

Now, we can compute $D^{[2]} = D_{[1]2}D_{[0]1}$:

$$D^{[2]}(\lambda) = \left(\frac{2\kappa_2(\kappa_1\cos\varphi - \kappa_2\cos\psi)}{M}D_{[0]1} - \frac{(\kappa_1^2 - \kappa_2^2)}{M}D_{[0]2}\right)D_{[0]1}. \tag{63}$$

Using a general property of the Clifford product of vectors

$$vw = \langle v\mid w\rangle + v\wedge w \tag{64}$$

(where the wedge denotes the skew product) we get (43). □

Corollary 1. *The symmetric form of two-fold Darboux transformation can be considered as yet another proof of Bianchi's permutability theorem* [5].

7. Seed Solutions

In order to produce exact solution by iterating the Darboux transformation we need some starting point: a seed solution. Below we give two simple examples.

7.1. The Trivial Background (Plane)

The data $\vartheta = 0$, $k_1 = k_2 = 0$ correspond to the trivial background, i.e., to the plane. The linear system (3) assumes the form

$$\Psi_{,u} = \frac{1}{2}\lambda \mathbf{e}_1 \mathbf{e}_5 \Psi, \qquad \Psi_{,v} = \frac{1}{2}\lambda \mathbf{e}_2 \mathbf{e}_4. \tag{65}$$

Hence

$$\Psi = e^{\frac{1}{2}\lambda u \mathbf{e}_1 \mathbf{e}_5} e^{\frac{1}{2}\lambda v \mathbf{e}_2 \mathbf{e}_4} \tag{66}$$

Finally

$$2\Psi^{-1}\Psi_{,\lambda}|_{\lambda=0} = u\mathbf{e}_1\mathbf{e}_5 + v\mathbf{e}_2\mathbf{e}_4, \tag{67}$$

Performing the projection (13) we get

$$\mathbf{r} = P(\Psi^{-1}\Psi_{,\lambda}|_{\lambda=0}) = u\mathbf{e}_1 + v\mathbf{e}_2, \tag{68}$$

7.2. Cylinder

One can easily see that $\vartheta = 0$, $k_2 = 0$, $k_1 \equiv k = \text{const}$ satisfy the system (2). The linear system (3) assumes the form

$$\Psi_{,u} = \tfrac{1}{2}\lambda \mathbf{e}_1 \mathbf{e}_5 \Psi,$$

$$\Psi_{,v} = \tfrac{1}{2}\mathbf{e}_2(\lambda \mathbf{e}_4 - k\mathbf{e}_3).$$

$\mathbf{e}_1\mathbf{e}_5$ commutes with $\lambda \mathbf{e}_2\mathbf{e}_4 - k\mathbf{e}_2\mathbf{e}_3$ and they do not depend on u, v. Therefore Ψ can be easily computed

$$\Psi = e^{\frac{1}{2}\lambda u \mathbf{e}_1 \mathbf{e}_5} e^{\frac{1}{2}v(\lambda \mathbf{e}_2 \mathbf{e}_4 - k\mathbf{e}_2\mathbf{e}_3)} \tag{69}$$

Then

$$(\lambda \mathbf{e}_2\mathbf{e}_4 - k\mathbf{e}_2\mathbf{e}_3)^2 = -(\lambda^2 + k^2)$$

Therefore

$$\Psi = \left(\cosh\frac{\lambda u}{2} + \mathbf{e}_1\mathbf{e}_5 \sinh\frac{\lambda u}{2}\right)\left(\cos\frac{v\sqrt{\lambda^2+k^2}}{2} + \frac{\mathbf{e}_2(\mathbf{e}_4\lambda - \mathbf{e}_3 k)}{\sqrt{\lambda^2+k^2}}\sin\frac{v\sqrt{\lambda^2+k^2}}{2}\right)$$

$$\Psi^{-1}(0) = \cos\frac{kv}{2} + \mathbf{e}_2\mathbf{e}_3 \sin\frac{kv}{2}$$

$$\Psi_{,\lambda}(0) = \frac{1}{2}u\mathbf{e}_1\mathbf{e}_5\left(\cos\frac{kv}{2} - \mathbf{e}_2\mathbf{e}_3 \sin\frac{kv}{2}\right) + \frac{1}{k}\mathbf{e}_2\mathbf{e}_4 \sin\frac{kv}{2}$$

Then

$$2\Psi^{-1}\Psi_{,\lambda}|_{\lambda=0} = u\mathbf{e}_1\mathbf{e}_5 + \frac{2}{k}\sin\frac{kv}{2}\cos\frac{kv}{2}\mathbf{e}_2\mathbf{e}_4 - \frac{2}{k}\mathbf{e}_3\mathbf{e}_4 \sin^2\frac{kv}{2}$$

Finally, using the projection (13), we get the cylinder immersed in $\mathbb{R}^3$

$$\mathbf{r} = u\mathbf{e}_1 - \frac{1}{k}\mathbf{e}_3 + \frac{1}{k}(\mathbf{e}_3 \cos kv + \mathbf{e}_2 \sin kv) \tag{70}$$

8. Conclusions

We constructed an iterated Darboux transformation for isothermic surfaces using the Clifford algebra approach. Our main result is a symmetric representation of two-fold Darboux transformation (Theorem 4). Thus we made some progress in the direction of constructing symmetric compact formulas for "multi-soliton" isothermic surfaces, what reduces to transforming $\Psi^{[K]}$ (given by (38)) into a form that is explicitly invariant with respect to permutations of real eigenvalues $\kappa_1, \ldots, \kappa_K$. Another open problem, more challenging, is to find analogous formulas in a direct way and with more general set of complex eigenvalues. We also expect to extend our approach on related multidimensional problems (see, e.g., [32]).

Author Contributions: conceptualization, J.L.C.; methodology, J.L.C.; formal analysis, J.L.C. and Z.H.; investigation, J.L.C. and Z.H.; writing—original draft preparation, J.L.C.; writing—review and editing, J.L.C. All authors have read and agreed to the published version of the manuscript.

Funding: This research received no external funding.

Conflicts of Interest: The authors declare no conflict of interest.

References

1. Lamé, G. Mémoire sur les surfaces isothermes dans les corps solides homogènes en équilibre de température. *J. Math. Pures Appl.* **1837**, *2*, 147–183.
2. Bertrand, J. Mémoire sur les surfaces isothermes orthogonales. *J. Math. Pures Appl.* **1844**, *9*, 117–130.
3. Klimczewski, P.; Nieszporski, M.; Sym, A. Luigi Bianchi, Pasquale Calapso and solitons. *Rend. Sem. Mat. Messina (Atti del Congresso Internazionale in onore di Pasquale Calapso)* **2000**, 223–240.
4. Darboux, G. Sur les surfaces isothermiques. *C. R. Acad. Sci. Paris* **1899**, *128*, 1299–1305. [CrossRef]
5. Bianchi, L. Ricerche sulle superficie isoterme e sulle deformazione delle quadriche. *Ann. Matem.* **1905**, *11*, 93–157. [CrossRef]
6. Novikov, S.; Manakov, S.V.; Pitaevskii, L.P.; Zakharov, V.E. *Theory of Solitons. The Inverse Sattering Method*; Springer: New York, NY, USA, 1984.
7. Cieśliński, J.; Goldstein, P.; Sym, A. Isothermic surfaces in E^3 as soliton surfaces. *Phys. Lett. A* **1995**, *205*, 37–43. [CrossRef]
8. Bobenko, A.; Pinkall, U. Discrete isothermic surfaces. *J. Reine Angew. Math.* **1996**, *475*, 187–208.
9. Burstall, F.; Hertrich-Jeromin, U.; Pedit, F.; Pinkall, U. Curved flats and isothermic surfaces. *Math. Z.* **1997**, *225*, 199–209. [CrossRef]
10. Hertrich-Jeromin, U.; Pedit, F. Remarks on the Darboux transform of isothermic surfaces. *Doc. Math.* **1997**, *2*, 313–333.
11. Musso, E.; Nicolodi, L. Special isothermic surfaces and solitons. *Contemp. Math.* **2001**, *288*, 129–148.
12. Burstall, F.E. Isothermic surfaces: Conformal geometry, Clifford algebras and integrable systems. *AMS/IP Stud. Adv. Math.* **2006**, *36*, 1–82.
13. Burstall, F.E.; Hertrich-Jeromin, U. The Ribaucour transformation in Lie sphere geometry. *Differ. Geom. Appl.* **2006**, *24*, 503–520. [CrossRef]
14. Burstall, F.; Hertrich-Jeromin, U.; Müller, C.; Rossman, W. Semi-discrete isothermic surfaces. *Geom. Dedicata* **2016**, *183*, 43–58. [CrossRef]
15. Tafel, J. Covariant Description of Isothermic Surfaces. *Rep. Math. Phys.* **2016**, *78*, 295–303. [CrossRef]
16. Hertrich-Jeromin, U.; Honda, A. Minimal Darboux transformations. *Beitr. Algebra Geom.* **2017**, *58*, 81–91. [CrossRef]
17. Cieśliński, J.L.; Kobus, A. Group interpretation of the spectral parameter. The case of isothermic surfaces. *J. Geom. Phys.* **2017**, *113*, 28–37. [CrossRef]
18. Fuchs, A. Transformations and singularities of polarized curves. *Ann. Glob. Anal. Geom.* **2019**, *55*, 529–553. [CrossRef]
19. Corro, A.M.V.; Ferro, M.L. New Isothermic surfaces. *arXiv* **2020**, arXiv:2011.07941v1.
20. Cieśliński, J.L. A class of linear spectral problems in Clifford algebras. *Phys. Lett. A* **2000**, *267*, 251–255. [CrossRef]
21. Cieśliński, J.L. The Darboux-Bäcklund transformation without using a matrix representation. *J. Phys. A Math. Gen.* **2000**, *33*, L363–L368. [CrossRef]
22. Bobenko, A.I.; Hertrich-Jeromin, U.J. Orthogonal nets and Clifford algebras. *arXiv* **1998**, arXiv:math/9802126.
23. Cieśliński, J. The Darboux-Bianchi transformation for isothermic surfaces. Classical results versus the soliton approach. *Diff. Geom. Appl.* **1997**, *7*, 1–28. [CrossRef]
24. Sym, A. Soliton surfaces and their applications. In *Geometric Aspects of the Einstein Equations and Integrable Systems*; Lecture Notes in Physics; Martini, R., Ed.; Springer: Berlin, Germany, 1985; Volume 239, pp. 154–231.
25. Lounesto, P. *Clifford Algebras and Spinors*; Cambridge University Press: Cambridge, UK, 1997.
26. Vaz, J.; da Rocha, R. *An Introduction to Clifford Algebras and Spinors*; Oxford University Press: Oxford, UK, 2016.
27. Cieśliński, J. An algebraic method to construct the Darboux matrix. *J. Math. Phys.* **1995**, *36*, 5670–5706. [CrossRef]
28. Gu, C.H. Bäcklund Transformations and Darboux Transformations. In *Soliton Theory and Its Applications*; Springer: Berlin, Germany, 1995; pp. 122–151..
29. Mikhailov, A.V. The reduction problem and the inverse scattering method. *Phys. D Nonlinear Phenom.* **1981**, *3*, 73–117. [CrossRef]
30. Cieśliński, J.L. Algebraic construction of the Darboux matrix revisited. *J. Phys. A Math. Theor.* **2009**, *42*, 404003. [CrossRef]
31. Biernacki, W.; Cieśliński, J.L. A compact form of the Darboux-Bäcklund transformation for some spectral problems in Clifford algebras. *Phys. Lett. A* **2001**, *288*, 167–172. [CrossRef]
32. Cieśliński, J.L. Geometry of submanifolds derived from Spin-valued spectral problems. *Theor. Math. Phys.* **2003**, *137*, 1394–1403. [CrossRef]
33. Matveev, V.B.; Salle, M.A. *Darboux Transformations and Solitons*; Springer: Berlin/Heidelberg, Germany, 1991.

34. Rogers, C.; Schief, W.K. *Bäcklund and Darboux Transformations. Geometry and Modern Applications in Soliton Theory*; Cambridge University Press: Cambridge, UK, 2002.
35. Sakhnovich, A.L. Generalized Bäcklund–Darboux Transformation: Spectral Properties and Nonlinear Equations. *J. Math. Anal. Appl.* **2001**, *262*, 274–306. [CrossRef]
36. Sakhnovich, A.L.; Sakhnovich, L.A.; Roitberg, I.Y. *Inverse Problems and Nonlinear Evolution Equations*; De Gruyter: Berlin, Germany, 2013.
37. Cieśliński, J.L.; Biernacki, W. A new approach to the Darboux-Bäcklund transformation versus the standard dressing method. *J. Phys. A Math. Gen.* **2005**, *38*, 9491–9501. [CrossRef]

Article

Singularities in Euler Flows: Multivalued Solutions, Shockwaves, and Phase Transitions

Valentin Lychagin [1] and Mikhail Roop [1,2,*]

[1] V.A. Trapeznikov Institute of Control Sciences of Russian Academy of Sciences, 65 Profsoyuznaya Str., 117997 Moscow, Russia; valentin.lychagin@uit.no
[2] Faculty of Physics, Lomonosov Moscow State University, Leninskie Gory, 119991 Moscow, Russia
* Correspondence: mihail_roop@mail.ru

Abstract: In this paper, we analyze various types of critical phenomena in one-dimensional gas flows described by Euler equations. We give a geometrical interpretation of thermodynamics with a special emphasis on phase transitions. We use ideas from the geometrical theory of partial differential equations (PDEs), in particular symmetries and differential constraints, to find solutions to the Euler system. Solutions obtained are multivalued and have singularities of projection to the plane of independent variables. We analyze the propagation of the shockwave front along with phase transitions.

Keywords: Euler equations; shockwaves; phase transitions; symmetries

Citation: Lychagin, V.; Roop, M. Singularities in Euler Flows: Multivalued Solutions, Shockwaves, and Phase Transitions. *Symmetry* **2021**, *13*, 54. https://doi.org/10.3390/sym13010054

Received: 28 November 2020
Accepted: 30 December 2020
Published: 31 December 2020

Publisher's Note: MDPI stays neutral with regard to jurisdictional claims in published maps and institutional affiliations.

Copyright: © 2020 by the authors. Licensee MDPI, Basel, Switzerland. This article is an open access article distributed under the terms and conditions of the Creative Commons Attribution (CC BY) license (https://creativecommons.org/licenses/by/4.0/).

1. Introduction

Various types of critical phenomena, such as singularities, discontinuities, wave fronts and phase transitions, have always been of interest from both mathematical [1–3] and practical [4] viewpoints. In the context of gases, discontinuous solutions to the Euler system, describing their motion, are usually treated as *shockwaves*. In the past decades, such phenomena have widely been studied (see, e.g., [5] for the case of Chaplygin gases [6,7], where the weak shocks are considered). It is also worth mentioning the works in [8,9], where the influence of turbulence on shocks and detonations is emphasized.

This paper can be seen as a natural continuation of the work in [10], where have considered the case of ideal gas flows. Here, we use the van der Waals model of gases, which is more complicated and at the same time more interesting from the singularity theory viewpoint. The van der Waals model is known to be one of the most popular in the description of phase transitions. Thus, singularities of shockwave type that can be viewed as in some sense singular solutions to the Euler system are analyzed together with singularities of purely thermodynamic nature, phase transitions. Our approach to finding and investigating such phenomena is essentially based on the geometric theory of PDEs [11–15]. Namely, we find a class of multivalued solutions to the Euler system (see also [16]), and singularities of their projection to the plane of independent variables are exactly what drives the appearance of the shockwave [17]. Similar ideas are used in a series of works [18–20], where multivalued solutions to filtration equations are obtained along with analysis of shocks. To find such solutions, we use the idea of adding a differential constraint to the original PDE in such a way that the resulting overdetermined system of PDEs is compatible [21]. The same concepts were also used by Schneider [22], who found a general solution to the Hunter–Saxton equation; LY1 [23], who considered the two-dimensional Euler system; and LY2 [24], who applied this approach to the Khokhlov–Zabolotskaya equation.

The paper is organized as follows. Section 2 presents the preliminary concepts, where we describe the necessary concepts from thermodynamics. In Section 3, we analyze a multivalued solution to Euler equations and its singularities, including shockwaves and phase transitions. In the last section, we discuss the results. The essential computations for this paper were made with the DifferentialGeometry package [25] in Maple.

2. Thermodynamics

In this section, we give necessary concepts from thermodynamics. As shown below, geometrical interpretation of thermodynamic states allows one to use Arnold's ideas from the theory of Legendrian and Lagrangian singularities [1–3], which are crucial in description of phase transitions. The geometrical approach to thermodynamics was already initiated by Gibbs [26]. It was further developed, for example, by the authors of [27,28] and, more recently, by Lychagin [29]. For more detailed analysis regarding the geometrical methods in thermodynamics, we also refer to [30].

2.1. Legendrian and Lagrangian Manifolds

Consider the contact space $(\mathbb{R}^5, \theta)$ with coordinates (s, e, ρ, p, T) standing for specific entropy, specific inner energy, density, pressure and temperature. The contact structure θ is given by

$$\theta = T^{-1}de - ds - pT^{-1}\rho^{-2}d\rho. \tag{1}$$

Then, a *thermodynamic state* is a Legendrian manifold $\widehat{L} \subset (\mathbb{R}^5, \theta)$, i.e., $\theta|_{\widehat{L}} = 0$ and $\dim \widehat{L} = 2$. From the physical viewpoint, this means that the first law of thermodynamics holds on $\widehat{L}$. Due to (1), it is natural to choose (e, ρ) as coordinates on $\widehat{L}$. Then, a two-dimensional manifold $\widehat{L} \subset (\mathbb{R}^5, \theta)$ is given by

$$\widehat{L} = \left\{ s = S(e, \rho),\ T = \frac{1}{S_e},\ p = -\rho^2 \frac{S_\rho}{S_e} \right\}, \tag{2}$$

where the function $S(e, \rho)$ specifies the dependence of the specific entropy on e and ρ.

Note that determining a Legendrian manifold $\widehat{L}$ by means of (2) requires the knowledge of $S(e, \rho)$, while in experiments one usually obtains relations among pressure, density and temperature. Thus, we get rid of the specific entropy s by means of projection $\pi \colon \mathbb{R}^5 \to \mathbb{R}^4$, $\pi(s, e, \rho, p, T) = (e, \rho, p, T)$ and consider an immersed Lagrangian manifold $\pi(\widehat{L}) = L \subset (\mathbb{R}^4, \Omega)$ in a symplectic space $(\mathbb{R}^4, \Omega)$, where the structure symplectic form Ω is

$$\Omega = d\theta = d(T^{-1}) \wedge de - d(pT^{-1}\rho^{-2}) \wedge d\rho.$$

Then, one can treat thermodynamic state manifolds as Lagrangian manifolds $L \subset (\mathbb{R}^4, \Omega)$, i.e., $\Omega|_L = 0$. In coordinates (T, ρ), a thermodynamic Lagrangian manifold L is given by two functions

$$L = \{p = P(T, \rho),\ e = E(T, \rho)\}. \tag{3}$$

Since $\Omega|_L = 0$, the functions $P(T, \rho)$ and $E(T, \rho)$ are not arbitrary, but are related by

$$[p - P(T, \rho), e - E(T, \rho)]|_L = 0, \tag{4}$$

where $[f, g]$ is the Poisson bracket of functions f and g on $(\mathbb{R}^4, \Omega)$ uniquely defined by the relation

$$[f, g]\, \Omega \wedge \Omega = df \wedge dg \wedge \Omega.$$

Equation (4) forces the following relation between $P(T, \rho)$ and $E(T, \rho)$: $(-\rho^{-2}T^{-1}P)_T = (T^{-2}E)_\rho$, and therefore the following theorem is valid:

Theorem 1. *The Lagrangian manifold L is given by means of the Massieu–Planck potential $\phi(\rho, T)$*

$$p = -\rho^2 T \phi_\rho, \quad e = T^2 \phi_T. \tag{5}$$

Remark 1. *Having given the Lagrangian manifold L by means of (3), one can find the entropy function $S(e, \rho)$ solving the overdetermined system*

$$T = \frac{1}{S_e}, \quad p = -\rho^2 \frac{S_\rho}{S_e}$$

with compatibility condition (4).

2.2. Riemannian Structures, Singularities, Phase Transitions

There is one more important structure arising, as shown in [29], from measurement approach to thermodynamics. Indeed, if one considers equilibrium thermodynamics as a theory of measurement of random vectors, whose components are inner energy and volume $v = \rho^{-1}$, one drives to the universal quadratic form on $(\mathbb{R}^4, \Omega)$ of signature $(2,2)$:

$$\kappa = d(T^{-1}) \cdot de - \rho^{-2} d(pT^{-1}) \cdot d\rho,$$

where $\cdot$ is the symmetric product of differential forms, and areas on L, where the restriction $\kappa|_L$ of κ to L is negative, are those where the variance of a random vector $(e, v = \rho^{-1})$ is positive [29,31]. Using (5), we get

$$\kappa|_L = -(2T^{-1}\phi_T + \phi_{TT})dT \cdot dT + (2\rho^{-1}\phi_\rho + \phi_{\rho\rho})d\rho \cdot d\rho, \quad (6)$$

and, taking into account (5), we conclude that the condition of positive variance is satisfied at points on L, where

$$e_T > 0, \quad p_\rho > 0,$$

which is known as the condition of the thermodynamic stability.

Let us now explore singularities of Lagrangian manifolds. We are interested in the singularities of their projection to the plane of intensive variables (p, T), i.e., points where the form $dp \wedge dT$ degenerates. We assume that extensive variables (e, ρ) may serve as global coordinates on L, i.e., the form $de \wedge d\rho$ is non-degenerate everywhere. The set where $dp \wedge dT = 0$ coincides with that where $2\rho^{-1}\phi_\rho + \phi_{\rho\rho} = 0$, or, equivalently, where the from $\kappa|_L$ degenerates. A manifold L turns out to be divided into submanifolds L_i, where both (e, ρ) and (p, T) may serve as coordinates, or, equivalently, the form (6) is non-degenerate. Such L_i are called *phases*. Additionally, those of L_i, where (6) is negative, are called *applicable phases*. Thus, we end up with the observation that singularities of projection of thermodynamic Lagrangian manifolds are related with the theory of phase transitions. Indeed, by a *phase transition* of the first order, we mean a jump from one applicable state to another, governed by the conservation of intensive variables p and T and specific Gibbs potential

$$\gamma = e - Ts + p/\rho,$$

which in terms of the Massieu–Planck potential is expressed as $\gamma = -T(\phi + \rho\phi_\rho)$ [30]. Consequently, to find the points of phase transition, one needs to solve the system

$$p = -\rho_1^2 T \phi_\rho(T, \rho_1), \quad p = -\rho_2^2 T \phi_\rho(T, \rho_2), \quad \phi(T, \rho_1) + \rho_1 \phi_\rho(T, \rho_1) = \phi(T, \rho_2) + \rho_2 \phi_\rho(T, \rho_2), \quad (7)$$

where p and T are the pressure and temperature of the phase transition and ρ_1 and ρ_2 are the densities of gas and liquid phases.

Example 1 (Ideal gas). *The simplest example of a gas is an ideal gas model. In this case, the Legendrian manifold is given by*

$$\widehat{L} = \left\{ p = R\rho T, \, e = \frac{n}{2}RT, \, s = R \ln\left(\frac{T^{n/2}}{\rho}\right) \right\}, \quad (8)$$

where R is the universal gas constant and n is the degree of freedom. The differential quadratic form $\kappa|_L$ is

$$\kappa|_L = -\frac{Rn}{2}\frac{dT^2}{T^2} - R\rho^{-2}d\rho^2.$$

It is negative definite on the entire $\widehat{L}$, and there are neither phase transitions nor singularities of projection of $\widehat{L}$ to the $p - T$ plane.

Example 2 (van der Waals gas). *To define the Legendrian manifold for van der Waals gases, we use reduced state equations:*

$$\widehat{L} = \left\{ p = \frac{8T\rho}{3-\rho} - 3\rho^2,\ e = \frac{4nT}{3} - 3\rho,\ s = \ln\left(T^{4n/3}(3\rho^{-1} - 1)^{8/3}\right) \right\}. \qquad (9)$$

The differential quadratic form $\kappa|_L$ is

$$\kappa|_L = -\frac{4n}{3T^2} dT^2 + \frac{6(\rho^3 - 6\rho^2 - 4T + 9\rho)}{\rho^2 T(\rho - 3)^2} d\rho^2.$$

In this case, it changes its sign; the manifold $\widehat{L}$ has a singularity of cusp type. The singular set of $\widehat{L}$, called also caustic, and the curve of phase transition are shown in Figure 1.

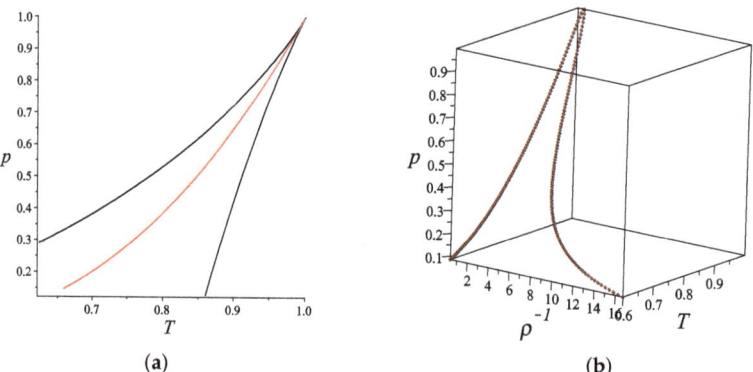

(a) (b)

Figure 1. Singularities of the van der Waals Legendrian manifold: caustic (black line) and phase transition curve (red line) in coordinates (p, T) (**a**); and the curve of phase transition in (p, ρ, T) (**b**). Points of the phase transition curve with the same values of pressure p and temperature T and different values of density $\rho_2 > \rho_1$ correspond to the liquid phase and the gas phase, respectively, while points between ρ_1 and ρ_2 correspond to wet steam.

3. Euler Equations

In this paper, we study non-stationary, one-dimensional flows of gases, described by the following system of differential equations:

- Conservation of momentum:

$$\rho(u_t + u u_x) = -p_x \qquad (10)$$

- Conservation of mass:

$$\rho_t + (\rho u)_x = 0 \qquad (11)$$

- Conservation of entropy along the flow:

$$s_t + u s_x = 0 \qquad (12)$$

Here, $u(t, x)$ is the flow velocity, $\rho(t, x)$ is the density of the medium, and $s(t, x)$ is the specific entropy. System (10)–(12) is incomplete. It becomes complete once extended by equations of thermodynamic state (2). We are interested in *homentropic flows*, i.e., those with $s(t, x) = s_0$. On the one hand, this assumption satisfies (12) identically. On the other hand, it allows us to express all the thermodynamic variables in terms of ρ. Indeed, the entropy s has the following expression in terms of the Massieu–Planck potential $\phi(T,\rho)$: $s = \phi + T\phi_T$ [30]. Putting $s = s_0$, we get the equation $s_0 = \phi + T\phi_T$, which determines $T(\rho)$ uniquely, since the derivative of its right-hand side with respect to T is positive due to the negativity of $\kappa|_L$. Substituting $T(\rho)$ into (3), one gets $p = p(\rho)$. Thus, we end up with the following two-component system of PDEs:

$$u_t + uu_x + A(\rho)\rho_x = 0, \quad \rho_t + (\rho u)_x = 0, \tag{13}$$

where $A(\rho) = p'(\rho)/\rho$.

We do not specify the function $A(\rho)$ yet; we do this while solving (13).

3.1. Finding Solutions

To find solutions to system (13), we use the idea of adding a differential constraint to (13), compatible with the original system. It is worth mentioning that a solution is an integral manifold of the Cartan distribution on (13) (see [11–13] for details). This geometrical interpretation of a solution to a PDE allows finding ones in the form of manifolds, which, in general, may not be globally given by functions. This approach gives rise to investigation of singularities in a purely geometrical manner, which is shown in this paper.

In general, finding differential constraints is not a trivial problem. However, having found ones, the problem of finding solutions is reduced to the integration of a completely integrable Cartan distribution of the resulting compatible overdetermined system. In rgw case the Cartan distribution has a solvable transversal symmetry algebra, whose dimension equals the codimension of the Cartan distribution, we are able to get explicit solutions in quadratures by applying the Lie–Bianchi theorem (for details, see [11–13]).

We look for a differential constraint compatible with (13) in the form of a quasilinear equation

$$u_x - \rho_x(\alpha(\rho)u + \beta(\rho)) = 0, \tag{14}$$

where functions $\alpha(\rho)$ and $\beta(\rho)$ are to be determined. We denote system (13) and (14) by $\mathcal{E}$.

Theorem 2. *System (13) and (14) is compatible if*

$$\alpha(\rho) = -\frac{1}{\rho(C_3\rho - 1)}, \quad \beta(\rho) = \frac{C_2}{\rho(C_3\rho - 1)}, \quad A(\rho) = C_1 + \frac{C_5}{\rho^3}\left(C_3 + \frac{C_7}{\rho}\right)^{C_6}, \tag{15}$$

where C_i are constants.

The proof of Theorem 2 is more technical rather than conceptual. First, we lift system (13) and (14) to the space of 3-jets $J^3(\mathbb{R}^2)$ by applying total derivatives

$$\begin{aligned}
D_t &= \partial_t + u_t\partial_u + \rho_t\partial_\rho + u_{tt}\partial_{u_t} + \rho_{tt}\partial_{\rho_t} + \ldots, \\
D_x &= \partial_x + u_x\partial_u + \rho_x\partial_\rho + u_{xx}\partial_{u_x} + \rho_{xx}\partial_{\rho_x} + \ldots.
\end{aligned}$$

to equations of $\mathcal{E}$ the required number of times, consequently. The resulting system $\mathcal{E}_3 \subset J^3(\mathbb{R}^2)$, consisting of equations only of the third order, contains nine equations for eight variables of purely third order: $u_{ttt}, u_{xxx}, u_{txx}, u_{ttx}, \rho_{ttt}, \rho_{xxx}, \rho_{txx}$ and ρ_{ttx}. Eliminating them from $\mathcal{E}_3$, we get seven relations (six obtained by lifting $\mathcal{E}$ to $J^2(\mathbb{R}^2)$ plus one remaining from eliminations of third-order variables). Again, we eliminate all the variables of the second order and we get four relations of the first order. Eliminating u_x, u_t and ρ_t, we end up with an expression of the form $\rho_x^3 G(\rho, u) = 0$, where $G(\rho, u)$ is a polynomial in u, whose coefficients are ordinary differential equations (ODEs) on $\alpha(\rho)$, $\beta(\rho)$ and $A(\rho)$, solving which we get (15). It is worth stating that these computations are algebraic and well suited for computer algebra systems.

Remark 2. Using (8) and (9), one can show that the function $A(\rho) = p'(\rho)/\rho$ given in (15) corresponds to that of:

- ideal gas in the case of

$$C_1 = C_3 = 0, \quad C_5 = R\left(1 + \frac{2}{n}\right)\exp\left(\frac{2s_0}{Rn}\right), \quad C_6 = -2 - \frac{2}{n}, \quad C_7 = 1;$$

- van der Waals gas in the case of

$$C_1 = -6, \quad C_3 = -1, \quad C_5 = 24\left(1 + \frac{2}{n}\right)\exp\left(\frac{3s_0}{4n}\right), \quad C_6 = -2 - \frac{2}{n}, \quad C_7 = 3. \tag{16}$$

The case of ideal gases was thoroughly investigated by LR2 [10]. Here, we are interested in the case of van der Waals gases.

Summarizing, we have a compatible overdetermined system of PDEs

$$\mathcal{E} = \{F_1 = u_t + uu_x + A(\rho)\rho_x = 0, \; F_2 = \rho_t + (\rho u)_x = 0, \; F_3 = u_x - \rho_x(\alpha(\rho)u + \beta(\rho)) = 0\} \subset J^1(\mathbb{R}^2),$$

where functions $\alpha(\rho)$, $\beta(\rho)$ and $A(\rho)$ are specified in (15). This system is a smooth manifold $\mathcal{E}$ in the space of 1-jets $J^1(\mathbb{R}^2)$ of functions on $\mathbb{R}^2$. Since $\dim J^1(\mathbb{R}^2) = 8$, and $\mathcal{E}$ consists of three relations on $J^1(\mathbb{R}^2)$, $\dim \mathcal{E} = 5$. The dimension of the Cartan distribution $\mathcal{C}_\mathcal{E}$ on $\mathcal{E}$ equals 2, therefore $\mathrm{codim}\, \mathcal{C}_\mathcal{E} = 3$. Let us choose (t, x, u, ρ, ρ_x) as internal coordinates on $\mathcal{E}$. Then, the Cartan distribution $\mathcal{C}_\mathcal{E}$ is generated by differential 1-forms

$$\omega_1 = du - u_x dx - u_t dt, \tag{17}$$
$$\omega_2 = d\rho - \rho_x dx - \rho_t dt, \tag{18}$$
$$\omega_3 = d\rho_x - \rho_{xx} dx - \rho_{xt} dt, \tag{19}$$

where $\rho_{xx}, \rho_{xt}, u_t, u_x, \rho_t$ are expressed due to $\mathcal{E}$ and its prolongation $\mathcal{E}_2 = \{D_t(F_1) = 0, D_t(F_2) = 0, D_t(F_3) = 0, D_x(F_1) = 0, D_x(F_2) = 0, D_x(F_3) = 0\}$:

$$\rho_{xx} = \frac{\rho_x^2(\rho(C_3\rho - 1)^3 A' + (C_3\rho - 1)^2 A + 3C_3(C_2 - u)^2)}{(C_3\rho - 1)((C_2 - u)^2 - A\rho(C_3\rho - 1)^2)}, \quad \rho_t = \frac{\rho_x(C_3\rho u + C_2 - 2u)}{1 - C_3\rho}, \tag{20}$$

$$u_x = \frac{\rho_x(C_2 - u)}{\rho(C_3\rho - 1)}, \quad u_t = -\frac{\rho_x(A\rho(C_3\rho - 1) + u(C_2 - u))}{\rho(C_3\rho - 1)}, \tag{21}$$

$$\rho_{xt} = \frac{\rho_x^2}{\rho(C_3\rho - 1)^2(A\rho(C_3\rho - 1)^2 - (C_2 - u)^2)}\Big(\rho^2(C_3\rho - 1)^3(C_3\rho u + C_2 - 2u)A' + \rho A(C_3\rho - 1)^2(C_3\rho u + 3C_2 - 4u) + (C_2 - u)^2(3C_3^2\rho^2 u + 3C_3\rho(C_2 - 2u) - 2C_2 + 2u)\Big), \tag{22}$$

where $A(\rho)$ is given by (15). We look for integrals of the distribution (17)–(22), which give us an (implicit) solution to (13) and (14).

Theorem 3. The distribution (17)–(22) is a completely integrable distribution with a three-dimensional Lie algebra $\mathfrak{g}$ of transversal infinitesimal symmetries generated by vector fields

$$X_1 = t\partial_t + x\partial_x - \rho_x\partial_{\rho_x}, \quad X_2 = \partial_t, \quad X_3 = \partial_x$$

with brackets $[X_1, X_3] = -X_3$, $[X_1, X_2] = -X_2$, $[X_2, X_3] = 0$.
The Lie algebra $\mathfrak{g}$ is solvable, and its sequence of derived algebras is

$$\mathfrak{g} = \langle X_1, X_2, X_3 \rangle \supset \langle X_2, X_3 \rangle \supset 0.$$

Thus, the Lie–Bianchi theorem [11–13] can be applied to integrate (17)–(22).

Let us choose another basis $\langle \varkappa_1, \varkappa_2, \varkappa_3 \rangle$ in $\mathcal{C}_\mathcal{E}$ by the following way:

$$\begin{pmatrix} \varkappa_1 \\ \varkappa_2 \\ \varkappa_3 \end{pmatrix} = \begin{pmatrix} \omega_1(X_1) & \omega_1(X_2) & \omega_1(X_3) \\ \omega_2(X_1) & \omega_2(X_2) & \omega_2(X_3) \\ \omega_3(X_1) & \omega_3(X_2) & \omega_3(X_3) \end{pmatrix}^{-1} \begin{pmatrix} \omega_1 \\ \omega_2 \\ \omega_3 \end{pmatrix}.$$

Due to the structure of the symmetry Lie algebra $\mathfrak{g}$, the form $\varkappa_1$ is closed [11,12], and therefore locally exact, i.e., $\varkappa_1 = dQ_1$, where $Q_1 \in C^\infty(J^1)$, while restrictions $\varkappa_2|_{M_1}$ and $\varkappa_3|_{M_1}$ to the manifold $M_1 = \{Q_1 = \text{const}\}$ are closed and locally exact too. Integrating the differential 1-form $\varkappa_1$ we observe that variables u, ρ, t, x can be chosen as local coordinates on M_1 and

$$M_1 = \left\{ \rho_x = \frac{\alpha_1 \rho^2 (C_3 \rho - 1)}{\rho A (C_3 \rho - 1)^2 - (C_2 - u)^2} \right\},$$

where α_1 is a constant. Integrating restrictions $\varkappa_2|_{M_1}$ and $\varkappa_3|_{M_1}$, we get two more relations that give us a solution to (13) and (14) implicitly:

$$t + \alpha_2 + \frac{C_2 - u}{\alpha_1 \rho} + \frac{C_3 u}{\alpha_1} = 0, \tag{23}$$

and

$$0 = x + \alpha_3 + \frac{1}{\alpha_1} \left(C_1 \ln \rho - C_1 C_3 \rho + \frac{C_3 u^2}{2} + \frac{u(C_2 - u)}{\rho} - C_5 \left(C_3 + \frac{C_7}{\rho} \right)^{C_6 + 1} \right.$$
$$\left. \cdot \frac{2\rho^2 C_3^2 - C_7^2 (C_6 + 1)(C_3 \rho (C_6 + 3) - C_6 - 2) + C_3 C_7 \rho (C_3 \rho (C_6 + 3) - 2C_6 - 2)}{(C_6 + 1)(C_6 + 2)(C_6 + 3) C_7^3 \rho^2} \right), \tag{24}$$

where we have already substituted $A(\rho)$ from (15), and α_2, α_3 are constants. The graph of a multivalued solution for the density is shown in Figure 2. We used substitution (16), where $C_5 = 240, n = 3$, together with $C_2 = 1, \alpha_1 = 1, \alpha_2 = 2, \alpha_3 = 1$.

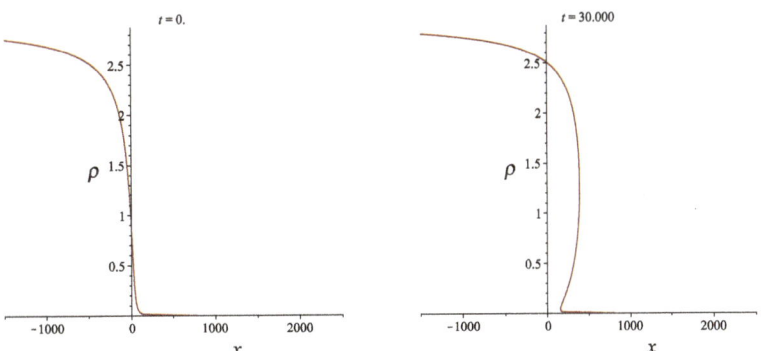

Figure 2. Graph of the density in case of $n = 3$ for time moments $t = 0, t = 30$.

3.2. Caustics and Shockwaves

We can see that solution given by (23) and (24) is, in general, multivalued. To figure out where the two-dimensional manifold N given by (23) and (24) has singularities of projection to the plane of independent variables, one needs to find zeroes of the two-form

$dt \wedge dx$. Condition $(dt \wedge dx)|_N = 0$ gives us a curve in the plane $\mathbb{R}^2(t,x)$ called *caustic*. Choosing ρ as a coordinate on the caustic, we get its equations in a parametric form:

$$x(\rho) = -\frac{1}{2\alpha_1}\left(2C_1 \ln\rho + C_1(C_3^3\rho^3 - 4\rho^2 C_3^2 + 3C_3\rho - 2) + C_3 C_7^2 + 2\alpha_1\alpha_3\right) \pm$$
$$\pm \frac{C_2(C_3\rho - 1)^2}{\alpha_1\rho^2}\sqrt{C_1\rho^3 + C_5\left(C_3 + \frac{C_7}{\rho}\right)^{C_6}} - \frac{C_5\left(C_3 + \frac{C_7}{\rho}\right)^{C_6}}{2(C_6+2)(C_6+3)C_7^3\alpha_1(C_6+1)\rho^3} \cdot$$
$$\cdot \left(C_3^3(-4 + C_7^3(C_6^3 + 6C_6^2 + 11C_6 + 6) + (-2C_6 - 6)C_7)\rho^3 - \right.$$
$$- 2C_7((2(C_6^3 + 6C_6^2 + 11C_6 + 6))C_7^2 + (-C_6^2 - 3C_6)C_7 - C_6)C_3^2\rho^2 +$$
$$\left. + C_7^2(C_6+1)((C_6+3)(5C_6+12)C_7 - 2C_6)C_3\rho - 2C_7^3(C_6+4)(C_6+2)(C_6+1)\right), \quad (25)$$

$$t(\rho) = -\alpha_2 - \frac{C_2 C_3}{\alpha_1} \pm \frac{(C_3\rho - 1)^2}{\alpha_1\rho^2}\sqrt{C_1\rho^3 + C_5\left(C_3 + \frac{C_7}{\rho}\right)^{C_6}}. \quad (26)$$

To construct a discontinuous solution from the multivalued one given by (23) and (24), we use the mass conservation law. Equation (11) with the velocity u found from (23) in terms of t and ρ takes the form:

$$\rho_t + \left(\rho \frac{\alpha_1\rho(t+\alpha_2) + C_2}{1 - C_3\rho}\right)_x = 0,$$

and therefore the conservation law is

$$\Theta = \rho dx - \rho\frac{\alpha_1\rho(t+\alpha_2) + C_2}{1 - C_3\rho} dt.$$

Its restriction $\Theta|_N$ to the manifold N given by (23) and (24) is a closed form, locally $\Theta|_N = dH$, and the potential $H(\rho, t)$ equals

$$H(\rho,t) = \frac{\rho}{2\alpha_1(C_3\rho-1)^2}\left(C_1 C_3^3 \rho^3 - 4C_1 C_3^2 \rho^2 + \rho\left(C_2^2 C_3^2 + (2C_2(t+\alpha_2)\alpha_1 + 5C_1)C_3 + \alpha_1^2(t+\alpha_2)^2\right) - 2C_1\right) -$$
$$- \frac{C_5\left(C_3 + \frac{C_7}{\rho}\right)^{C_6}}{(C_6+2)\alpha_1 C_7^2(C_6+1)\rho^2}(C_3\rho + C_7)(C_3(1+(C_6+2)C_7)\rho - (C_6+1)C_7).$$

The discontinuity line, or a shockwave front, is found from the system of equations

$$H(\rho_1,t) = H(\rho_2,t), \quad x(\rho_1,t) = x(\rho_2,t),$$

where $x(\rho,t)$ is obtained from (23) and (24) by eliminating u. Caustics along with the shockwave front are shown in Figure 3. Note that the picture is similar to that in the case of phase transitions.

The final result here is the expression for the time interval, within which the solution (23) and (24) is smooth.

Theorem 4. *The solution given by (23) and (24) is smooth and unique in the time interval $t \in [0, t^*)$, where*

$$t^* = \frac{1}{\alpha_1}\left(-C_2 C_3 - \alpha_1\alpha_2 + (C_3-3)^2\sqrt{\frac{C_1}{27} + C_5(C_3+3C_7)^{C_6}}\right),$$

and in the case of (16), where $C_5 = 240$, $n = 3$, together with $C_2 = 1$, $\alpha_1 = 1$, $\alpha_2 = 2$, $\alpha_3 = 1$ approximately $t^* = 12.53$.

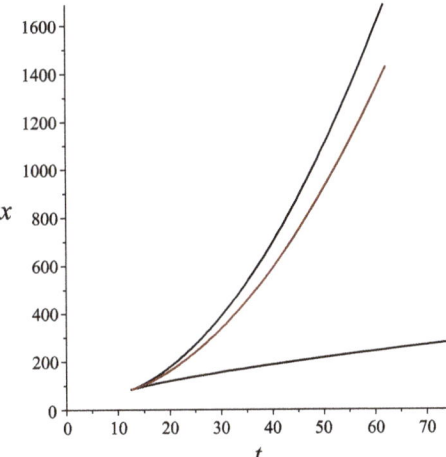

Figure 3. Caustic (black) and shockwave front (red) for $n = 3$.

3.3. Phase Transitions

Having a solution, one can remove the phase transition curve from the space of thermodynamic variables to $\mathbb{R}^2(t, x)$. Indeed, on the one hand, we have all the thermodynamic parameters as functions of (t, x). On the other hand, we have conditions on phase transitions (7) in the space of thermodynamic variables. In combination, they give us a curve of phase transitions in (t, x) plane. Phase transitions together with the shockwave are presented in Figure 4. We use substitution (16), where $C_5 = 240$, $n = 3$, together with $C_2 = 1$, $\alpha_1 = 1$, $\alpha_2 = 2$, $\alpha_3 = 1$.

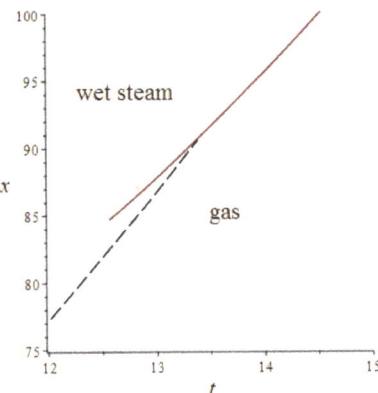

Figure 4. Phase transition curve (dash line) and shockwave front (red line).

4. Discussion

In the present work, we analyze critical phenomena in gas flows of purely thermodynamic nature, which are phase transitions and shockwaves arising from singularities of solutions to the Euler system. To obtain such solutions, we use a differential constraint compatible with the original system. In this work, it is found in a purely computational way, and how to get it in a more constructive way seems interesting. One possible way to find such constraints is using differential invariants. Then, constraints can be found constructively by solving quotient PDEs (see [32] for details), which was successfully realized by Schneider [22]. We hope to make use of this method in future research. The analysis of phase transitions shows that sometimes shockwaves can be accompanied with phase transitions, which is shown in Figure 4, since the phase transition curve intersects the shockwave front, and on the one side of the discontinuity curve we observe a pure gas phase, while on the other side we can see a wet steam.

Author Contributions: Conceptualization, V.L.; Formal analysis, M.R.; Investigation, V.L. and M.R.; and Writing—original draft, M.R. All authors have read and agreed to the published version of the manuscript.

Funding: This work was partially supported by the Russian Foundation for Basic Research (project 18-29-10013) and by the Foundation for the Advancement of Theoretical Physics and Mathematics "BASIS" (project 19-7-1-13-3).

Conflicts of Interest: The authors declare no conflict of interest.

References

1. Arnold, V. *Singularities of Caustics and Wave Fronts*; Springer: Dordrecht, The Netherlands, 1990.
2. Arnold, V. *Catastrophe Theory*; Springer: Berlin/Heidelberg, Germany, 1984.
3. Arnold, V.; Gusein-Zade, S.; Varchenko, A. *Singularities of Differentiable Maps*; Birkhäuser: Basel, Switzerland, 1985.
4. Zeldovich, A.; Kompaneets, I. *Theory of Detonation*; Academic Press: Cambridge, MA, USA, 1960.
5. Huang, S.J.; Wang, R. On blowup phenomena of solutions to the Euler equations for Chaplygin gases. *Appl. Math. Comput.* **2013**, *219*, 4365–4370. [CrossRef]
6. Rosales, R.; Tabak, E. Caustics of weak shock waves. *Phys. Fluids* **1997**, *10*, 206–222. [CrossRef]
7. Chaturvedi, R.; Gupta, P.; Singh, L.P. Evolution of weak shock wave in two-dimensional steady supersonic flow in dusty gas. *Acta Astronaut.* **2019**, *160*, 552–557. [CrossRef]
8. Poludnenko, A.; Oran, E. The interaction of high-speed turbulence with flames: Global properties and internal flame structure. *Combust. Flame* **2010**, *157*, 995–1011. [CrossRef]
9. Poludnenko, A.; Gardiner, T.; Oran, E. Spontaneous Transition of Turbulent Flames to Detonations in Unconfined Media. *Phys. Rev. Lett.* **2011**, *107*, 054501. [CrossRef] [PubMed]
10. Lychagin, V.; Roop, M. Shock waves in Euler flows of gases. *Lobachevskii J. Math.* **2020**, *41*, 2466–2472.
11. Kushner, A.; Lychagin, V.; Rubtsov, V. *Contact Geometry and Nonlinear Differential Equations*; Cambridge University Press: Cambridge, UK, 2007.
12. Vinogradov, A.; Krasilshchik, I. (Eds.) *Symmetries and Conservation Laws for Differential Equations of Mathematical Physics*; Factorial: Moscow, Russia, 1997.
13. Vinogradov, A.; Krasilshchik, I.; Lychagin, V. *Geometry of Jet Spaces and Nonlinear Partial Differential Equations*; Gordon and Breach: New York, NY, USA, 1996.
14. Ovsiannikov, L. *Group Analysis of Differential Equations*; Academic Press: Cambridge, MA, USA, 1982.
15. Olver, P. *Applications of Lie Groups to Differential Equations*; Springer: New York, NY, USA, 1986.
16. Tunitsky, D. On multivalued solutions of equations of one-dimensional gas flow. In Proceedings of the 12th International Conference "Management of Large-Scale System Development" (MLSD), Moscow, Russia, 1–3 October 2019.
17. Lychagin, V. Singularities of multivalued solutions of nonlinear differential equations, and nonlinear phenomena. *Acta Appl. Math.* **1985**, *3*, 135–173. [CrossRef]
18. Akhmetzyanov, A.; Kushner, A.; Lychagin, V. Control of displacement front in a model of immiscible two-phase flow in porous media. *Dokl. Math.* **2016**, *94*, 378–381. [CrossRef]
19. Akhmetzyanov, A.; Kushner, A.; Lychagin, V. Integrability of Buckley-Leverett's filtration model. *IFAC PapersOnLine* **2016**, *49*, 1251–1254.
20. Akhmetzyanov, A.; Kushner, A.; Lychagin, V. Shock waves in initial boundary value problem for filtration in two-phase 2-dimensional porous media. *Glob. Stoch. Anal.* **2016**, *3*, 41–46.
21. Kruglikov, B.; Lychagin, V. Compatibility, Multi-Brackets and Integrability of Systems of PDEs. *Acta Appl. Math.* **2010**, *109*, 151–196. [CrossRef]
22. Schneider, E. Solutions of second-order PDEs with first-order quotients. *arXiv* **2020**, arXiv:2005.06794.
23. Lychagin, V.; Yumaguzhin, V. On Geometric Structures of 2-Dimensional Gas Dynamics Equations. *Lobachevskii J. Math.* **2009**, *30*, 327–332. [CrossRef]
24. Lychagin, V.; Yumaguzhin, V. Minkowski Metrics on Solutions of the Khokhlov-Zabolotskaya Equation. *Lobachevskii J. Math.* **2009**, *30*, 333–336. [CrossRef]
25. Anderson, I.; Torre, C.G. The Differential Geometry Package. Downloads. 2016, Paper 4. Available online: http://digitalcommons.usu.edu/dg_downloads/4 (accessed on 27 November 2020).
26. Gibbs, J.W. A Method of Geometrical Representation of the Thermodynamic Properties of Substances by Means of Surfaces. *Trans. Conn. Acad.* **1873**, *1*, 382–404.

27. Mrugala, R. Geometrical formulation of equilibrium phenomenological thermodynamics. *Rep. Math. Phys.* **1978**, *14*, 419–427. [CrossRef]
28. Ruppeiner, G. Riemannian geometry in thermodynamic fluctuation theory. *Rev. Mod. Phys.* **1995**, *67*, 605–659. [CrossRef]
29. Lychagin, V. Contact Geometry, Measurement, and Thermodynamics. In *Nonlinear PDEs, Their Geometry and Applications*; Kycia, R., Schneider, E., Ulan, M., Eds.; Birkhäuser: Cham, Switzerland, 2019; pp. 3–52.
30. Lychagin, V.; Roop, M. Critical phenomena in filtration processes of real gases. *Lobachevskii J. Math.* **2020**, *41*, 382–399. [CrossRef]
31. Kushner, A.; Lychagin, V.; Roop, M. Optimal Thermodynamic Processes for Gases. *Entropy* **2020**, *22*, 448. [CrossRef] [PubMed]
32. Kruglikov, B.; Lychagin, V. Global Lie-Tresse theorem. *Selecta Math.* **2016**, *22*, 1357–1411. [CrossRef]

Article
Differential Invariants of Linear Symplectic Actions

Jørn Olav Jensen and Boris Kruglikov *

Institute of Mathematics and Statistics, UiT the Arctic University of Norway, 90-37 Tromsø, Norway; jje041@post.uit.no
* Correspondence: boris.kruglikov@uit.no

Received: 20 October 2020; Accepted: 2 December 2020; Published: 7 December 2020

Abstract: We consider the equivalence problem for symplectic and conformal symplectic group actions on submanifolds and functions of symplectic and contact linear spaces. This is solved by computing differential invariants via the Lie-Tresse theorem.

Keywords: differential invariants; invariant derivations; symplectic; contact spaces

MSC: 53A55; 14L24; 37J15; 15A72

1. Introduction

Differential invariants of various groups play an important role in applications [1–3]. Classical curvatures of submanifolds in Euclidean space arise as differential invariants of the orthogonal group. The corresponding problem for symplectic spaces was initiated in [4]. Further works in this direction include [5–10]. In this paper we consider the linear symplectic group action and compute the corresponding algebra of differential invariants. We will use either the standard representation or its trivial extension; other actions were also considered in the literature [11] and we comment on the relations of the above cited works to ours at the conclusion of the paper.

Let $V = \mathbb{R}^{2n}(x,y)$ be equipped with the standard symplectic form $\omega = \sum_1^n dx_i \wedge dy_i$. Every infinitesimal symplectic transformation of V is given by the Hamiltonian function $H \in C^\infty(V)$ and has the form $X_H = \omega^{-1} dH$, and the Lie bracket of vector fields corresponds to the Poisson bracket of functions. By the Darboux-Givental theorem, the action of $\text{Symp}(V, \omega)$ has no local invariants. However these arise when we restrict to finite-dimensional subalgebras/subgroups. Namely, functions H quadratic in x, y form a subalgebra isomorphic to $\mathfrak{sp}(2n, \mathbb{R})$. For functions of degree ≤ 2 we get the affine symplectic algebra $\mathfrak{sp}(2n, \mathbb{R}) \ltimes \mathbb{R}^{2n}$. We will concentrate on the linear case and compute the algebra of differential invariants for submanifolds and functions on V.

It turns out that for curves and hypersurfaces one can describe the generators for all n that we provide, while for the case of dimension and codimension greater than one, this becomes more complicated. Of those, we consider in details only the case of surfaces in $\mathbb{R}^4$. Generators of the algebra of differential invariants will be presented in the Lie-Tresse form as functions and derivations, and for lower dimensions, we also compute the differential syzygies. We will mainly discuss the geometric coordinate-free approach. The explicit formulae are rather large and will be shown in the Appendix A only for $n = 2$.

We also consider the space $W = \mathbb{R}^{2n+1}(x, y, z)$ equipped with the standard contact form $\alpha = dz - \sum_1^n y_i dx_i$. Every infinitesimal contact transformation of W is given by the contact Hamiltonian $H \in C^\infty(W)$ via $\alpha(X_H) = H$, $X_H(\alpha) = \partial_u(H)$, and the Lie bracket of vector fields corresponds to the Lagrange bracket of functions. Again, the action of $\text{Cont}(W, [\alpha])$ has no local invariants, however, these arise when we restrict to finite-dimensional subalgebras/subgroups. Namely, functions H quadratic in x, y, z with weights $w(x_i) = 1 = w(y_i)$, $w(z) = 2$ form a subalgebra isomorphic to

$\mathfrak{csp}(2n,\mathbb{R})$. For functions of degree ≤ 2 we get the affine extension $(\mathbb{R} \oplus \mathfrak{sp}(2n,\mathbb{R})) \ltimes \mathfrak{heis}(2n+1)$ by the Heisenberg algebra. For simplicity, we will concentrate on the action of $\mathfrak{sp}(2n,\mathbb{R})$, and then comment how to extend to the conformally symplectic algebra and include the translations.

It is interesting to remark on the computational aspect of the results. There are two approaches to compute differential invariants. The infinitesimal method is based on the defining Lie equations and works universally for arbitrary Lie algebras of vector fields. The moving frame method is based on elimination of group parameters and is dependent on explicit parametrization of the Lie group (or pseudogroup in infinite-dimensional situation) and its action. In MAPLE, these in turn rely on pdsolve and eliminate commands or some algorithmically optimized versions of those (via Gröbner basis or similar). For the problem at hand, we can use both since one can locally parametrize the group $\mathrm{Sp}(2n,\mathbb{R})$ and its linear action. The Lie algebra method works well in dimension 2 (symplectic case $n=1$) and fails further. The Lie group method works well in dimension 3 (contact case $n=1$) and fails further. Computational difficulties obstruct finishing calculations already in dimension 4 with these straightforward approaches. We show, however, how other geometric methods allow to proceed further.

This paper is partially based on the results of [12], extending and elaborating it in several respects. Some applications will be briefly discussed at the end of the paper. The paper is organized as follows. In the next section, we recall the basics. Then, we describe in turn differential invariants of functions, curves and hypersurfaces in symplectic vector spaces, and also discuss the particular case of surfaces in $\mathbb{R}^4$. Then, we briefly discuss the invariants in contact vector spaces and demonstrate how to compute differential invariants for conformal and affine extensions from our preceding computations.

We present most computations explicitly. Some large formulae are delegated to the Appendix A, the other can be found as Supplementary Material in this article.

2. Recollections and Setup

We refer to [13] for details of the jet-formalism, summarizing the essentials here.

2.1. Jets

Let M be a smooth manifold. Two germs at $a \in M$ of submanifolds $N_1, N_2 \subseteq M$ of dimension n and codimension m are equivalent if they are tangent up to order k at a. The equivalence class $[N]_a^k$ is called the k-jet of N at a. Denote $J_a^k(M,n)$ the set of all k-jets at a and $J^k(M,n) = \cup_{a \in M} J_a^k(M,n)$ the space of k-jets of n-submanifolds. This is a smooth manifold of dimension $n + m\binom{n+k}{k}$ and there are natural bundle projections $\pi_{k,l} : J^k(M,n) \to J^l(M,n)$ for $k > l \geq 0$. Note that $J^0(M,n) = M$ and $J^1(M,n) = \mathrm{Gr}_n(TM)$, while $\pi_{k,k-1} : J^k(M,n) \to J^{k-1}(M,n)$ are affine bundles for $k > 1$.

Since functions $f \in C^\infty(M)$ can be identified with their graphs $\Sigma_f \subset M \times \mathbb{R}$, the space of k-jets of functions $J^k M$ is defined as the space of k-jets of hypersurfaces $\Sigma \subset M \times \mathbb{R}$ transversal to the fibers of the projection to M. This jet space embeds as an open subset into $J^k(M \times \mathbb{R}, n)$, where $n = \dim M$ (and $m = 1$) and so its dimension is $n + \binom{n+k}{k}$.

Sometimes, we denote spaces $J^k M$ and $J^k(M,n)$ simply by J^k. The inverse limit along projections $\pi_{k,k-1}$ yields the space $J^\infty = \varprojlim J^k$.

In local coordinates (x,y) on M a submanifold N can be written as $y^j = y^j(x^i)$, $i = 1, \ldots, n$, $j = 1, \ldots, m$. Then the jet-coordinates are given by $x^i([N]_a^k) = a^i$, $y_\sigma^j([N]_a^k) = \frac{\partial^{|\sigma|} y^j}{\partial x^\sigma}(a)$ for a multi-index $\sigma = (i_1, \ldots, i_n)$ of length $|\sigma| = \sum_1^n i_s \leq k$.

For the jets of functions $u = u(x)$ we use the jet-coordinates $x^i([u]_a^k) = a^i$, $u_\sigma([u]_a^k) = \frac{\partial^{|\sigma|} u}{\partial x^\sigma}(a)$. We sometimes also write u instead of u_0, and we often lower indices for the base coordinates, like x_i instead of x^i etc, if no summation suffers.

2.2. Prolongations

A Lie group action on a manifold M is a homomorphism $\Phi : G \to \text{Diff}(M)$. Any $g \in G$ determines a point transformation $\Phi_g(a) = g \cdot a$, $a \in M$. This induces an action on germs of submanifolds, hence on jets of submanifolds, namely

$$\Phi_g^{(k)}([N]_a^k) = [\Phi_g(N)]_{\Phi_g(a)}^k.$$

Similarly, if X is a vector field on M, corresponding to the Lie algebra $\mathfrak{g} = \text{Lie}(G)$, the prolongation or lift gives a vector field $X^{(k)}$ on J^k. If (x, u) are local coordinates on M (with x^i interpreted as independent and u^j as dependent variables) and the vector field is given as $X = a^i \partial_{x^i} + b^j \partial_{u^j}$, then its prolongation has the form

$$X^{(k)} = a^i \mathcal{D}_{x^i}^{(k+1)} + \sum_{|\sigma| \leq k} \mathcal{D}_\sigma(\varphi^j) \partial_{u_\sigma^j},$$

where $\varphi = (\varphi^1, \ldots, \varphi^m)$ and $\varphi^j = b^j - a^i u_i^j$ is the generating vector-function, $\mathcal{D}_{x^i} = \partial_{x^i} + \sum_{j,\tau} u_{\tau+1_i}^j \partial_{u_\tau^j}$ is the total derivative, $\mathcal{D}_{x^i}^{(k+1)}$ its truncation (restriction to $(k+1)$-jets: $|\tau| \leq k$) and $\mathcal{D}_\sigma = \mathcal{D}_{x^1}^{i_1} \cdots \mathcal{D}_{x^n}^{i_n}$ for $\sigma = (i_1, \ldots, i_n)$ is the iterated total derivative.

2.3. Differential Invariants

A differential invariant of order k is a function I on J^k, which is constant on the orbits of $\Phi^{(k)}$ action of G. If the Lie group G is connected this is equivalent to $L_{X^{(k)}} I = 0$ for all $X \in \mathfrak{g}$ (some care should be taken with this statement, mostly related to usage of local coordinate charts in jets, see the first example in [14]).

The space of k-th order differential invariants forms a commutative algebra over $\mathbb{R}$, denoted by $\mathcal{A}_k$. The injection $\pi_{k+1,k}^*$ induces the embedding $\mathcal{A}_k \subset \mathcal{A}_{k+1}$, and in the inductive limit we get the algebra of differential invariants $\mathcal{A} \subseteq C^\infty(J^\infty)$, namely

$$\mathcal{A} = \varinjlim \mathcal{A}_k.$$

Denote by $G_a = \{g \in G : g \cdot a = a\}$ the stabilizer of $a \in M$. This subgroup of G acts on J_a^k. The prolonged action of G is called algebraic if the prolongation $G_a^{(k)}$ is an algebraic group acting algebraically on $J_a^k \ \forall a \in M$. For our problem, the action of G on M is almost transitive and algebraic, so by [14] the invariants $I \in \mathcal{A}$ can be taken as *rational functions* in jet-variables u_σ^j; moreover they may be chosen polynomial starting from some jet-order. This will be assumed in what follows.

In our situation $\mathcal{A}$ is not finitely generated in the usual sense since the number of independent invariants is infinite. We will use the Lie–Tresse theorem [14] that guarantees that $\mathcal{A}$ is generated by a finite set of differential invariants and invariant derivations.

Recall that an invariant derivation is such a horizontal (or Cartan) derivation $\nabla : \mathcal{A} \to \mathcal{A}$ (obtained by a combination of total derivatives) that it commutes with the action of the group: $\forall g \in G$ we have $g_*^{(k+1)} \nabla = \nabla g_*^{(k)}$ for $k \geq k_0$, where k_0 is the order of ∇, which can be identified with the highest order of coefficients in the decomposition $\nabla = \sum_i a^i(x, u_\sigma^j) \mathcal{D}_{x^i}$. Equivalently we can write $\forall X \in \mathfrak{g} : L_{X^{(k+1)}} \nabla = \nabla L_{X^{(k)}}$ for $k \geq k_0$. This implies $\nabla : \mathcal{A}_k \to \mathcal{A}_{k+1}$ in the same range.

Invariant derivations form a submodule $\mathcal{CD}^G \subseteq \mathcal{CD}(J^\infty)$ in the space of all horizontal derivations. It is a finitely generated $\mathcal{A}$ module: any $\nabla \in \mathcal{CD}^G$ has the form $\nabla = I^i \nabla_i$ for a fixed set ∇_i and $I^i \in \mathcal{A}$. By ([14] Theorem 21), the number of derivations ∇_i is n.

We compose iterated operators $\nabla_J : \mathcal{A}_k \to \mathcal{A}_{k+|J|}$ for multi-indices J, and then $\mathcal{A}$ is generated by $\nabla_J I_i$ for a finite set of I_i.

2.4. Counting the Invariants

An important part of our computations is a count of independent differential invariants. Denote the number of those on the level of k-jets by s_k. This number is equal to the transcendence degree of the field of differential invariants on J^k (when the elements of $\mathcal{A}_k$ are rational functions) and it coincides with the codimension of $G^{(k)}$ orbit in J^k.

Since in our case G is a (finite-dimensional) Lie group, the action becomes eventually free, i.e., $G^{(k)}_{a_k} = \mathrm{id}$ for sufficiently large $k \geq \ell$ and generic $a_k \in J^k$ cf. [1]. In this case, the orbit is diffeomorphic to G, in particular $s_k = \dim J^k - \dim G$ for $k \geq \ell$.

The number of "pure order" k differential invariants is $h_k = s_k - s_{k-1}$, so it becomes

$$h_k = \dim J^k - \dim J^{k-1} = m\binom{n+k-1}{k} \text{ for } k > \ell.$$

The Poincaré function $P(z) = \sum_{k=0}^{\infty} h_k z^k$ is rational in all local problems of analysis according to Arnold's conjecture [15]. In our case, this $P(z)$ differs from $m(1-z)^{-n}$ by a polynomial reflecting the action of G.

Note that by the eventual freeness of the action, the algebra $\mathcal{A}$ is generated by invariants and derivations at most from the jet-level ℓ.

2.5. The Equivalence Problem

The generators I_i $(1 \leq i \leq s)$, ∇_j $(1 \leq j \leq n)$ are not independent, i.e., the algebra $\mathcal{A}$ is not freely generated by them, in general. A differential syzygy is a relation among these generators. Such an expression has the form $F(\nabla_{J_1}(I_{i_1}), \ldots, \nabla_{J_t}(I_{i_t})) = 0$, where F is a function of t arguments and $J_1, \ldots, J_t$ are multi-indices. Choosing a generating set F_ν of differential syzygies, we express

$$\mathcal{A} = \langle I_i; \nabla_j \mid F_\nu \rangle.$$

This allows to solve the equivalence problem for submanifolds of functions with respect to G as follows. Consider the above Lie–Tresse type representation of $\mathcal{A}$. The collection of invariants $I_i, \nabla_j(I_i)$ (totally r functions) allows to restore the generators, while the relations F_ν constrain this collection. Any submanifold N (for function f given as the graph $\Sigma_f \simeq M$) canonically lifts to the jet-space J^∞: $N \ni a \mapsto [N]^\infty_a$. We thus map $\Psi : N \to \mathbb{R}^r$, $\Psi(a) = (I_i([N]^\infty_a), \nabla_j(I_i)([N]^\infty_a))$. Due to differential syzygies the image is contained in some algebraic subset $Q \subset \mathbb{R}^r$. Two generic submanifolds N_1, N_2 are G-equivalent iff $\Psi(N_1) = \Psi(N_2)$ as (un-parametrized) subsets.

2.6. Conventions

All differential invariants below are denoted by I with a subscript. The subscript consists of a number and a letter. The number reflects the order of an invariant, while the letter distinguishes invariants of the same order. If no letter is given, there is only one new (independent) invariant on the corresponding jet-space.

The symplectic Hamiltonian vector field in canonical coordinates on V has the form $X_H = \sum_i H_{y_i} \partial_{x_i} - H_{x_i} \partial_{y_i}$. The Poisson bracket given by $[X_f, X_g] = X_{\{f,g\}}$ is equal to

$$\{f, g\} = \sum_{i=1}^{n} \left(\frac{\partial f}{\partial x_i} \frac{\partial g}{\partial y_i} - \frac{\partial f}{\partial y_i} \frac{\partial g}{\partial x_i} \right).$$

A basis of quadratic functions $\langle x_i x_j, x_i y_j, y_i y_j \rangle \ni f$ gives a basis of vector fields X_f forming $\mathfrak{sp}(2n, \mathbb{R})$. This may be extended to $\mathfrak{csp}(2n, \mathbb{R})$ by adding the homothety $\zeta = \sum_i x_i \partial_{x_i} + y_i \partial_{y_i}$ that commutes with $\mathfrak{sp}(2n, \mathbb{R})$.

The contact Hamiltonian vector field in canonical coordinates on W has the form $X_H = H\partial_z + \sum_1^n \mathcal{D}_{x_i}^{(1)}(H)\partial_{y_i} - H_{y_i}\mathcal{D}_{x_i}^{(1)} = (H - \sum y_i H_{y_i})\partial_z + \sum_1^n (H_{x_i} + y_i H_z)\partial_{y_i} - H_{y_i}\partial_{x_i}$. The Lagrange bracket given by $[X_f, X_g] = X_{[f,g]}$ is equal to

$$[f, g] = \sum_{i=1}^{n} \left(\frac{\partial f}{\partial x_i} \frac{\partial g}{\partial y_i} - \frac{\partial g}{\partial x_i} \frac{\partial f}{\partial y_i} \right) + \sum_{i=1}^{n} y_i \left(\frac{\partial f}{\partial z} \frac{\partial g}{\partial y_i} - \frac{\partial g}{\partial z} \frac{\partial f}{\partial y_i} \right) + \left(f \frac{\partial g}{\partial z} - g \frac{\partial f}{\partial z} \right).$$

A basis of quadratic functions $\langle x_i x_j, x_i y_j, y_i y_j \rangle \ni f$ gives a basis of vector fields X_f forming $\mathfrak{sp}(2n,\mathbb{R})$. This may be extended to $\mathfrak{csp}(2n,\mathbb{R})$ by adding the homothety $X_f = \sum_i x_i \partial_{x_i} + y_i \partial_{y_i} + 2z \partial_z$ for $f = 2z - \sum_i x_i y_i$ that commutes with $\mathfrak{sp}(2n,\mathbb{R})$.

3. Functions on Symplectic Vector Spaces

The group $G = \mathrm{Sp}(2n,\mathbb{R})$ acts almost transitively on $V = \mathbb{R}^{2n}$ (one open orbit that complements the origin), and it is lifted to $J^0 V = V \times \mathbb{R}(u)$ with $I_0 = u$ being invariant. The prolonged action has orbits of codimension 2 on $J^1 V$ (one more invariant appears) and then the action becomes free on $J^2 V$.

An invariant on J^1 is due to the invariant 1-form du and the invariant (radial) vector field $\zeta = \sum_i x_i \partial_{x_i} + y_i \partial_{y_i}$: their contraction yields

$$I_1 = du(\zeta) = \sum_{i=1}^{n} x_i u_{x_i} + y_i u_{y_i}.$$

3.1. The Case of Dimension 2n = 2

Here $V = \mathbb{R}^2(x,y)$. To compute differential invariants of order k we solve the equation $\mathcal{L}_{X_i^{(k)}} I = 0$, $I \in C^\infty(J^k V)$, for a basis of the Lie algebra $\mathfrak{sp}(2,\mathbb{R}) = \mathfrak{sl}(2,\mathbb{R})$: $X_1 = x \partial_y$, $X_2 = x \partial_x - y \partial_y$, $X_3 = y \partial_x$. For $k = 2$, in addition to I_0 and I_1, we get

$$I_{2a} = x^2 u_{xx} + 2xy u_{xy} + y^2 u_{yy},$$
$$I_{2b} = x u_y u_{xx} - y u_x u_{yy} + (y u_y - x u_x) u_{xy},$$
$$I_{2c} = u_x^2 u_{yy} - 2 u_x u_y u_{xy} + u_y^2 u_{xx}.$$

These invariants are functionally (hence algebraically) independent.

To determine the invariant derivations, we solve its defining PDE. The invariant derivations of order $k = 1$ are linear combinations of

$$\nabla_1 = x \mathcal{D}_x + y \mathcal{D}_y, \quad \nabla_2 = u_x \mathcal{D}_y - u_y \mathcal{D}_x.$$

Let $\mathcal{A}$ denote the algebra of differential invariants, whose elements can be assumed polynomial in all jet-variables. Since the obtained invariants are quasi-linear in their respective top jet-variables, and this property is preserved by invariant derivations, the algebra $\mathcal{A}$ is generated by them.

To find a more compact description, note that $I_1 = \nabla_1(I_0)$ and

$$I_{2a} = \nabla_1^2(I_0) - \nabla_1(I_0), \quad I_{2b} = -\nabla_2 \nabla_1(I_0).$$

Thus only I_0 and I_{2c} suffice to generate $\mathcal{A}$.

To describe the differential syzygies, note that $\nabla_2(I_0) = 0$, and the commutator relation is

$$[\nabla_1, \nabla_2] = \frac{I_{2b}}{I_1} \nabla_1 + \frac{I_{2a} - I_1}{I_1} \nabla_2.$$

In addition, when applying ∇_1, ∇_2 to I_{2a}, I_{2b}, I_{2c} and using the commutator relation we get five different invariants of order 3, while there are only four independent 3-jet coordinates. Thus computing the symbols of the invariants and eliminating those coordinates we obtain the remaining syzygy:

$$(\nabla_2(I_{2b}) + \nabla_1(I_{2c})) I_1 - (3 I_{2a} - I_1) I_{2c} + 3 I_{2b}^2 = 0.$$

To summarize, define

$$\mathcal{R}_1 = \nabla_2(I_0),$$
$$\mathcal{R}_2 = I_1[\nabla_1, \nabla_2] - I_{2b}\nabla_1 - (I_{2a} - I_1)\nabla_2,$$
$$\mathcal{R}_3 = I_1\nabla_2(I_{2b}) + I_1\nabla_1(I_{2c}) - (3I_{2a} - I_1)I_{2c} + 3I_{2b}^2.$$

Then, the algebra of differential invariants is given by generators and relations as follows:

$$\mathcal{A} = \langle I_0, I_{2c} ; \nabla_1, \nabla_2 \mid \mathcal{R}_1, \mathcal{R}_2, \mathcal{R}_3 \rangle.$$

3.2. Another Approach for $n = 1$

We act similar to [16].

Note that ∇_1 corresponds to the radial vector field ζ and $\nabla_2 = \omega^{-1}\hat{d}u$, where $\hat{d}$ is the horizontal differential (in this case $\hat{d} = dx \otimes \mathcal{D}_x + dy \otimes \mathcal{D}_y$, so $\hat{d}u = u_x\,dx + u_y\,dy$). To find further invariants and derivations we consider the quadratic form

$$Q_2 = \hat{d}^2 u = u_{xx}dx^2 + 2u_{xy}dx\,dy + u_{yy}dy^2 \in \pi_2^* S^2 T^* V.$$

Lowering the indices with respect to the symplectic form (or partially contracting with $\omega^{-1} = \partial_x \wedge \partial_y$) we get the endomorphism

$$A = \omega^{-1}Q_2 = u_{yy}\partial_x \otimes dy - u_{xy}\partial_y \otimes dy + u_{xy}\partial_x \otimes dx - u_{xx}\partial_y \otimes dx.$$

This can be lifted to the Cartan distribution on J^∞ and thus applied to horizontal fields:

$$A\nabla_1 = (xu_{xy} + yu_{yy})\mathcal{D}_x - (xu_{xx} + yu_{xy})\mathcal{D}_y,$$
$$A\nabla_2 = (u_x u_{yy} - u_y u_{xy})\mathcal{D}_x - (u_x u_{xy} - u_y u_{xx})\mathcal{D}_y.$$

These are also invariant derivations and they can be expressed through the previous as follows:

$$A\nabla_1 = -\frac{I_{2b}}{I_1}\nabla_1 - \frac{I_{2a}}{I_1}\nabla_2, \quad A\nabla_2 = \frac{I_{2c}}{I_1}\nabla_1 + \frac{I_{2b}}{I_1}\nabla_2.$$

Note also that $I_{2a} = Q_2(\nabla_1, \nabla_1)$, $I_{2b} = -Q_2(\nabla_1, \nabla_2)$, $I_{2c} = Q_2(\nabla_2, \nabla_2)$, so that we can generate all the invariants uniformly.

3.3. The General Case

In general dimension $2n$ we still have the invariant derivations ∇_1 corresponding to the radial field ζ and $\nabla_2 = \omega^{-1}Q_1$ for $Q_1 = \hat{d}I_0$. Then, the horizontal field of endomorphisms $A = \omega^{-1}Q_2$ for $Q_2 = \hat{d}^2 I_0$ generates the rest: the invariant derivations $\nabla_{i+2} = A^i \nabla_2$ (alternatively $\nabla_{i+2} = A^i \nabla_1$) for $i = 1, \ldots, 2n - 2$ are independent (also with ∇_1, ∇_2) on a Zariski open subset in the space of jets. This gives a complete set of invariant derivations $\nabla_1, \ldots, \nabla_{2n}$.

Taking into account $I_1 = \nabla_1(I_0)$ the generating set of invariants can be taken I_0 and $I_{ij} = Q_2(\nabla_i, \nabla_j)$. By dimensional count and independence it is enough to restrict to $i = 1, 2$ and $1 \leq j \leq 2n$. We obtain:

Theorem 1. *The algebra of differential invariants of the G-action on $J^\infty(V)$ is*

$$\mathcal{A} = \langle I_0, I_{1i}, I_{2j}; \nabla_k \mid \mathcal{R}_l \rangle$$

for some finite set of differential syzygies $\mathcal{R}_l$.

This is a Lie-Tresse type of generation of $\mathcal{A}$. Note also the following (non-finite) generation of this algebra. The higher symmetric differentials $Q_k = d^k u \in \pi_k^* S^k T^* V$ can be contracted with invariant derivations to get k-th order differential invariants $Q_k(\nabla_{j_1}, \ldots, \nabla_{j_k})$. These clearly generate $\mathcal{A}$.

There is an algorithmic way of describing relations (syzygies) between these invariants similar to ([16], Section 4). We refer for explicit formulae of invariants to [12] for $n = 2$.

4. Curves in Symplectic Vector Spaces

Locally a curve in $\mathbb{R}^{2n}$ is given as $u = u(t)$ for $t = x_1$ and $u = (x_2, \ldots, x_n, y_1, \ldots, y_n)$ in the canonical coordinates $(x_1, x_2, \ldots, x_n, y_1, \ldots, y_n)$. The corresponding jet-space $J^k(V, 1)$ has coordinates u_l, $l \leq k$, where l stands for the l-tuple of t. For instance, $J^1(V, 1) = \mathbb{R}^{4n-1}(t, u, u_1)$. Note that $\dim J^k(V, 1) = 2n + k(2n - 1)$.

4.1. The Case of Dimension $2n = 2$

Let us again start with the simplest example $V = \mathbb{R}^2(x, y)$. The jet-space is $J^k(V, 1) = \mathbb{R}^{k+2}(x, y, y_1, \ldots, y_k)$. Here $G = \mathrm{Sp}(2, \mathbb{R})$ has an open orbit in $J^1(V, 1)$, and there is one new differential invariant in every higher jet-order k.

Let us indicate in this simple case how to verify algebraicity of the action (this easily generalizes to the other cases and will not be discussed further). The 1-prolonged action of $g = \begin{pmatrix} a & b \\ c & d \end{pmatrix} \in G$ is

$$\Phi_g^{(1)}(x, y, y_1) = \left(ax + by, cx + dy, \frac{dy_1 + c}{by_1 + a} \right).$$

Since the action is transitive on $J^0(V, 1) \setminus 0 = \mathbb{R}^2_\times$, choose $p = (1, 0)$ as a generic point. Its stabilizer is $G_p = \left\{ \begin{pmatrix} 1 & b \\ 0 & 1 \end{pmatrix} \right\} \subset G$. The action of this on the fiber $\pi_{1,0}^{-1}(p)$ is algebraic: $y_1 \mapsto \frac{y_1}{by_1 + 1}$.

Thus, the Lie-Tresse theorem [14] applies and the algebra of invariants $\mathcal{A}$ can be taken to consist of rational functions in jet-variables, which are polynomial in jets of order ≥ 2.

The first differential invariant is easily found from the Lie equation:

$$I_2 = \frac{y_2}{(xy_1 - y)^3}.$$

Similarly, solving the PDE for the coefficients of invariant derivation, we find

$$\nabla = \frac{1}{xy_1 - y} \mathcal{D}_x.$$

Now by differentiation, we get new differential invariants $I_3 = \nabla I_2$, $I_4 = \nabla^2 I_2$, etc. Since these are quasilinear differential operators, they generate the entire algebra. In other words, the algebra of differential invariants is free:

$$\boxed{\mathcal{A} = \langle I_2; \nabla \rangle.}$$

4.2. The Case of Dimension $2n = 4$

Let us use coordinates (t, x, y, z) on $V = \mathbb{R}^4$ with the symplectic form $\omega = dt \wedge dy + dx \wedge dz$. Note that $\dim J^k(V, 1) = 3k + 4$, and the jet-coordinates on J^k are $(t, x, y, z, \ldots, x_k, y_k, z_k)$. The action of $G = \mathrm{Sp}(4, \mathbb{R})$ on $J^k(V, 1)$ has orbits of dimensions $4, 7, 9, 10$ for $k = 0, 1, 2, 3$ respectively. Thus the first differential invariant appears already in jet-order 2, then two more appear in jet-order 3, and then $h_k = 3$ new invariants in every jet-order $k \geq 4$.

The infinitesimal and moving frame methods fail to produce enough invariants here, so we apply more geometric considerations.

We exploit that G preserves the symplectic form on V, but also the fact that the action is linear, so the vector space structure of V is preserved as well. In particular, the origin is preserved, so we can form a vector from the origin to any point $p = (t, x, y, z) \in J^0(V, 1)$. Denote the corresponding vector by

$$v_0 = (t, x, y, z) \equiv t\partial_t + x\partial_x + y\partial_y + z\partial_z.$$

Consider the space of 1-jets of unparametrized curves $J^1(V, 1)$. For a parameterization of the curve $c = (t, x(t), y(t), z(t))$ the tangent vector at any point of this curve can be computed as $w_1 = \mathcal{D}_t^{(1)} = \partial_t + x_1 \partial_x + y_1 \partial_y + z_1 \partial_z$, which is rescaled $v_1 = \beta w_1$ upon a change of parametrization. To make v_1 invariant we fix β by the condition $\omega(v_0, v_1) = 1$. This normalization $\beta = 1/(ty_1 + xz_1 - x_1 z - y)$ gives a canonical horizontal (that is tangent to the curve) vector field, which can be interpreted as an invariant derivative

$$\nabla = \frac{1}{(ty_1 + xz_1 - x_1 z - y)} \mathcal{D}_t.$$

The further approach is as follows. On every step there is a freedom associated to a parameterization of a given curve. Fixing it in a canonical way via evaluation with the symplectic form, we obtain invariantly defined vectors and henceforth invariants.

On the first step, changing the parameterization $c = c(t)$ to another parameterization $c = c(\tau)$ results in a change of the tangent vector by the chain rule:

$$\frac{dc}{dt} = \frac{d\tau}{dt} \frac{dc}{d\tau}.$$

This can be written as $w_1 = k_1 v_1$, for $d\tau/dt = k_1$. The vector w_1, associated with a specific choice of parameterization, is not canonical but convenient for computations. The above normalization $k_1 = 1/\beta$ makes v_1 a canonical choice.

The change of parameterization on 2-jets gives

$$\frac{d^2 c}{dt^2} = \frac{d^2 c}{d\tau^2} \left(\frac{d\tau}{dt}\right)^2 + \frac{dc}{d\tau} \frac{d^2 \tau}{dt^2}.$$

Denote $v_2 = d^2 c/d\tau^2$, $w_2 = d^2 c/dt^2$ and $d^2\tau/dt^2 = k_2$. The equation becomes

$$w_2 = v_2 k_1^2 + v_1 k_2.$$

In the parameterization $c = c(t)$ the acceleration is $w_2 = (0, x_2, y_2, z_2)$. We solve for v_2 as

$$v_2 = \frac{w_2 - v_1 k_2}{k_1^2}.$$

Then, k_2 can be fixed by $\omega(v_0, v_2) = 0$. This uniquely determines v_2, which can now be used to find the first differential invariant. In fact, $I_2 = \omega(v_1, v_2)$ is a differential invariant of order 2. In coordinates

$$I_2 = \omega(v_1, v_2) = \frac{x_1 z_2 - z_1 x_2 + y_2}{(ty_1 + xz_1 - zx_1 - y)^3}.$$

There are 2 independent third order invariants by our dimension count. The first can be obtained as $\nabla(I_2)$, to find the second we exploit the above normalization method on 3-jets. The change of parameterization is

$$\frac{d^3c}{dt^3} = \frac{d^3c}{d\tau^3}\left(\frac{d\tau}{dt}\right)^3 + 3\frac{d^2c}{d\tau^2}\frac{d\tau}{dt}\frac{d^2\tau}{dt^2} + \frac{dc}{d\tau}\frac{d^3\tau}{dt^3}.$$

Again, rewrite it in simpler notations as

$$w_3 = v_3 k_1^3 + 3k_1 k_2 v_2 + k_3 v_1.$$

Here, $w_3 = (0, x_3, y_3, z_3)$ and the unknown k_3 can be fixed by the condition $\omega(v_0, v_3) = 0$, where

$$v_3 = \frac{w_3 - 3k_1 k_2 v_2 - k_3 v_1}{k_1^3}.$$

This uniquely determines v_3, which allows the computation of two new differential invariants:

$$I_{3a} = \omega(v_1, v_3), \quad I_{3b} = \omega(v_2, v_3).$$

The invariants I_{3a} and I_{3b} are independent, but I_{3a} can be expressed through $\nabla(I_2)$, so it is not required in what follows.

Finally, we explore the forth order chain rule

$$\frac{d^4c}{dt^4} = \frac{d^4c}{d\tau^4}\left(\frac{d\tau}{dt}\right)^4 + 6\frac{d^3c}{d\tau^3}\left(\frac{d\tau}{dt}\right)^2 \frac{d^2\tau}{dt^2} + \frac{d^2c}{d\tau^2}\left(4\frac{d\tau}{dt}\frac{d^3\tau}{dt^3} + 3\left(\frac{d^2\tau}{dt^2}\right)^2\right) + \frac{dc}{d\tau}\frac{d^4\tau}{dt^4}$$

that can be written as

$$w_4 = v_4 k_1^4 + 6v_3 k_1^2 k_2 + v_2\left(4k_1 k_3 + 3k_2^2\right) + v_1 k_4$$

with $w_4 = (0, x_4, y_4, z_4)$. Find k_4 by $\omega(v_0, v_4) = 0$. This uniquely determines v_4, then the invariants of order 4 are found by the formulae

$$I_{4a} = \omega(v_1, v_4), \quad I_{4b} = \omega(v_2, v_4), \quad I_{4c} = \omega(v_3, v_4).$$

These are independent, but I_{4a} and I_{4b} can be expressed by the invariants of order 3 and the invariant derivation, so they will not be required in what follows.

This gives the necessary invariants to generate the entire algebra of differential invariants. To summarize, if we denote $I_3 = I_{3b}$ and $I_4 = I_{4c}$, then the algebra of differential invariants is freely generated as follows

$$\mathcal{A} = \langle I_2, I_3, I_4; \nabla \rangle.$$

The explicit coordinate formulae of invariants are shown in the Appendix A.

4.3. The General Case

In dimension $\dim V = 2n$ the following dimensional analysis readily follows from the normalization procedure developed above.

Jet-level k	dim $J^k(V,1)$	G-orbit dimension	# new invariants h_k
0	$2n$	$2n$	0
1	$4n-1$	$2n+(2n-1)=4n-1$	0
2	$6n-2$	$(4n-1)+(2n-2)=6n-3$	1
3	$8n-3$	$(6n-3)+(2n-3)=8n-6$	2
4	$10n-4$	$(8n-6)+(2n-4)=10n-10$	3
...	...	...	...
k	$2n+k(2n-1)$	$2(k+1)n-\binom{k+1}{2}$	$k-1$
...	...	...	...
$2n-1$	$(2n-1)^2+2n$	$\binom{2n+1}{2}$	$2n-2$
$2n$	$4n^2$	$\dashrightarrow$ stabilized	$2n-1$

In particular, the number of pure order k differential invariants is $h_k = k-1$ for $1 \leq k \leq 2n$ and $h_k = 2n-1$ for $k > 2n$.

If the canonical coordinates in $\mathbb{R}^{2n}$ are (t,x,y,z), where x and z and $(n-1)$-dimensional vectors, then the invariant derivation is equal to

$$\nabla = \frac{1}{(ty_1-y+xz_1-x_1z)}\mathcal{D}_t.$$

We also obtain the first differential invariant of order 2

$$I_2 = \frac{(x_1z_2-x_2z_1+y_2)}{(ty_1-y+xz_1-x_1z)^3}.$$

Then, we derive the differential invariant $\nabla(I_2)$ and add to it another differential invariant I_3 of order 3. Then, we derive the differential invariants $\nabla^2(I_2), \nabla(I_3)$ and add another differential invariant I_4 of order 4. We continue obtaining new invariants by using the higher order chain rule and normalization via the symplectic form up to order $2n$.

In summary, we obtain $2n-1$ independent differential invariants $I_2,\ldots,I_{2n}$ of orders $2,\ldots,2n$ respectively.

Theorem 2. *The algebra of differential invariants of the G-action on $J^\infty(V,1)$ is freely generated as follows:*

$$\mathcal{A} = \langle I_2,\ldots,I_{2n}\,;\,\nabla\rangle.$$

5. Hypersurfaces in Symplectic Vector Spaces

Since hypersurfaces in $\mathbb{R}^2$ are curves, the first new case come in dimension 4. We consider this first and then discuss the general case.

5.1. The Case of Dimension $2n=4$.

Let $V = \mathbb{R}^4$, denote its canonical coordinates by (x,y,z,u) with $\omega = dx \wedge dz + dy \wedge du$. Hypersurfaces can be locally identified as graphs $u = u(x,y,z)$ and this gives parametrization of an open chart in $J^k(V,3)$. We use the usual jet-coordinates u_x, u_{xx}, u_{yz}, etc.

As is the cases above, straightforward computations become harder. Maple is not able to compute all required invariants and derivations, so we again rely on a more geometric approach. Before going through the method, we investigate the count of invariants.

The group $G = \text{Sp}(4,\mathbb{R})$ acts with an open orbit on $J^0(V,3)$. On the space of 1-jets the dimension of the orbit is $7 = \dim J^1(V,3)$, hence there are no invariants. The orbit stabilization is reached on $J^2(V,3)$, where the action is free. The rank of the action is 10 and $\dim J^2(V,3) = 13$, so there are $h_2 = 3$ independent differential invariants. For $k > 2$, the number of new differential invariants is $h_k = \binom{k+2}{2}$. In particular, $h_3 = 10$.

The number of independent invariant derivations is 3, so these and 3 invariants of order 2 generate a total number of 9 invariants of order 3. In addition, commutators of invariant derivations $[\nabla_i, \nabla_j] = I_{ij}^k \nabla_k$ give up to 9 more differential invariants of order 3. We will confirm that the totality of these 18 contain 10 independent invariants of order 3, and hence suffice to generate also the differential invariants of higher order.

The 0-jet $p = (x, y, z, u) \in J^0(V, 3)$ can be identified with the vector from the origin to this point, which we denote by
$$v_0 = (x, y, z, u) \equiv x\partial_x + y\partial_y + z\partial_z + u\partial_u.$$

The 1-jet of a hypersurface $\Sigma = \{u = u(x, y, z)\}$ can be identified with its tangent space
$$T_p\Sigma = \langle \partial_x + u_x\partial_u, \partial_y + u_y\partial_u, \partial_z + u_z\partial_u \rangle = \langle \mathcal{D}_x^{(1)}, \mathcal{D}_y^{(1)}, \mathcal{D}_z^{(1)} \rangle.$$

The orthogonal complement to $T_p\Sigma$ with respect to ω is generated by
$$w_1 = \partial_y - u_z\partial_x + u_x\partial_z + u_y\partial_u,$$

that is $T_p\Sigma^{\perp\omega} = \langle w_1 \rangle$. The vector w_1 is determined up to scale, which we fix via the symplectic form so: $v_1 = k_1 w_1$ must satisfy $\omega(v_0, v_1) = 1$. This normalization gives $k_1 = 1/(xu_x + yu_y + zu_z - u)$, so the canonical vector v_1 is equal to
$$v_1 = \frac{1}{xu_x + yu_y + zu_z - u}(\partial_y - u_z\partial_x + u_x\partial_z + u_y\partial_u).$$

This vector field is tangent to the hypersurface, so it is horizontal and can be rewritten in terms of the total derivative. This yields the first invariant derivation:
$$\nabla_1 = \frac{D_y - u_z D_x + u_x D_z}{xu_x + yu_y + zu_z - u}.$$

Let $q = -u + u(x, y, z)$ be a defining function of the hypersurface $\Sigma = \{q = 0\}$. We have $T_p\Sigma = \ker dq$. A change of the defining function $q' = fq$ of Σ, with $f \in C^\infty(V)$ such that $f|_\Sigma \neq 0$, has the following effect on the differential: $dq' = q\, df + f\, dq$. Therefore at $p \in \Sigma$ we have $d_p q' = f(p) d_p q$ and so $T_p\Sigma = \ker dq'$.

Next we compute the second symmetric differential $d^2 q$ of the defining function for Σ. A change of the defining function $q' = fq$ has the following effect on the second differential:
$$d^2 q' = d(d(fq)) = d(q\, df + f\, dq) = q\, d^2 f + 2\, df\, dq + f\, d^2 q.$$

At the points $p \in \Sigma$ this simplifies to
$$d_p^2 q' = 2\, d_p f\, d_p q + f(p)\, d_p^2 q.$$

Restricting to the tangent space of Σ gives
$$d^2 q'|_{T_p\Sigma} = f(p) d^2 q|_{T_p\Sigma}.$$

Thus, the defining differential dq and the quadratic form $d^2 q$ are defined up to the same scale. We fix it again via the symplectic form: $d_p q' = k_2 d_p q$ must satisfy $d_p q'(v_0) = 1$, i.e., $k_2 = 1/dq(v_0)$ for generic 1-jets. This normalization gives the quadratic form $d^2q'|_{T_p\Sigma} = k_2 d^2 q|_{T_p\Sigma} = d^2 q|_{T_p\Sigma}/dq(v_0)$. In coordinates, with $q = -u + u(x,y,z)$, we get the expression

$$Q = d^2q'|_{T_p\Sigma} = \frac{u_{xx}dx^2 + 2u_{xy}dxdy + 2u_{xz}dxdz + u_{yy}dy^2 + 2u_{yz}dydz + u_{zz}dz^2}{xu_x + yu_y + zu_z - u}.$$

The first invariant is then computed by

$$I_{2a} = Q(v_1, v_1) = \frac{u_x^2 u_{zz} - 2u_x u_z u_{xz} + u_z^2 u_{xx} + 2u_x u_{yz} - 2u_z u_{xy} + u_{yy}}{(xu_x + yu_y + zu_z - u)^3}.$$

Let us summarize the geometric data encoding the 2-jet that we obtained and which are supported on the 3-dimensional tangent space $T_p\Sigma$: the invariant vector v_1, the symmetric 2-form Q of general rank, the skew 2-form $\omega|_{T_p\Sigma}$ of rank 2 (v_1 spans its kernel), and 1-form $\alpha = \omega(v_0, \cdot)$. These data give a canonical splitting of the tangent space $T_p\Sigma = \langle v_1 \rangle \oplus \Pi$, where $\Pi = \text{Ker}(\alpha)$. Indeed, $v_1 \notin \text{Ker}(\alpha)$ because $\omega(v_0, v_1) = 1$ by the normalization. Using this data, we can construct 2 more invariant derivations.

Choose a nonzero $w_3 \in \Pi$, $Q(v_1, w_3) = 0$. Then, choose $w_2 \in \Pi$, $Q(w_2, w_3) = 0$. For generic 2-jet, the vectors w_2, w_3 are defined up to scale that we fix so: $v_2 \in \langle w_2 \rangle$, $v_3 \in \langle w_3 \rangle$ must satisfy $Q(v_1, v_2) = 1$, $\omega(v_2, v_3) = 1$.

Since $v_2, v_3 \in T_p\Sigma$ are horizontal, they generate two invariant derivations ∇_2, ∇_3. Additionally we get 2 differential invariants:

$$I_{2b} = Q(v_2, v_2), \quad I_{2c} = Q(v_3, v_3).$$

A calculation of the rank of the corresponding Jacobi matrix shows that these are independent, and moreover, that the data generate all differential invariants of order 3. Then, by independence of $\nabla_1, \nabla_2, \nabla_3$ all higher order invariants can be derived, so for a finite set of differential syzygies $\mathcal{R}_l$ we get:

$$\boxed{\mathcal{A} = \langle I_{2a}, I_{2b}, I_{2c}\, ;\, \nabla_1, \nabla_2, \nabla_3 \mid \mathcal{R}_l \rangle}$$

The coordinate formulae can be found in [12] (note that renumeration $v_2 \leftrightarrow v_3$ and a different normalization is taken here).

5.2. The General Case

Now, we consider jets of hypersurfaces $\Sigma \subset V = \mathbb{R}^{2n}$ for general n and compute their differential invariants with respect to $G = \text{Sp}(2n, \mathbb{R})$.

By the Lie-Tresse theorem [14] the algebra $\mathcal{A}$ can be assumed to consist of rational functions on $J^\infty(V, 2n-1)$, which are polynomial in jet-variables of order ≥ 2.

The dimensional count easily generalizes to give $h_0 = h_1 = 0$, $h_2 = 2n - 1$ and $h_k = \binom{2n-2+k}{k}$ for $k > 2$. There will be $2n - 1$ independent invariant derivations ∇_j, and as before these together with second order invariants I_{2s} ($1 \leq s \leq 2n - 1$) and the structure coefficients l_{ij}^k of the horizontal frame ∇_j will suffice to generate all invariants.

We again have the position vector v_0, the tangent vector v_1 normalized by $\omega(v_0, v_1) = 1$, and the quadratic form Q on $T_p\Sigma$. From this data in a Zariski open set of $J^2(V, 2n-1)$ of generic 2-jets we get a canonical basis $v_1, \ldots, v_{2n-1}$ by normalizing in turn via ω and Q as follows (we repeat steps 0 and 1 that are already performed).

Step 0: $T_p\Sigma = \langle v_1, \ldots, v_{2n-1} \rangle$.

Step 1: Choose v_1 by $\langle v_1 \rangle \perp_\omega \langle v_1, \ldots, v_{2n-1} \rangle$, $\langle v_2, \ldots, v_{2n-1} \rangle \perp_\omega \langle v_0 \rangle$. Normalize $\omega(v_0, v_1) = 1$.
Step 2: Choose v_2 by $\langle v_3, \ldots, v_{2n-1} \rangle \perp_Q \langle v_1 \rangle$, $\langle v_2 \rangle \perp_Q \langle v_3, \ldots, v_{2n-1} \rangle$. Normalize $Q(v_1, v_2) = 1$.
Step 3: Choose v_3 by $\langle v_3 \rangle \perp_\omega \langle v_3, \ldots, v_{2n-1} \rangle$, $\langle v_4, \ldots, v_{2n-1} \rangle \perp_\omega \langle v_2 \rangle$. Normalize $\omega(v_2, v_3) = 1$.
Step 4: Choose v_4 by $\langle v_5, \ldots, v_{2n-1} \rangle \perp_Q \langle v_3 \rangle$, $\langle v_4 \rangle \perp_Q \langle v_5, \ldots, v_{2n-1} \rangle$. Normalize $Q(v_3, v_4) = 1$.
Inductively, we get the interchangeable steps as follows.

Step $(2r-1)$: Choose v_{2r-1} by $\langle v_{2r-1} \rangle \perp_\omega \langle v_{2r-1}, \ldots, v_{2n-1} \rangle$, $\langle v_{2r}, \ldots, v_{2n-1} \rangle \perp_\omega \langle v_{2r-2} \rangle$. Normalize $\omega(v_{2r-2}, v_{2r-1}) = 1$.

Step $2r$: Choose v_{2r} by $\langle v_{2r+1}, \ldots, v_{2n-1} \rangle \perp_Q \langle v_{2r-1} \rangle$, $\langle v_{2r} \rangle \perp_Q \langle v_{2r+1}, \ldots, v_{2n-1} \rangle$. Normalize $Q(v_{2r-1}, v_{2r}) = 1$.

The procedure stops at step $(2n-1)$. The frame v_i is canonical:

$$\omega^{-1} = v_0 \wedge v_1 + v_2 \wedge v_3 + \cdots + v_{2n-2} \wedge v_{2n-1}.$$

The only non-constant entries of the Gram matrix of Q in the basis v_i are diagonal $Q(v_i, v_i) = I_{2,i}$ for $1 \leq i < 2n$. The Gram matrix consists of $(n-1)$ diagonal blocks of size 2×2 and 1 diagonal block of size 1×1 as follows:

Q	v_1	v_2	v_3	v_4	$\ldots$	v_{2n-3}	v_{2n-2}	v_{2n-1}
v_1	$I_{2,1}$	1	0	0	$\ldots$	0	0	0
v_2	1	$I_{2,2}$	0	0	$\ldots$	0	0	0
v_3	0	0	$I_{2,3}$	1	$\ldots$	0	0	0
v_4	0	0	1	$I_{2,4}$	$\ldots$	0	0	0
$\vdots$	$\vdots$	$\vdots$	$\vdots$	$\vdots$	$\ddots$	$\vdots$	$\vdots$	$\vdots$
v_{2n-3}	0	0	0	0	$\ldots$	$I_{2,2n-3}$	1	0
v_{2n-2}	0	0	0	0	$\ldots$	1	$I_{2,2n-2}$	0
v_{2n-1}	0	0	0	0	$\ldots$	0	0	$I_{2,2n-1}$

The horizontal vector fields v_j correspond to invariant derivations ∇_j, $1 \leq j \leq 2n-1$. To summarize, we obtain the following statement.

Theorem 3. *For the G-action on $J^\infty(V, 2n-1)$ the algebra $\mathcal{A}$ is generated by the differential invariants $I_{2,i}$ and the invariant derivations ∇_j, where $1 \leq i, j \leq 2n-1$.*

6. General Submanifolds in a Symplectic Vector Space

The case of submanifolds of dimension and codimension greater than 1 is more complicated, no straightforward computations work for $G = \mathrm{Sp}(2n, \mathbb{R})$ action on $J^\infty(V, m)$, $V = \mathbb{R}^{2n}$. Yet, the geometric methods applied above do generalize, and to illustrate this, we consider the simplest case $n = m = 2$ and then remark on the general case.

6.1. Surfaces in a Four-Dimensional Symplectic Space

The action has an open orbit in $J^1(V, 2)$, but becomes free on the level of 2-jets. Since $\dim J^2(V, 2) = 14$ we get $h_2 = 4$ differential invariants of order 2 and then at every higher order $k > 2$ there will be $h_k = 2(k+1)$ new invariants.

There will be two independent invariant derivations. Applying those to four differential invariants of the second order gives a total of 8 invariants of order 3. A direct computation shows that these are functionally (hence algebraic) independent. Since $h_3 = 8$ this is enough to generate all differential invariants.

In this case the algebra $\mathcal{A}$ of differential invariants can be chosen to consist of rational functions that are polynomial in jets-variables of order >2.

Having done the counting, we can proceed with the geometric approach. Choose canonical coordinates (t, s, x, y) on $V = (\mathbb{R}^4, \omega)$. Locally surfaces in V are given as $\Sigma = \{x = x(s,t), y = y(s,t)\}$.

Here s, t will be treated as independent and x, y as dependent variables, whence the coordinates on $J^\infty(V, 2)$.

The 0-jet $p = (t, s, x, y) \in J^0$ can be identified with the vector to that point from the origin $v_0 = t\partial_t + s\partial_s + x\partial_x + y\partial_y$. The 1-jet can be identified with the tangent space

$$T_p\Sigma = \langle \mathcal{D}_t^{(1)}, \mathcal{D}_s^{(1)} \rangle = \langle \partial_t + x_t\partial_x + y_t\partial_y, \partial_s + x_s\partial_x + y_s\partial_y \rangle.$$

Equivalently, if the surface is described by $\Sigma = \{f = 0, g = 0\}$ with $f = x - x(t, s)$ and $g = y - y(t, s)$, then $T_p\Sigma = \text{Ann}(d_p f, d_p g)$, where $d_p f = dx - x_t dt - x_s ds$ and $d_p g = dy - y_t dt - y_s ds$.

The restriction of ω to $T_p\Sigma$ has rank 2 on generic 1-jets, so $T_p\Sigma$ is a symplectic subspace of dimension 2 and $T_p V = T_p\Sigma \oplus T_p\Sigma^{\perp\omega}$.

Denote by $\pi_1 : T_p V \to T_p\Sigma$ and $\pi_2 : T_p V \to T_p\Sigma^{\perp\omega}$ the natural projections with respect to this decomposition. Further for $v \in T_p V$ denote $v = v^\| + v^\perp$, where $v^\| = \pi_1(v) \in T_p\Sigma$ and $v^\perp = \pi_2(v) \in T_p\Sigma^{\perp\omega}$.

Thus, 1-jet $[\Sigma]_p^1$ is entirely encoded by $(T_p\Sigma, \omega|_{T_p\Sigma}, v_0^\|)$ and $(T_p\Sigma^{\perp\omega}, \omega|_{T_p\Sigma^{\perp\omega}}, v_0^\perp)$. Note also that $\text{Ann}(T_p\Sigma)$ is identified with $T_p\Sigma^{\perp\omega}$ by the symplectic form ω.

Moving on to 2-jets there is more structure on the tangent space. The defining functions f, g can be changed to $F = \alpha f + \beta g$, $G = \gamma f + \delta g$, where $\alpha, \beta, \gamma, \delta$ are arbitrary functions that satisfy $\alpha\delta - \beta\gamma \neq 0$ along Σ. Then $\Sigma = \{F = 0, G = 0\}$ and the tangent space can be described as the annihilator of the differentials of the new defining functions at $p \in \Sigma$:

$$d_p F = \alpha(p) d_p f + \beta(p) d_p g,$$
$$d_p G = \gamma(p) d_p f + \delta(p) d_p g.$$

Next, compute the second symmetric differential of f, g and restrict to $T_p\Sigma$. Doing the same for F, G results in

$$d_p^2 F = \alpha(p) d_p^2 f + \beta(p) d_p^2 g,$$
$$d_p^2 G = \gamma(p) d_p^2 f + \delta(p) d_p^2 g.$$

This gives a 2-dimensional space $\mathcal{Q} = \langle d_p^2 f|_{T_p\Sigma}, d_p^2 g|_{T_p\Sigma} \rangle = \langle d_p^2 F|_{T_p\Sigma}, d_p^2 G|_{T_p\Sigma} \rangle$ of quadratic forms, and the above formulae show that there is a natural isomorphism between $\text{Ann}(T_p\Sigma) \subset T_p^* V$ and $\mathcal{Q}$. Our goal is to find a canonical basis Q_1, Q_2 in this space.

Let $Q_1 \in \mathcal{Q}$ be given by the condition $Q_1(v_0^\|, v_0^\|) = 0$. This ensures that Q_1 has a Lorentzian signature or is degenerate, and for a generic 2-jet we get that Q_1 is non-degenerate. The vector $v_0^\|$ becomes null-like vector for Q_1 that is yet defined up to scale. A Lorentzian metric on the plane has two independent null-like vectors and this gives a way to fix Q_1 and a vector $w^\| \in T_p\Sigma$ complementary to $v_0^\|$ as follows:

$$\omega(v_0^\|, w^\|) = 1, \quad Q_1(w^\|, w^\|) = 0, \quad Q_1(v_0^\|, w^\|) = 1.$$

Note that this does not involve square roots, but only linear algebra. Indeed, the first condition fixes the second null-like vector up to change $w^\| \mapsto w^\| + k v_0^\|$. The second condition fixes k and the last normalizes Q_1.

The quadratic form Q_1 corresponds to a 1-form $\sigma_1 \in \text{Ann}(T_p\Sigma)$ such that the symmetric differential of an extension of σ_1 to a section of $\text{Ann}(T\Sigma)$, restricted to $T_p\Sigma$ equals $Q_1 = d_p^{\text{sym}} \sigma_1$. Then, fix $w^\perp \in T_p\Sigma^{\perp\omega}$ uniquely by the conditions $\sigma_1(w^\perp) = 0$, $\omega(v_0^\perp, w^\perp) = 1$ (for a generic 2-jet $\sigma_1(v_0^\perp) \neq 0$).

Then, define $\sigma_2 \in \text{Ann}(T_p\Sigma)$ by the conditions $\sigma_2(v_0^\perp) = 0$, $\sigma_2(w^\perp) = 1$. This gives a unique 1-form independent of σ_1. It in turn corresponds to a quadratic form $Q_2 = d_p^{\text{sym}} \sigma_2$.

The remaining evaluations yield differential invariants

$$I_{2a} = \sigma_1(v_0^\perp), \quad I_{2b} = Q_2(v_0^\|, v_0^\|), \quad I_{2c} = Q_2(v_0^\|, w^\|), \quad I_{2d} = Q_2(w^\|, w^\|).$$

The vectors $v_0^\|$ and $w^\|$ are tangent vectors to Σ (horizontal) so they correspond to the invariant derivations ∇_1, ∇_2 and we conclude:

Theorem 4. *For the G-action on $J^\infty(V,2)$ the algebra $\mathcal{A}$ is generated by the differential invariants $I_{2a}, I_{2b}, I_{2c}, I_{2d}$ and invariant derivations ∇_1, ∇_2.*

The explicit form of these generators in jet-coordinates can be found in [12].

6.2. A Remark about the General Case

In general, it is easy to check that $G = \mathrm{Sp}(2n, \mathbb{R})$ acts with one open orbit in $J^1(V, m)$, $V = \mathbb{R}^{2n}$, so there are no first order invariants. However, there are always second order invariants. Their number is at least $\dim J^2(V, m) - n(2n+1)$, but this can be non-positive for $m \ll n$.

Thus, combining the ideas on differentials and quadratic forms with ω-orthogonal complements, one can get some of the invariants. If they are not sufficient, third and higher symmetric powers $d^r f$ of the defining functions f should be explored.

From the investigated cases, we cannot observe a pattern and hence cannot universally describe all differential invariants of $G = \mathrm{Sp}(2n, \mathbb{R})$ action on $J^\infty(V, m)$, $V = \mathbb{R}^{2n}$.

7. Note on Extension of the Group

One can also consider invariants of functions and submanifolds in symplectic $V = \mathbb{R}^{2n}$ with respect to conformal symplectic group $\mathrm{CSp}(2n, \mathbb{R}) = \mathrm{Sp}(2n, \mathbb{R}) \times \mathbb{R}_+$, the affine symplectic group $\mathrm{ASp}(2n, \mathbb{R}) = \mathrm{Sp}(2n, \mathbb{R}) \ltimes \mathbb{R}^{2n}$ and affine conformal symplectic group $\mathrm{ACSp}(2n, \mathbb{R}) = \mathrm{CSp}(2n, \mathbb{R}) \ltimes \mathbb{R}^{2n}$. Denote a group in this list by H.

Since our G is a subgroup of H, the algebras of differential invariants $\mathcal{A}_H$ for each of the cases are subalgebras in the algebra $\mathcal{A}_G$ that we previously computed (enhanced notations should be self-evident). One imposes the homogeneity assumption or translation-invariance or both on a general combination of invariants.

Let us discuss how to do this in all three cases. For brevity of exposition, we restrict to the case $n = 1$ (functions and curves on symplectic plane), the general case is similar.

7.1. Conformal Symplectic Group Action: Functions

Consider functions on the conformal symplectic plane, $H = \mathrm{CSp}(2n, \mathbb{R})$. For $n = 1$ observe $H \simeq \mathrm{GL}(2, \mathbb{R})$. We recall the invariants from Section 3.1 and note that all of them are homogeneous with respect to scaling $\xi = x\partial_x + y\partial_y$, corresponding to the center of $\mathfrak{h} = \mathfrak{gl}(2, \mathbb{R})$. Restricting to invariants and derivations of weight 0 we obtain the algebra of differential $\mathfrak{h}$-invariants.

The invariants I_0, I_1 have weight 0, and the invariants I_{2a}, I_{2b}, I_{2c} have weights $0, -2, -4$ respectively. Therefore, for the new algebra $\mathcal{A}_H$ there are two independent invariants of order ≤ 1 and two additional invariants of order 2, namely I_0, I_1, I_{2a} and $I'_{2b} = I_{2b}^{-2} I_{2c}$ in the notations of Section 3.1.

The invariant derivations are ∇_1, ∇_2 of weights $0, -2$ respectively. Therefore we obtain two invariant derivations with respect to $\mathfrak{h}$: ∇_1 and $\nabla'_2 = I_{2b}^{-1} \nabla_2$.

Now a straightforward verification shows that $\nabla_1(I_{2a}), \nabla_1(I'_{2b}), \nabla'_2(I_{2a}), \nabla'_2(I'_{2b})$ are independent in 3-jets, which implies that the algebra $\mathcal{A}_H$ of differential invariants is generated by I_0, I'_{2b} and ∇_1, ∇'_2. Note that $I_1 = \nabla_1(I_0)$ and $I_{2a} = \nabla_1(I_1) - I_1$.

To complete the picture, here are the differential syzygies: $\nabla'_2(I_0) = 0$, $\nabla'_2(I_1) = -1$ and

$$[\nabla_1, \nabla'_2] = \frac{1}{I_1}\nabla_1 + \left(\frac{I_{2a}}{I_1} + \nabla'_2(I_{2a})\right)\nabla'_2.$$

Denote these by $\mathcal{R}_1, \mathcal{R}_2, \mathcal{R}_3$. There is also a forth order differential syzygy $\mathcal{R}_4$:

$$\nabla'_2(I_{3a}) + \frac{1}{2I'_{2b}}\nabla'_2(I_{3b}) - \frac{1}{2I'_{2b}}\nabla_1(I_{3c}) - \frac{1}{2I_1^2 I'_{2b}}\left[(I'_{2b}I_{3b} - 3I_{3b}I_{3c} - I_{3c})I_1^2\right.$$
$$\left. + \left(((3I_{3b}+4)I_{2a} - 5I_{3a})I'_{2b} - 4I_{2a}I_{3c} - 4I_{3b} - 4\right)I_1 + 6I_{2a}^2 I'_{2b} - 6I_{2a}\right] = 0,$$

where $I_{3a} = \nabla_1(I_{2a})$, $I_{3b} = \nabla'_2(I_{2a})$, $I_{3c} = \nabla_1(I'_{2b})$ and $I_{3d} = \nabla'_2(I'_{2b})$. With this we obtain a complete description of the algebra of differential H-invariants:

$$\mathcal{A}_H = \langle I_0, I'_{2b}; \nabla_1, \nabla'_2 \mid \mathcal{R}_1, \mathcal{R}_2, \mathcal{R}_3, \mathcal{R}_4 \rangle.$$

7.2. Conformal Symplectic Group Action: Curves

Now, we discuss differential invariants of curves with respect to the same H as in Section 7.1. Consider the invariants from Section 4.1 and note that all of them are homogeneous with respect to scaling $\xi = x\partial_x + y\partial_y$, corresponding to the center of $\mathfrak{h}$. Again, we have to restrict to invariants and derivations of weight 0 to describe the algebra $\mathcal{A}_H$.

The invariant I_2 has weight -4 and the derivation ∇ weight -2. Thus the derived invariants $I_{k+2} = \nabla^k(I_2)$ have weights $-2(k+2)$ for $k \geq 0$. In particular, $I'_3 = I_3^2/I_2^3$ has weight 0 and similar for $\nabla' = I_2 I_3^{-1}\nabla$ in the notations of Section 4.1. Therefore, these freely generate the algebra of differential H-invariants:

$$\mathcal{A}_H = \langle I'_3; \nabla' \rangle.$$

7.3. Affine Symplectic Group Action: Functions

Consider differential invariants of functions on the affine symplectic plane, $H = \mathrm{ASp}(2n, \mathbb{R})$. For $n = 1$ observe $H = \mathrm{SAff}(2, \mathbb{R})$. We recall the generating invariants from Section 3.1, and note that they indeed depend explicitly on x, y except for I_0 and I_{2c}.

To single out invariants in $\mathcal{A}_G$ that are x, y-independent eliminate x, y from the system $\{I_1 = c_1, I_{2a} = c_2, I_{2b} = c_3\}$ to get a translation-invariant polynomial on J^2 that depends parametrically on c_1, c_2, c_3. Taking the coefficients of this expression with respect to those parameters, we obtain the invariants I_{2c} and $I'_2 = u_{xx}u_{yy} - u_{xy}^2 = \mathrm{Hess}(u)$. Then substituting the obtained expressions for x, y into the invariant derivative ∇_1 and simplifying modulo the obtained invariants (note that ∇_2 is already H-invariant) we get new invariant derivative

$$\nabla'_1 = (u_x u_{yy} - u_y u_{xy})\mathcal{D}_x - (u_x u_{xy} - u_y u_{xx})\mathcal{D}_y.$$

Note that $\nabla'_1(I_0) = I_{2c}$ so the latter generator can be omitted. The commutator of invariant derivations is

$$[\nabla'_1, \nabla_2] = -\frac{\nabla_2(I_{2c})}{I_{2c}}\nabla'_1 + \left(\frac{\nabla'_1(I_{2c})}{I_{2c}} - 2I'_2\right)\nabla_2.$$

Denote this relation and the relation $\nabla_2(I_0) = 0$ by $\mathcal{R}_1, \mathcal{R}_2$. Note that $\nabla_1'(I_0)$ is a second order invariant, and application of ∇_1', ∇_2 to it and I_2' gives four third-order invariants. Further differentiation gives six fourth-order invariants, whence the syzyzy $\mathcal{R}_3$:

$$-I_{2c}\nabla_2(I_{3b}) + \nabla_1'(I_{3c}) + I_2'\nabla_2(I_{3d})$$
$$-\frac{1}{I_{2c}}\left[12I_2'^2 I_{2c}^2 - 10I_2' I_{2c} I_{3c} + 3I_2'^2 I_{3d}^2 + 3I_{2c}^2 I_{3a} - 3I_{2c} I_{3b} I_{3d} + 3I_{3c}^2\right] = 0,$$

where $I_{3a} = \nabla_1'(I_2')$, $I_{3b} = \nabla_2(I_2')$, $I_{3c} = \nabla_1'(I_{2c})$ and $I_{3d} = \nabla_2(I_{2c})$. Therefore the algebra of differential H-invariants is

$$\mathcal{A}_H = \langle I_0, I_2'; \nabla_1', \nabla_2 \mid \mathcal{R}_1, \mathcal{R}_2, \mathcal{R}_3 \rangle.$$

7.4. Affine Symplectic Group Action: Curves

Now, we discuss the case of curves on the conformal symplectic plane, with the same H as in Section 7.3. Consider the invariants from Section 4.1 and note that $I_{k+2} = \nabla^k I_2$ are not translationally invariant. However, using the elimination of parameters trick as above we arrive to micro-local differential invariant and invariant derivation

$$I_4' = \sqrt[3]{y_2}(3y_2^{-2}y_4 - 5y_2^{-3}y_3^2), \quad \nabla' = \frac{1}{\sqrt[3]{y_2}}D_x.$$

In other words, these are invariants with respect to $\mathfrak{h}$ but not with respect to H. Indeed, by the global Lie-Tresse theorem [14] we know that the invariants should be rational. To get generators we therefore pass to

$$I_4'' = (I_4')^3 \quad \text{and} \quad \nabla'' = I_4' \nabla'.$$

Consequently these freely generate the algebra of differential H-invariants:

$$\mathcal{A}_H = \langle I_4''; \nabla'' \rangle.$$

7.5. Affine Conformal Symplectic Group Action: Functions

Let us discuss differential invariants of functions on the affine conformal symplectic plane, $H = \text{ACSp}(2n, \mathbb{R})$. For $n = 1$ observe $H = \text{Aff}(2, \mathbb{R})$. We can combine the approaches of the previous two sections, for instance by taking the affine symplectic differential invariants and restricting to those of weight 0 with respect to the scaling by the center action.

Referring to the notations of Section 7.3 we get that the weights of I_0, I_2' are $0, -4$, while that of ∇_1', ∇_2 are $-4, -2$ respectively. Therefore, the algebra of differential invariants $\mathcal{A}_H$ is generated by the invariant derivations

$$\nabla_1'' = \frac{1}{I_2'}\nabla', \quad \nabla_2'' = \frac{I_2'}{\nabla_2(I_2')}\nabla_2$$

and the differential invariants (derived invariants $\nabla_1'' I_0, (\nabla_1'')^2 I_0, \nabla_2'' \nabla_1'' I_0$ are omitted)

$$I_0, \quad I_{3a}'' = \frac{\nabla_1'(I_2')}{(I_2')^2}, \quad I_{3b}'' = \frac{(\nabla_2 I_2')^2}{(I_2')^3}.$$

Denote by $\mathcal{R}_l$ the unknown differential syzygies. Then the algebra of differential invariants is

$$\mathcal{A}_H = \langle I_0, I''_{3a}, I''_{3b}; \nabla''_1, \nabla''_2 \mid \mathcal{R}_l \rangle.$$

7.6. Affine Conformal Symplectic Group Action: Curves

Similarly for the case of curves on the conformal symplectic plane, with the same H as in Section 7.5, we get in the notations of Section 7.4 that the weights of I''_4 is -4 and that of ∇'' is -2. Therefore, the algebra of differential invariants $\mathcal{A}_H$ is generated by

$$I_5 = \frac{(\nabla'' I''_4)^2}{(I''_4)^3} \quad \text{and} \quad \nabla''' = \frac{I''_4}{\nabla''(I''_4)} \nabla''.$$

In fact, it is a free differential algebra

$$\mathcal{A}_H = \langle I_5; \nabla''' \rangle.$$

8. Differential Invariants in Contact Spaces

Let W be a contact space that is a contactification of the symplectic vector space V. In coordinates, $W = \mathbb{R}^{2n+1}(x, y, z)$ is equipped with the contact form $\alpha = dz - y\, dx$ such that its differential $d\alpha = dx \wedge dy$ descends to the symplectic form on $V = \mathbb{R}^{2n}(x, y)$.

As the equivalence group, we take either $G = \mathrm{Sp}(2n, \mathbb{R})$ lifted to an action on W from the standard linear action on V, or its central extension $\hat{G} = \mathrm{CSp}(2n, \mathbb{R})$ corresponding to the scaling $(x, y, z) \mapsto (\lambda x, \lambda y, \lambda^2 z)$. (One can also consider the affine extensions, as was done in Section 7 but we skip doing this.)

Note that the group G does not have an open orbit on W because $I_0 = 2z - xy$ is an invariant. This gives a way to carry over the results on the algebra of differential invariants in V to that in W (for both functions and submanifolds; note that the formulae from the symplectic case enter through a change of variables, which is due to the lift of Hamiltonian vector fields X_H to contact Hamiltonian fields).

Then, we can single out the subalgebra $\mathcal{A}_{\hat{G}} \subset \mathcal{A}_G$ as the space of functions of weight 0 with respect to the scaling above (or its infinitesimal field). In particular, as I_0 has weight 2, it is not a scaling invariant, and in fact, the action of $\hat{G}$ on W is almost transitive.

Below, we demonstrate this two-stage computation in the simplest case $n = 1$. Note that the action of $\hat{G} = \mathrm{GL}(2, \mathbb{R}) \supset G = \mathrm{SL}(2, \mathbb{R})$ on $W = \mathbb{R}^3(x, y, z)$ has the formula

$$\Phi_A(x, y, z) = \big(ax + by, cx + dy, (ad - bc)(z - \tfrac{1}{2}xy) + \tfrac{1}{2}(ax + by)(cx + dy)\big).$$

with $A = \begin{pmatrix} a & b \\ c & d \end{pmatrix} \in \hat{G}$. This explicit parametrization is a base for an application of the moving frame method, which involves normalization of the group parameters via elimination. (This was already exploited in Sections 7.3 and 7.4.) This algorithm (we refer for details to [1]; an elaborated version of it, the method of equivariant moving frame, was further developed in the works by Peter Olver and co-authors) allows to carry the computations below; however, for $n > 1$ it would meet the complexity issues. Yet, the method we propose works for arbitrary $n > 1$ as a straightforward generalization.

8.1. Differential Invariants: Curves

We begin with the group $G = \mathrm{Sp}(2,\mathbb{R}) = \mathrm{SL}(2,\mathbb{R})$. Its action on W has a base invariant

$$I_0 = 2z - xy.$$

The curves will be represented as $y = y(x)$, $z = z(x)$ and the projection to $\mathbb{R}^2(x,y)$ restores the symplectic action. We note that invariants from Section 4.1 are still G-invariants in the contact action, and we will use them:

$$I_{2a} = \frac{y_2}{(xy_1 - y)^3}, \qquad \nabla = \frac{1}{xy_1 - y}\mathcal{D}_x.$$

Differential invariants of order ≤ 2 are generated by I_0, $I_1 = \nabla(I_0)$, I_{2a} and $I_{2b} = \nabla(I_1)$. Of course, in Lie-Tresse generating set we omit the derived invariants I_1, I_{2b}, namely

$$\mathcal{A}_G = \langle I_0, I_{2a}; \nabla \rangle.$$

However these derived invariants are useful in generating the algebra of $\hat{G}$-invariants. Indeed, with respect to the action of the center $\xi = x\partial_x + y\partial_y + 2z\partial_z$, the weights of I_0, I_1, I_{2a}, I_{2b} are $2, 0, -4, -2$ and the weight of ∇ is -2. Thus, in order to obtain $\hat{G}$-invariants we pass to weight 0 combinations (I_1 is already invariant)

$$I'_{2a} = I_0^2 I_{2a}, \quad I'_{2b} = I_0 I_{2b}, \quad \nabla' = I_0 \nabla.$$

Explicitly after simplifications $I_1 \mapsto \tfrac{1}{2}(I_1 + 1)$, $I'_{2b} \mapsto \tfrac{1}{2}I'_{2b}$ we get:

$$I_1 = \frac{z_1 - y}{xy_1 - y}, \qquad \nabla' = \frac{2z - xy}{xy_1 - y}\mathcal{D}_x,$$

$$I'_{2a} = \frac{(2z - xy)^2}{(xy_1 - y)^3} y_2,$$

$$I'_{2b} = \frac{2z - xy}{(xy_1 - y)^3}\Big(x(y - z_1)y_2 - (xy_1 - y)(y_1 - z_2)\Big).$$

The count of invariants is $h_0 = 0$, $h_1 = 1$ and $h_k = 2$ for $k \geq 2$. We conclude:

Theorem 5. *The algebra of differential invariants of the $\hat{G}$-action on $J^\infty(W,1)$ is freely generated as follows:*

$$\mathcal{A}_{\hat{G}} = \langle I_1, I'_{2a}; \nabla' \rangle.$$

8.2. Differential Invariants: Surfaces

Now we consider the action of G and $\hat{G}$ on surfaces given as $z = z(x,y)$. Since projection to $\mathbb{R}^2(x,y)$ gives the symplectic plane, the G-computations can be derived from Section 3.1 with substitution $u = 2z - xy$. This gives us the following differential invariants and invariant derivations with respect to G:

$$I_0 = 2z - xy, \qquad \nabla_1 = x\mathcal{D}_x + y\mathcal{D}_y, \qquad \nabla_2 = (x - 2z_y)\mathcal{D}_x + (2z_x - y)\mathcal{D}_y,$$

and with the notations $I_1 = \frac{1}{2}\nabla_1(I_0)$, $I_{2a} = \nabla_1(I_1) - I_1$, $I_{2b} = -\frac{1}{2}(\nabla_2(I_1) + I_{2a} - I_1)$ the following first and second order invariants

$$I_1 = xz_x + yz_y - xy,$$
$$I_{2a} = x^2 z_{xx} + 2xyz_{xy} + y^2 z_{yy} - xy,$$
$$I_{2b} = x(z_y - x)z_{xx} - yz_x z_{yy} + (y(z_y - x) - xz_x)z_{xy} + xz_x,$$
$$I_{2c} = z_x^2 z_{yy} - 2z_x(z_y - x)z_{xy} + (z_y - x)^2 z_{xx} + z_x(z_y - x).$$

Now to obtain $\hat{G}$-invariants note that $I_0, I_1, I_{2a}, I_{2b}, I_{2c}$ all have weight 2 with respect to ξ, while ∇_1, ∇_2 are already invariant. Thus the invariants are

$$I_1' = I_0^{-1} I_1, \ I_{2a}' = I_0^{-1} I_{2a}, \ I_{2b}' = I_0^{-1} I_{2b}, \ I_{2c}' = I_0^{-1} I_{2c}.$$

We have:

$$I_{2a}' = \nabla_1(I_1') + 2(I_1')^2 - I_1', \quad I_{2b}' = -\frac{1}{2}\nabla_1(I_1') - \frac{1}{2}\nabla_2(I_1') - (I_1')^2 + I_1',$$

so these can be omitted from the list of generators.

The count of $\hat{G}$-invariants is $h_0 = 0$, $h_1 = 1$ and $h_k = k + 1$ for $k \geq 2$.

Applying the derivations to the generating invariants and counting the relations, we find that beside the commutation relation

$$[\nabla_1, \nabla_2] + \frac{\nabla_2(I_1')}{I_1'} \nabla_1 - \left(\frac{\nabla_1(I_1')}{I_1'} + 2(I_1' - 1)\right) \nabla_2 = 0$$

there is one more relation generating the module of differential syzygies

$$\nabla_1^2(I_1') + 2\nabla_1\nabla_2(I_1') + \nabla_2^2(I_1') - 4\nabla_1(I_{2c}')$$
$$- 3(I_1')^{-1}\left(\nabla_1(I_1')^2 + 2\nabla_1(I_1')\nabla_2(I_1') - 4\nabla_1(I_1')I_{2c}' + \nabla_2(I_1')^2\right)$$
$$- 2(I_1' - 1)(3\nabla_1(I_1') + 4\nabla_2(I_1') - 8I_{2c}') - 4I_1'(I_1' - 1)(2I_1' - 1) = 0.$$

Denote these syzygies by $\mathcal{R}_1$ and $\mathcal{R}_2$.

Let us summarize the results.

Theorem 6. *The algebra of differential invariants of the $\hat{G}$-action on $J^\infty(W, 2)$ is generated as follows:*

$$\mathcal{A}_{\hat{G}} = \langle I_1', I_{2c}'; \nabla_1, \nabla_2 \mid \mathcal{R}_1, \mathcal{R}_2 \rangle.$$

8.3. Differential Invariants: Functions

Skipping the intermediate computation with the group G let us directly pass to the description of invariants on $J^\infty(W)$ with respect to the group $\hat{G}$. Fix the coordinates as follows: $W = \mathbb{R}^3(x, y, z)$ with the contact form $\alpha = dz - y\, dx$ as before, $J^0 = W \times \mathbb{R}(u)$ and for the jet-coordinates we use the numbered multi-index notations u_σ.

The count of the number of differential invariants is as follows: $h_0 = 1$, $h_1 = 2$ and $h_k = \binom{k+2}{2}$ for $k \geq 2$.

The zero and first order invariants are

$$I_0 = u, \quad I_{1a} = (xy - 2z)u_3, \quad I_{1b} = xu_1 + y(u_2 + xu_3).$$

Next we obtain the invariant derivations

$$\nabla_1 = x\mathcal{D}_x + y\mathcal{D}_y + 2z\mathcal{D}_z,$$
$$\nabla_2 = (xy - 2z)\mathcal{D}_z,$$
$$\nabla_3 = (xu_3 + u_2)(xy - 2z)\mathcal{D}_x - u_1(xy - 2z)\mathcal{D}_y - xu_1(xy - 2z)\mathcal{D}_z.$$

Note that $I_{1a} = \nabla_2(I_0)$, $I_{1b} = (\nabla_1 + \nabla_2)(I_0)$ and $\nabla_3(I_0) = 0$. The latter is the first differential syzygy, denoted

$$\mathcal{R}_1 = \nabla_3(I_0).$$

Second order differential invariants $\nabla_i(I_{1a})$, $\nabla_i(I_{1b})$ contain only 5 independent. We find the remaining 1 differential invariant via the the method of moving frames and get

---

Second Order Differential Invariants

$I_{2a} = y^2 u_{2,2} + y(4zu_{2,3} + 2xu_{1,2} + u_2) + 4z^2 u_{3,3}$
$\quad + z(4xu_{1,3} + 4u_3) + x(xu_{1,1} + u_1)$

$I_{2b} = (xy - 2z)(yu_{2,3} + 2zu_{3,3} + xu_{1,3} + 2u_3)$

$I_{2c} = -(xy - 2z)(x^2(u_1 u_{1,3} - u_3 u_{1,1}) + x(u_1(yu_{2,3} + 2zu_{3,3} + u_3 + u_{1,2})$
$\quad - yu_3 u_{1,2} - 2zu_3 u_{1,3} - u_2 u_{1,1}) + u_1(yu_{2,2} + 2zu_{2,3})$
$\quad - u_2(yu_{1,2} + 2zu_{1,3}))$

$I_{2d} = (xy - 2z)(-2u_3 + (xy - 2z)u_{3,3})$

$I_{2e} = -(xy - 2z)(x^2 y(u_1 u_{3,3} - u_3 u_{1,3}) + x(y(u_1 u_{2,3} - u_3^2 - u_2 u_{1,3})$
$\quad + u_1(-2zu_{3,3} - u_3) + 2zu_3 u_{1,3}) - yu_2 u_3 - 2z(u_1 u_{2,3} - u_2 u_{1,3}))$

$I_{2f} = x^4 y^2 u_1^2 u_{3,3} - 2x^4 y^2 u_1 u_3 u_{1,3} + x^4 y^2 u_3^2 u_{1,1} + 2x^3 y^2 u_1^2 u_{2,3}$
$\quad - x^3 y^2 u_1 u_3^2 - 2x^3 y^2 u_1 u_3 u_{1,2} - 2x^3 y^2 u_1 u_2 u_{1,3} + 2x^3 y^2 u_2 u_3 u_{1,1}$
$\quad - 4x^3 yzu_1^2 u_{3,3} + 8x^3 yzu_1 u_3 u_{1,3} - 4x^3 yzu_3^2 u_{1,1} + x^2 y^2 u_1^2 u_{2,2}$
$\quad - x^2 y^2 u_1 u_2 u_3 - 2x^2 y^2 u_1 u_2 u_{1,2} + x^2 y^2 u_2^2 u_{1,1} - 8x^2 yzu_1^2 u_{2,3}$
$\quad + 4x^2 yzu_1 u_3^2 + 8x^2 yzu_1 u_3 u_{1,2} + 8x^2 yzu_1 u_2 u_{1,3} - 8x^2 yzu_2 u_3 u_{1,1}$
$\quad + 4x^2 z^2 u_1^2 u_{3,3} - 8x^2 z^2 u_1 u_3 u_{1,3} + 4x^2 z^2 u_3^2 u_{1,1} - 4xyzu_1^2 u_{2,2}$
$\quad + 4xyzu_1 u_2 u_3 + 8xyzu_1 u_2 u_{1,2} - 4xyzu_2^2 u_{1,1} + 8xz^2 u_1^2 u_{2,3}$
$\quad - 4xz^2 u_1 u_3^2 - 8xz^2 u_1 u_3 u_{1,2} - 8xz^2 u_1 u_2 u_{1,3} + 8xz^2 u_2 u_3 u_{1,1}$
$\quad + 4z^2 u_1^2 u_{2,2} - 4z^2 u_1 u_2 u_3 - 8z^2 u_1 u_2 u_{1,2} + 4z^2 u_2^2 u_{1,1}$

Note that $I_{2a}, I_{2b}, I_{2c}, I_{2d}, I_{2e}$ can be expressed through I_0, I_{1a}, I_{1b} and invariant derivations. Thus they need not enter the set of generators.

All the differential syzygies coming from the commutators are

$$\mathcal{R}_2 = [\nabla_1, \nabla_2],$$
$$\mathcal{R}_3 = (I_{1a} + I_{1b})[\nabla_1, \nabla_3] + I_{2c}(\nabla_1 + \nabla_2) - (I_{2a} + I_{2b})\nabla_3,$$
$$\mathcal{R}_4 = (I_{1a} + I_{1b})[\nabla_2, \nabla_3] - (I_{1b}(I_{1a} + I_{1b}) - I_{2e})\nabla_1 + (I_{1a}(I_{1a} + I_{1b}) + I_{2e})\nabla_2$$
$$\quad - (I_{2b} + I_{2d} - 2(I_{1a} + I_{1b}))\nabla_3.$$

The remaining differential syzygies are found by the symbolic method: find a relation between the symbols of differentiated invariants, get a linear combination of lower order and express it through the invariants established earlier.

$$\mathcal{R}_5 = (I_{1a} + I_{1b})(\nabla_3(I_{2b}) - \nabla_1(I_{2e})) - (I_{2c} - I_{2e})I_{2b} + I_{2a}I_{2e} - I_{2c}I_{2d},$$
$$\mathcal{R}_6 = (I_{1a} + I_{1b})(\nabla_3(I_{2c}) - \nabla_1(I_{2f})) - 3I_{2c}^2 - (I_{1a}^2 + I_{1a}I_{1b} + 3I_{2e})I_{2c} + 3I_{2f}(I_{2a} + I_{2b}),$$
$$\mathcal{R}_7 = (I_{1a} + I_{1b})(-\nabla_3(I_{2e}) + \nabla_2(I_{2f})) - I_{1b}^4 - 4I_{1a}I_{1b}^3 - (5I_{1a}^2 + 2I_{2c})I_{1b}^2$$
$$- (2I_{1a}^3 + (2I_{2c} - 3I_{2e})I_{1a} + 4I_{2f})I_{1b} + 3I_{2e}I_{1a}^2 + 4I_{2f}I_{1a} + 3I_{2e}^2$$
$$+ 3I_{2c}I_{2e} - 3I_{2f}(I_{2b} + I_{2d}).$$

Theorem 7. *The algebra of differential invariants of the $\hat{G}$-action on $J^\infty(W)$ is generated as follows:*

$$\mathcal{A}_{\hat{G}} = \langle I_0, I_{2f}\,;\, \nabla_1, \nabla_2, \nabla_3 \mid \mathcal{R}_i = 0,\, i = 1\ldots 7\rangle.$$

9. Conclusions

In this paper, we computed the algebra of differential invariants for various geometric objects on symplectic spaces with several choices of the equivalence group and touched upon a relation between the invariants of the pair (group, subgroup) action.

For most of the text we worked with the linear symplectic group, but we demonstrated how to extend the results for conformal symplectic and affine symplectic groups, treated in other publications. Some of the objects were also investigated by different authors, namely jets of curves [5,10] and hypersurfaces [9], yet the technique and the description of the algebras are quite distinct. Surfaces in four-dimensional symplectic space were also studied in [6–8], but they considered Lagrangian surfaces while our focus was on symplectic (generic) submanifolds.

Other geometric objects appeared in [11], which intersects with our work by studying functions on the symplectic spaces. Again the approaches differ significantly: in [11] the infinite number of generators were computed (with a nontrivial change of variables) while our method uses the Lie-Tresse finite type presentation of the algebra (in the original jet-coordinates). This latter allows, in particular, to solve the equivalence problem via a finite-dimensional signature variety.

The work [11] also described invariants in the adjoint bundle, and one can consider other geometric spaces on which the symplectic group acts. For instance, [17] was devoted to four-fold surfaces in 6-dimensional Lagrangian Grassmanian, satisfying the integrability condition. It would be worth characterizing those via symplectic invariants.

Finally note that one can approach the equivalence problem of geometric objects via discretizations, with more algebraic methods, see [18].

Supplementary Materials: The following are available online at http://www.mdpi.com/2073-8994/12/12/2023/s1 .

Author Contributions: Both authors are equally responsible for all results in this paper. All authors have read and agreed to the published version of the manuscript.

Funding: The publication charges for this article have been funded by a grant from the publication fund of UiT The Arctic University of Norway.

Acknowledgments: J.O.J. thanks Fredrik Andreassen, Eivind Schneider and Henrik Winther for useful discussions. B.K. thanks Peter Olver and Niky Kamran for helpful correspondence.

Conflicts of Interest: The authors declare no conflict of interest.

Appendix A. Differential Invariants of Curves in 4-Dimensions

Here are explicit expressions of the differential invariants of curves $x = x(t), y = y(t), z = z(t)$, as derived in Section 4.2. These as well as other long formulae resulting from our calculations can be found in Supplementary Materials. Below $\gamma = 1/(ty_1 + xz_1 - x_1z - y)$ is the factor of $\mathcal{D}_t$ in ∇. The jet notations are $x_t = x_1, x_{tt} = x_2, x_{ttt} = x_3$ etc, likewise for y and z.

Differential invariants that together with ∇ generate $\mathcal{A}$
$I_2 = \gamma^3(x_1z_2 - z_1x_2 + y_2)$
$I_{3b} = -\gamma^6(tx_1y_2z_3 - tx_1y_3z_2 - tx_2y_1z_3 + tx_2y_3z_1 + tx_3y_1z_2 - tx_3y_2z_1$
$\quad -xy_2z_3 + xy_3z_2 + x_2yz_3 - x_2y_3z - x_3yz_2 + x_3y_2z)$
$I_{4c} = -\gamma^{10}(t^3x_1y_1^2y_3z_4 - t^3x_1y_1^2y_4z_3 - 3t^3x_1y_1y_2^2z_4 + 4t^3x_1y_1y_2y_3z_3$
$\quad +3t^3x_1y_1y_2y_4z_2 - 4t^3x_1y_1y_3^2z_2 + 3t^3x_1y_2^3z_3 - 3t^3x_1y_2^2y_3z_2 + 3t^3x_2y_1^2y_2z_4$
$\quad -4t^3x_2y_1^2y_3z_3 - 3t^3x_2y_1y_2^2z_3 - 3t^3x_2y_1y_2y_4z_1 + 4t^3x_2y_1y_3^2z_1$
$\quad +3t^3x_2y_2^2y_3z_1 - t^3x_3y_1^3z_4 + 4t^3x_3y_1^2y_3z_2 + t^3x_3y_1^2y_4z_1 + 3t^3x_3y_1y_2^2z_2$
$\quad -4t^3x_3y_1y_2y_3z_1 - 3t^3x_3y_2^2z_1 + t^3x_4y_1^3z_3 - 3t^3x_4y_1^2y_2z_2 - t^3x_4y_1^2y_3z_1$
$\quad +3t^3x_4y_1y_2^2z_1 - 3t^2xx_1y_1y_2z_2z_4 + 4t^2xx_1y_1y_2z_3^2 + 2t^2xx_1y_1y_3z_1z_4$
$\quad -4t^2xx_1y_1y_3z_2z_3 - 2t^2xx_1y_1y_4z_1z_3 + 3t^2xx_1y_1y_4z_2^2 - 3t^2xx_1y_2^2z_1z_4$
$\quad +6t^2xx_1y_2^2z_2z_3 + 4t^2xx_1y_2y_3z_1z_3 - 6t^2xx_1y_2y_3z_2^2 + 3t^2xx_1y_2y_4z_1z_2$
$\quad -4t^2xx_1y_3^2z_1z_2 + 3t^2xx_2y_1^2z_2z_4 - 4t^2xx_2y_1^2z_3^2 + 3t^2xx_2y_1y_2z_1z_4$
$\quad -6t^2xx_2y_1y_2z_2z_3 - 3t^2xx_2y_1y_4z_1z_2 + 6t^2xx_2y_2y_3z_1z_2 - 3t^2xx_2y_2y_4z_1^2$
$\quad +4t^2xx_2y_3^2z_1^2 - 2t^2xx_3y_1^2z_1z_4 + 4t^2xx_3y_1^2z_2z_3 - 4t^2xx_3y_1y_2z_1z_3$
$\quad +6t^2xx_3y_1y_2z_2^2 + 4t^2xx_3y_1y_3z_1z_2 + 2t^2xx_3y_1y_4z_1^2 - 6t^2xx_3y_2^2z_1z_2$
$\quad -4t^2xx_3y_2y_3z_1^2 + 2t^2xx_4y_1^2z_1z_3 - 3t^2xx_4y_1^2z_2^2 - 2t^2xx_4y_1y_3z_1^2$
$\quad +3t^2xx_4y_2^2z_1^2 - 2t^2x_1^2y_1y_3zz_4 + 2t^2x_1^2y_1y_4zz_3 + 3t^2x_1^2y_2^2zz_4$
$\quad -4t^2x_1^2y_2y_3zz_3 - 3t^2x_1^2y_2y_4zz_2 + 4t^2x_1^2y_3^2zz_2 + 4t^2x_1x_2y_1y_3zz_3$
$\quad -3t^2x_1x_2y_1y_4zz_2 - 6t^2x_1x_2y_2^2zz_3 + 6t^2x_1x_2y_2y_3zz_2 + 3t^2x_1x_2y_2y_4zz_1$
$\quad -4t^2x_1x_2y_3^2zz_1 + 2t^2x_1x_3y_1^2zz_4 - 4t^2x_1x_3y_1y_2zz_3 - 2t^2x_1x_3y_1y_4zz_1$
$\quad +4t^2x_1x_3y_2y_3zz_1 - 2t^2x_1x_4y_1^2zz_3 + 3t^2x_1x_4y_1y_2zz_2 + 2t^2x_1x_4y_1y_3zz_1$
$\quad -3t^2x_1x_4y_2^2zz_1 - 3t^2x_2^2y_1^2zz_4 + 6t^2x_2^2y_1y_2zz_3 + 3t^2x_2^2y_1y_4zz_1$
$\quad -6t^2x_2^2y_2y_3zz_1 + 4t^2x_2x_3y_1^2zz_3 - 6t^2x_2x_3y_1y_2zz_2 - 4t^2x_2x_3y_1y_3zz_1$
$\quad +6t^2x_2x_3y_2^2zz_1 + 3t^2x_2x_4y_1^2zz_2 - 3t^2x_2x_4y_1y_2zz_1 - 4t^2x_3^2y_1^2zz_2$
$\quad +4t^2x_3^2y_1y_2zz_1 - 3tx^2x_1y_2z_1z_2z_4 + 4tx^2x_1y_2z_1z_3^2 + 3tx^2x_1y_2z_2^2z_3$
$\quad +tx^2x_1y_3z_1^2z_4 - 4tx^2x_1y_3z_1z_2z_3 - 3tx^2x_1y_3z_2^3 - tx^2x_1y_4z_1^2z_3$
$\quad +3tx^2x_1y_4z_1z_2^2 + 3tx^2x_2y_1z_1z_2z_4 - 4tx^2x_2y_1z_1z_3^2 - 3tx^2x_2y_1z_2^2z_3$
$\quad +4tx^2x_2y_3z_1^2z_3 + 3tx^2x_2y_3z_1z_2^2 - 3tx^2x_2y_4z_1^2z_2 - tx^2x_3y_1z_1^2z_4$
$\quad +4tx^2x_3y_1z_1z_2z_3 + 3tx^2x_3y_1z_2^3 - 4tx^2x_3y_2z_1^2z_3 - 3tx^2x_3y_2z_1z_2^2 + tx^2x_3y_4z_1^3$
$\quad +tx^2x_4y_1z_1^2z_3 - 3tx^2x_4y_1z_1z_2^2 + 3tx^2x_4y_2z_1^2z_2 - tx^2x_4y_3z_1^3 + 3txx_1^2y_2zz_2z_4$
$\quad -4txx_1^2y_2zz_3^2 - 2txx_1^2y_3zz_1z_4 + 4txx_1^2y_3zz_2z_3 + 2txx_1^2y_4zz_1z_3 - 3txx_1^2y_4zz_2^2$
$\quad -3txx_1x_2y_1zz_2z_4 + 4txx_1x_2y_1zz_3^2 + 3txx_1x_2y_2zz_1z_4 - 6txx_1x_2y_2zz_2z_3$
$\quad -4txx_1x_2y_3zz_1z_3 + 6txx_1x_2y_3zz_2^2 + 2txx_1x_3y_1zz_1z_4 - 4txx_1x_3y_1zz_2z_3$
$\quad +4txx_1x_3y_3zz_1z_2 - 2txx_1x_3y_4zz_1^2 - 2txx_1x_4y_1zz_1z_3 + 3txx_1x_4y_1zz_2^2$
$\quad -3txx_1x_4y_2zz_1z_2 + 2txx_1x_4y_3zz_1^2 - 3txx_2^2y_1zz_1z_4 + 6txx_2^2y_1zz_2z_3$
$\quad -6txx_2^2y_3zz_1z_2 + 3txx_2^2y_4zz_1^2 + 4txx_2x_3y_1zz_1z_3 - 6txx_2x_3y_1zz_2^2$
$\quad +6txx_2x_3y_2zz_1z_2 - 4txx_2x_3y_3zz_1^2 + 3txx_2x_4y_1zz_1z_2 - 3txx_2x_4y_2zz_1^2$
$\quad -4txx_3^2y_1zz_1z_2 + 4txx_3^2y_2zz_1^2 + tx_1^3y_3z^2z_4 - tx_1^3y_4z^2z_3 - 3tx_1^2x_2y_2z^2z_4$
$\quad +3tx_1^2x_2y_4z^2z_2 - tx_1^2x_3y_1z^2z_4 + 4tx_1^2x_3y_2z^2z_3 - 4tx_1^2x_3y_3z^2z_2 + tx_1^2x_3y_4z^2z_1$
$\quad +tx_1^2x_4y_1z^2z_3 - tx_1^2x_4y_3z^2z_1 + 3tx_1x_2^2y_1z^2z_4 + 3tx_1x_2^2y_2z^2z_3 - 3tx_1x_2^2y_3z^2z_2$
$\quad -3tx_1x_2^2y_4z^2z_1 - 4tx_1x_2x_3y_1z^2z_3 + 4tx_1x_2x_3y_3z^2z_1 - 3tx_1x_2x_4y_1z^2z_2$
$\quad +3tx_1x_2x_4y_2z^2z_1 + 4tx_1x_3^2y_1z^2z_2 - 4tx_1x_3^2y_2z^2z_1 - 3tx_2^3y_1z^2z_3 + 3tx_2^3y_3z^2z_1$
$\quad +3tx_2^2x_3y_1z^2z_2 - 3tx_2^2x_3y_2z^2z_1 - t^2xy_1^2y_3z_4 + t^2xy_1^2y_4z_3 + 3t^2xy_1y_2^2z_4$
$\quad -4t^2xy_1y_2y_3z_3 - 3t^2xy_1y_2y_4z_2 + 4t^2xy_1y_3^2z_2 - 3t^2xy_2^3z_3 + 3t^2xy_2^2y_3z_2$
$\quad -2t^2x_1yy_1y_3z_4 + 2t^2x_1yy_1y_4z_3 + 3t^2x_1yy_2^2z_4 - 4t^2x_1yy_2y_3z_3 - 3t^2x_1yy_2y_4z_2$

$$\begin{aligned}
&+4t^2x_1yy_3^2z_2 - 6t^2x_2yy_1y_2z_4 + 8t^2x_2yy_1y_3z_3 + 3t^2x_2yy_2^2z_3 + 3t^2x_2yy_2y_4z_1\\
&-4t^2x_2yy_3^2z_1 + 3t^2x_2y_1y_2y_4z - 4t^2x_2y_1y_3^2z - 3t^2x_2y_2^2y_3z + 3t^2x_3yy_1^2z_4\\
&-8t^2x_3yy_1y_3z_2 - 2t^2x_3yy_1y_4z_1 - 3t^2x_3yy_2^2z_2 + 4t^2x_3yy_2y_3z_1\\
&-t^2x_3y_1^2y_4z + 4t^2x_3y_1y_2y_3z + 3t^2x_3y_3^2z - 3t^2x_4yy_1^2z_3 + 6t^2x_4yy_1y_2z_2\\
&+2t^2x_4yy_1y_3z_1 - 3t^2x_4yy_2^2z_1 + t^2x_4y_1^2y_3z - 3t^2x_4y_1y_2^2z + 3tx^2y_1y_2z_2z_4\\
&-4tx^2y_1y_2z_3^2 - 2tx^2y_1y_3z_1z_4 + 4tx^2y_1y_3z_2z_3 + 2tx^2y_1y_4z_1z_3 - 3tx^2y_1y_4z_2^2\\
&+3tx^2y_2^2z_1z_4 - 6tx^2y_2^2z_2z_3 - 4tx^2y_2y_3z_1z_3 + 6tx^2y_2y_3z_2 - 3tx^2y_2y_4z_1z_2\\
&+4tx^2y_3^2z_1z_2 + 3txx_1yy_2z_2z_4 - 4txx_1yy_2z_3^2 - 2txx_1yy_3z_1z_4 + 4txx_1yy_3z_2z_3\\
&+2txx_1yy_4z_1z_3 - 3txx_1yy_4z_2^2 + 2txx_1y_1y_3zz_4 - 2txx_1y_1y_4zz_3 - 3txx_1y_2^2zz_4\\
&+4txx_1y_2y_3zz_3 + 3txx_1y_2y_4zz_2 - 4txx_1y_3^2zz_2 - 6txx_2yy_1z_2z_4 + 8txx_2yy_1z_3^2\\
&-3txx_2yy_2z_1z_4 + 6txx_2yy_2z_2z_3 + 3txx_2yy_4z_1z_2 - 3txx_2y_1y_2zz_4 - 4txx_2y_1y_3zz_3\\
&+6txx_2y_1y_4zz_2 + 6txx_2y_2^2zz_3 - 12txx_2y_2y_3zz_2 + 3txx_2y_2y_4zz_1 - 4txx_2y_3^2zz_1\\
&+4txx_3yy_1z_1z_4 - 8txx_3yy_1z_2z_3 + 4txx_3yy_2z_1z_3 - 6txx_3yy_2z_2^2 - 4txx_3yy_3z_1z_2\\
&-2txx_3yy_4z_1^2 + 8txx_3y_1y_2zz_3 - 4txx_3y_1y_3zz_2 - 2txx_3y_1y_4zz_1 + 6txx_3y_2^2zz_2\\
&+4txx_3y_2y_3zz_1 - 4txx_4yy_1z_1z_3 + 6txx_4yy_1z_2^2 + 2txx_4yy_3z_1^2 - 3txx_4y_1y_2zz_2\\
&+2txx_4y_1y_3zz_1 - 3txx_4y_2^2zz_1 + 2tx_1^2yy_3zz_4 - 2tx_1^2yy_4zz_3 - 4tx_1x_2yy_3zz_3\\
&+3tx_1x_2yy_4z_2z_2 - 3tx_1x_2y_1y_2y_4z^2 + 4tx_1x_2y_3^2z^2 - 4tx_1x_3yy_1zz_4 + 4tx_1x_3yy_2zz_3\\
&+2tx_1x_3yy_4zz_1 + 2tx_1x_3y_1y_4z^2 - 4tx_1x_3y_2y_3z^2 + 4tx_1x_4yy_1zz_3 - 3tx_1x_4yy_2zz_2\\
&-2tx_1x_4yy_3zz_1 - 2tx_1x_4y_1y_3z^2 + 3tx_1x_4y_2^2z^2 + 6tx_2^2yy_1zz_4 - 6tx_2^2yy_2zz_3\\
&-3tx_2^2yy_4zz_1 - 3tx_2^2y_1y_4z^2 + 6tx_2^2y_2y_3z^2 - 8tx_2x_3yy_1zz_3 + 6tx_2x_3yy_2zz_2\\
&+4tx_2x_3yy_3zz_1 + 4tx_2x_3y_1y_3z^2 - 6tx_2x_3y_2^2z^2 - 6tx_2x_4yy_1zz_2 + 3tx_2x_4yy_2zz_1\\
&+3tx_2x_4y_1y_2z^2 + 8tx_3^2yy_1zz_2 - 4tx_3^2yy_2zz_1 - 4tx_3^2y_1y_2z^2 + 3x^3y_2z_1z_2z_4\\
&-4x^3y_2z_1z_3^2 - 3x^3y_2z_2^2z_3 - x^3y_3z_1^2z_4 + 4x^3y_3z_1z_2z_3 + 3x^3y_3z_2^3 + x^3y_4z_1^2z_3\\
&-3x^3y_4z_1z_2^2 - 3x^2x_1y_2z_2z_4 + 4x^2x_1y_2z_3^2 + 2x^2x_1y_3z_1z_4 - 4x^2x_1y_3z_2z_3\\
&-2x^2x_1y_4zz_1z_3 + 3x^2x_1y_4z_2^2 - 3x^2x_2y_2z_1z_2z_4 + 4x^2x_2y_2z_1z_3^2 + 3x^2x_2y_2z_2^2z_3\\
&-3x^2x_2y_2zz_1z_4 + 6x^2x_2y_2z_2z_2z_3 - 4x^2x_2y_3zz_1z_3 - 9x^2x_2y_3z_2^2 + 6x^2x_2y_4zz_1z_2\\
&+x^2x_3yz_1^2z_4 - 4x^2x_3yz_1z_2z_3 - 3x^2x_3yz_2^3 + 8x^2x_3y_2zz_1z_3 + 3x^2x_3y_2zz_2^2\\
&-4x^2x_3y_3zz_1z_2 - x^2x_3y_4zz_1^2 - x^2x_4yz_1^2z_3 + 3x^2x_4yz_1z_2^2 - 3x^2x_4y_2zz_1z_2\\
&+x^2x_4y_3zz_1^2 - xx_1^2y_3z^2z_4 + xx_1^2y_4z^2z_3 + 3xx_1x_2yzz_2z_4 - 4xx_1x_2yzz_3^2\\
&+3xx_1x_2y_2y_2z^2z_4 + 4xx_1x_2y_3z^2z_3 - 6xx_1x_2y_4z^2z_2 - 2xx_1x_3yzz_1z_4\\
&+4xx_1x_3yzz_2z_3 - 8xx_1x_3y_2z^2z_3 + 4xx_1x_3y_3z^2z_2 + 2xx_1x_3y_4z^2z_1\\
&+2xx_1x_4yzz_1z_3 - 3xx_1x_4yzz_2^2 + 3xx_1x_4y_2z^2z_2 - 2xx_1x_4y_3z^2z_1\\
&+3xx_2^2yzz_1z_4 - 6xx_2^2yzz_2z_3 - 3xx_2^2y_2z^2z_3 + 9xx_2^2y_3z^2z_2 - 3xx_2^2y_4z^2z_1\\
&-4xx_2x_3yzz_1z_3 + 6xx_2x_3yzz_2^2 - 6xx_2x_3y_2z^2z_2 + 4xx_2x_3y_3z^2z_1\\
&-3xx_2x_4yzz_1z_2 + 3xx_2x_4y_2z^2z_1 + 4xx_3^2yzz_1z_2 - 4xx_3^2y_2z^2z_1 + x_1^2x_3yz^2z_4\\
&-x_1^2x_3y_4z^3 - x_1^2x_4yz^2z_3 + x_1^2x_4y_3z^3 - 3x_1x_2^2yz^2z_4 + 3x_1x_2^2y_4z^3\\
&+4x_1x_2x_3yz^2z_3 - 4x_1x_2x_3y_3z^3 + 3x_1x_2x_4yz^2z_2 - 3x_1x_2x_4y_2z^3 - 4x_1x_3^2yz^2z_2\\
&+4x_1x_3^2y_2z^3 + 3x_2^3yz^2z_3 - 3x_2^3y_3z^3 - 3x_2^2x_3yz^2z_2 + 3x_2^2x_3y_2z^3\\
&+2txyy_1y_3z_4 - 2txyy_1y_4z_3 - 3txyy_2^2z_4 + 4txyy_2y_3z_3 + 3txyy_2y_4z_2\\
&-4txyy_3^2z_2 + tx_1y^2y_3z_4 - tx_1y^2y_4z_3 + 3tx_2y^2y_2z_4 - 4tx_2y^2y_3z_3\\
&-3tx_2yy_2y_4z + 4tx_2yy_3^2z - 3tx_3y^2y_1z_4 + 4tx_3y^2y_3z_2 + tx_3y^2y_4z_1z\\
&+2tx_3yy_1y_4z - 4tx_3yy_2y_3 + 3tx_4y^2y_1z_3 - 3tx_4y^2y_2z_2 - tx_4y^2y_3z_1\\
&-2tx_4yy_1y_3z + 3tx_4yy_2^2z - 3x^2yy_2z_2z_4 + 4x^2yy_2z_3^2 + 2x^2yy_3z_1z_4\\
&-4x^2yy_3z_2z_3 - 2x^2yy_4z_1z_3 + 3x^2yy_4z_2^2 - 2xx_1yy_3zz_4 + 2xx_1yy_4zz_3\\
&+3xx_2y^2z_2z_4 - 4xx_2y^2z_3^2 + 3xx_2yy_2zz_4 + 4xx_2yy_3zz_3 - 6xx_2yy_4zz_2\\
&-2xx_3y^2z_1z_4 + 4xx_3y^2z_2z_3 - 8xx_3yy_2zz_3 + 4xx_3yy_3zz_2 + 2xx_3yy_4zz_1\\
&+2xx_4y^2z_1z_3 - 3xx_4y^2z_2^2 + 3xx_4yy_2zz_2 - 2xx_4yy_3zz_1 + 2x_1x_3y^2zz_4\\
&-2x_1x_3yy_4z^2 - 2x_1x_4y^2zz_3 + 2x_1x_4yy_3z^2 - 3x_2^2y^2zz_4 + 3x_2^2yy_4z^2\\
&+4x_2x_3y^2zz_3 - 4x_2x_3yy_3z^2 + 3x_2x_4y^2zz_2 - 3x_2x_4yy_2z^2 - 4x_3^2y^2zz_2\\
&+4x_3^2yy_2z^2 - xy^2y_3z_4 + xy^2y_4z_3 + x_3y^3z_4 - x_3y^2y_4z - x_4y^3z_3 + x_4y^2y_3z).
\end{aligned}$$

References

1. Olver, P.J. *Equivalence, Invariants and Symmetry*; Cambridge University Press: Cambridge, UK, 1995.
2. Ovsiannikov, L.V. *Group Analysis of Differential Equations*; Academic Press: Cambridge, MA, USA, 1982.
3. Thomas, T.Y. *The Differential Invariants of Generalized Spaces*; Cambridge University Press: Cambridge, UK, 1934.
4. Chern, S.S.; Wang, H.C. Differential geometry in symplectic space. I. *Sci. Rep. Nat. Tsing Hua Univ.* **1947**, *4*, 453–477.
5. Kamran, N.; Olver, P.J.; Tenenblat, K. Local symplectic invariants for curves. *Commun. Contemp. Math.* **2009**, *11*, 165–183. [CrossRef]
6. McKay, B. Lagrangian submanifolds in affine symplectic geometry. *Differ. Geom. Appl.* **2006**, *24*, 670–689. [CrossRef]
7. Musso, E.; Hubert, E. Lagrangian curves in a 4-dimensional affine symplectic space. *Acta Appl. Math.* **2014**, *134*, 133–160. [CrossRef]
8. Musso, E.; Nicolodi, L. Symplectic applicability of Lagrangian surfaces. *SIGMA Symmetry Integr. Geom. Methods Appl.* **2009**, *5*, 67. [CrossRef]
9. Deconchy, V. Hypersurfaces in symplectic affine geometry. *Differ. Geom. Appl.* **2002**, *17*, 1–13. [CrossRef]
10. Valiquette, F. Geometric affine symplectic curve flows in $\mathbb{R}^4$. *Differ. Geom. Appl.* **2012**, *30*, 631–641. [CrossRef]
11. Xu, X. Differential invariants of classical groups. *Duke Math. J.* **1998**, *94*, 543–572. [CrossRef]
12. Jensen, J.O. Differential Invariants of Symplectic and Contact Lie Alebra Actions. Master's Thesis, UiT the Arctic University of Norway, Tromsø, Norway, June 2020.
13. Kruglikov, B.; Lychagin, V. Geometry of Differential equations. In *Handbook of Global Analysis*; Krupka, D., Saunders, D., Eds.; Elsevier: Amsterdam, The Netherlands, 2008; pp. 725–772.
14. Kruglikov, B.; Lychagin, V. Global Lie-Tresse theorem. *Selecta Math.* **2016**, *22*, 1357–1411. [CrossRef]
15. Kruglikov, B. Poincaré function for moduli of differential-geometric structures. *Moscow Math. J.* **2019**, *19*, 761–788. [CrossRef]
16. Kruglikov, B.; Lychagin, V. Differential invariants of the motion group actions. In *Variations, Geometry and Physics*; NOVA Scientific Publishers: Hauppauge, NY, USA, 2009; pp. 237–251.
17. Doubrov, B.; Ferapontov, E.; Kruglikov, B.; Novikov, V. On the integrability in Grassmann geometries: Integrable systems associated with fourfolds Gr(3,5). *Proc. Lond. Math. Soc.* **2018**, *116*, 1269–1300. [CrossRef]
18. Andreassen, F.; Kruglikov, B. Joint Invariants of Linear Symplectic Actions. *Symmetry* **2020**, *12*, 2020.

Publisher's Note: MDPI stays neutral with regard to jurisdictional claims in published maps and institutional affiliations.

© 2020 by the authors. Licensee MDPI, Basel, Switzerland. This article is an open access article distributed under the terms and conditions of the Creative Commons Attribution (CC BY) license (http://creativecommons.org/licenses/by/4.0/).

Article
Joint Invariants of Linear Symplectic Actions

Fredrik Andreassen and Boris Kruglikov *

Institute of Mathematics and Statistics, UiT the Arctic University of Norway, 90-37 Tromsø, Norway; fredrik.andreassen@gmail.com
* Correspondence: boris.kruglikov@uit.no

Received: 20 October 2020 ; Accepted: 26 November 2020; Published: 7 December 2020

Abstract: We review computations of joint invariants on a linear symplectic space, discuss variations for an extension of group and space and relate this to other equivalence problems and approaches, most importantly to differential invariants.

Keywords: polynomial and rational invariants; syzygy; free resolution; discretization

MSC: 15A72; 13A50; 53A55

1. Introduction

The classical invariant theory [1–3] investigates polynomial invariants of linear actions of a Lie group G on a vector space V, i.e., describes the algebra $(SV^*)^G$. For instance, the case of binary forms corresponds to $G = \mathrm{SL}(2,\mathbb{C})$ and $V = \mathbb{C}^2$; equivalently for $G = \mathrm{GL}(2,\mathbb{C})$ one studies instead the algebra of relative invariants. The covariants correspond to invariants in the tensor product $V \otimes W$ for another representation W. Changing to the Cartesian product $V \times W$ leads to joint invariants of G.

In this paper, we discuss joint invariants corresponding to the (diagonal) action of G on the iterated Cartesian product $V^{\times m}$ for increasing number of copies $m \in \mathbb{N}$. We will focus on the case $G = \mathrm{Sp}(2n,\mathbb{R})$, $V = \mathbb{R}^{2n}$ and discuss the conformal $G = \mathrm{CSp}(2n,\mathbb{R}) = \mathrm{Sp}(2n,\mathbb{R}) \times \mathbb{R}_+$ and affine $G = \mathrm{ASp}(2n,\mathbb{R}) = \mathrm{Sp}(2n,\mathbb{R}) \ltimes \mathbb{R}^{2n}$ versions later.

This corresponds to invariants of m-tuples of points in V, i.e., finite ordered subsets. By the Hilbert-Mumford [1] and Rosenlicht [4] theorems, the algebra of polynomial invariants (for the semi-simple G) or the field of rational invariants (in all other cases considered) can be interpreted as the space of functions on the quotient space $V^{\times m}/G$.

For $G = \mathrm{Sp}(2n,\mathbb{C})$ the algebra of invariants is known [5]. Generators and relations (syzygies) are described in the first and the second fundamental theorems, respectively. We review this in Theorem 1 (real version), and complement by explicit examples of free resolutions of the algebra. In addition, we describe the field of rational invariants.

We also discuss invariants with respect to the group $G = \mathrm{Sp}(2n,\mathbb{R}) \times S_m$, in which case considerably less is known. Another generalization we consider is the field of invariants for the conformal symplectic Lie group $G = \mathrm{CSp}(2n,\mathbb{R})$ on the contact space.

When approaching invariants of infinite sets, like curves or domains with smooth boundary, the theory of joint invariants is not directly applicable and the equivalence problem is solved via differential invariants [6]. In the case of a group G and a space V as above this problem was solved in [7]. We claim that the differential invariants from this reference can be obtained in a proper limit of joint invariants, i.e., via a certain discretization and quasiclassical limit, and demonstrate it explicitly in several cases.

In this paper, we focus on discussion of various interrelations of joint invariants. In particular, at the conclusion we note that joint invariants can be applied to the equivalence problem of binary forms. Since these have been studied also via differential invariants [2,8] a further link to the above symplectic discretization is possible.

The relation to binary forms mentioned above is based on the Sylvester theorem [9], which in turn can be extended to more general Waring decompositions, important in algebraic geometry [10]. Our computations should carry over to the general case. This note is partially based on the results of [11], generalized and elaborated in several respects.

2. Recollection: Invariants

We briefly recall the basics of invariant theory, referring to [3,12] for more details.

Let G be a Lie group acting on a manifold V. A point $x \in V$ is regular if a neighborhood of the orbit $G \cdot x$ is fibred by G-orbits. A point $x \in V$ is weakly regular, if its (not necessary G-invariant) neighborhood is foliated by the orbits of the Lie algebra $\mathfrak{g} = \mathrm{Lie}(G)$. In general, the action can lack regular points, but a generic point is weakly regular. For algebraic actions a Zariski open set of points is regular.

2.1. Smooth Invariants

If G and V are only smooth (and non-compact), there is little one can do to guarantee regularity a priori. An alternative is to look for local invariants, i.e., functions $I = I(x)$ in a neighborhood $U \subset V$ such that $I(x) = I(g \cdot x)$ as long as $x \in U$ and $g \in G$ satisfy $g \cdot x \in U$.

The standard method to search for such I is by elimination of group parameters, namely by computing quasi-transversals [3] or using normalization and moving frame [2]. Another way is to solve the linear PDE system $L_\xi(I) = 0$ for $\xi \in \mathfrak{g} = \mathrm{Lie}(G)$.

Given the space of invariants $\{I\}$ one can extend $U \subset V$ and address regularity. In our case the invariants are easy to compute and we do not rely on any of these methods; however instead we describe the algebra and the field of invariants depending on specification of the type of functions I.

2.2. Polynomial Invariants

If G is semi-simple and V is linear, then by the Hilbert-Mumford theorem generic orbits can be separated by polynomial invariants $I \in (SV^*)^G$, where $SV^* = \oplus_{k=0}^\infty S^k V^*$ is the algebra of homogeneous polynomials on V. With a choice of linear coordinates $x = (x_1, ..., x_n)$ on V we identify $SV^* = \mathbb{R}[x]$.

Moreover, by the Hilbert basis theorem, the algebra of polynomial invariants $\mathcal{A}_G = (SV^*)^G$ is Noetherian, i.e., finitely generated by some $a = (a_1, \ldots, a_s)$, $a_j = a_j(x) \in \mathcal{A}_G$.

Denote by $\mathcal{R} = \mathbb{R}[a]$ the free commutative $\mathbb{R}$-algebra generated by a. It forms a free module F_0 over itself. $\mathcal{A}_G$ is also an $\mathcal{R}$-module with surjective $\mathcal{R}$-homomorphism $\phi_0 : F_0 \to \mathcal{A}_G$, $\phi_0(a_j) = a_j(x)$. The first syzygy module $S_1 = \mathrm{Ker}(\phi_0)$ fits the exact sequence

$$0 \to S_1 \to F_0 \to \mathcal{A}_G \to 0.$$

A *syzygy* is an element of S_1, i.e., a relation $r = r(a)$ between the generators of $\mathcal{A}_G$ of the form $\sum_{p=1}^k r_{i_p} a_{j_p} = 0$, $r_{i_p} \in \mathcal{R}$.

The module S_1 is Noetherian, i.e., finitely generated by some $b = (b_1, \ldots, b_t)$. Denote the free $\mathcal{R}$-module generated by b by $F_1 = \mathcal{R}[b]$. The natural homomorphism $\phi_1 : F_1 \to S_1 \subset F_0$, $\phi_1(b_j) = b_j(a)$, defines the second syzygy module $S_2 = \mathrm{Ker}(\phi_1)$, and we can continue obtaining $S_2 \subset F_2 = \mathcal{R}[c]$, etc. This yields the exact sequence of $\mathcal{R}$-modules:

$$\cdots \xrightarrow{\phi_3} F_2 \xrightarrow{\phi_2} F_1 \xrightarrow{\phi_1} F_0 \xrightarrow{\phi_0} \mathcal{A}_G \to 0.$$

The Hilbert syzygy theorem states that q-th module of syzygies S_q is free for $q \geq s = \#a$. In particular, the minimal free resolution exists and has length $\leq s$, see [13].

To emphasize the generating sets, we depict free resolutions as follows:

$$\mathbb{R}[x] \supset \mathcal{A}_G \leftarrow \mathbb{R}[a] \leftarrow \mathcal{R}[b] \leftarrow \mathcal{R}[c] \leftarrow \cdots \leftarrow 0.$$

2.3. Rational Invariants

If G is algebraic, in particular reductive, then by the Rosenlicht theorem [4] generic orbits can be separated by rational invariants $I \in \mathcal{F}_G$. Here $\mathbb{R}(x)$ is the field of rational functions on V and $\mathcal{F}_G = \mathbb{R}(x)^G$.

Let d be the transcendence degree of $\mathcal{F}_G$. This means that there exist $(a_1, \ldots, a_d) = \bar{a}$, $a_j \in \mathcal{F}_G$, such that $\mathcal{F}_G$ is an algebraic extension of $\mathbb{R}(\bar{a})$. Then either $\mathcal{F}_G = \mathbb{R}(a)$ for $a = \bar{a}$ or $\mathcal{F}_G$ is generated by a set $a \supset \bar{a}$, which by the primitive element theorem can be assumed of cardinality $s = \#a = d + 1$, i.e., $a = (a_1, \ldots, a_d, a_{d+1})$. In the latter case there is one algebraic relation on a. Please note that $d \leq n$ because $\mathbb{R}(\bar{a}) \subset \mathbb{R}(x)$.

We adopt the following convention for depicting this:

$$\mathbb{R}(x) \supset \mathcal{F}_G \overset{\text{alg}}{\supset} \mathbb{R}(\bar{a}) \overset{d}{\supset} \mathbb{R}.$$

2.4. Our Setup

If the Lie group G acts effectively on V, then for some q it acts freely on $V^{\times q}$, and hence on all $V^{\times m}$ for $m \geq q$. The number of rational invariants separating a generic orbit in $V^{\times m}$ is equal to the codimension of the orbit.

It turns out that knowing all those invariants I on $V^{\times q}$ is enough to generate the invariants on $V^{\times m}$ for $m > q$. Indeed, let $\pi_{i_1, \ldots, i_q} : V^{\times m} \to V^{\times q}$ be the projection to the factors $(i_1, \ldots, i_q)$. Then the union of $\pi^*_{i_1, \ldots, i_q} I$ for I from the field $\mathcal{F}_G(V^{\times q})$ gives the generating set of the field $\mathcal{F}_G(V^{\times m})$, and similarly for the algebra of invariants.

Below we denote $\mathcal{A}_G^m = \mathcal{A}_G(V^{\times m})$ and $\mathcal{F}_G^m = \mathcal{F}_G(V^{\times m})$.

2.5. The Equivalence Problem

For a semi-simple Lie group G the field $\mathcal{F}_G$ is obtained from the ring $\mathcal{A}_G$ by localization (field of fractions): $\mathcal{F}_G = F(\mathcal{A}_G)$. Hence we discuss a solution to the equivalence problem through rational invariants.

Let $I_1, \ldots, I_s$ be a generating set of invariants of the action of G on $V^{\times q}$. If $s = d + 1$, this set of generators is subject to an algebraic condition, which constrains the generators to an algebraic set $\Sigma \subset \mathbb{R}^s$. If $s = d$ then $\Sigma = \mathbb{R}^d$. This Σ is the signature space, cf. [14].

Now the q-tuple of points $X = (x_1, \ldots, x_q)$ is mapped to $I_1(X), \ldots, I_s(X) \in \Sigma$. Denote this map by Ψ. Two generic configurations of points $X', X'' \in V^{\times q}$ are G-equivalent iff their signatures coincide $\Psi(X') = \Psi(X'')$.

3. Invariants on Symplectic Vector Spaces

Let $V = \mathbb{R}^{2n}(x^1, \ldots, x^n, y^1, \ldots, y^n)$ be equipped with the standard symplectic form $\omega = dx^1 \wedge dy^1 + \cdots + dx^n \wedge dy^n$. The group $G = \text{Sp}(2n, \mathbb{R})$ acts almost transitively on V, preserving the origin O. Thus, there are no continuous invariants of the action, $\mathcal{F}_G^1 = \mathbb{R}$. The first invariant occurs already for two copies of V. Namely for a pair of points $A_i, A_j \in V$ the double symplectic area of the triangle OA_iA_j is

$$a_{ij} = \omega(OA_i, OA_j) = x_i y_j - x_j y_i = \sum_{k=1}^{n} x_i^k y_j^k - x_j^k y_i^k.$$

3.1. The Case n = 1

Consider at first the case of dimension 2, where $V = \mathbb{R}^2(x,y)$, $\omega = dx \wedge dy$. The invariant $a_{12} = x_1y_2 - x_2y_1$ on $V \times V$ generates pairwise invariants a_{ij} on $V^{\times m}$ for $m \geq 2$ induced through the pull-back of the projection $\pi_{i,j} : V^{\times m} \to V \times V$ to the corresponding factors. Below we describe minimal free resolutions of $\mathcal{A}_G^m$ for $m \geq 2$.

3.1.1. $V \times V$

Here the algebra is generated by one element, whence the resolution:

$$\mathbb{R}[x_1, x_2, y_1, y_2] \supset \mathcal{A}_G^2 \leftarrow \mathbb{R}[a_{12}] \leftarrow 0$$

In other words, $\mathcal{A}_G^2 \simeq \mathcal{R} := \mathbb{R}[a_{12}]$. Please note that $\mathcal{F}_G^2 = \mathbb{R}(a_{12})$.

3.1.2. $V^{\times 3} = V \times V \times V$

Here the action is free on the level of $m = 3$ copies of V and we get $3 = \dim V^{\times 3} - \dim G$ independent invariants a_{12}, a_{13}, a_{23}. They generate the entire algebra, and we get the following minimal free resolution:

$$\mathbb{R}[x_1, x_2, x_3, y_1, y_2, y_3] \supset \mathcal{A}_G^3 \leftarrow \mathbb{R}[a_{12}, a_{13}, a_{23}] \leftarrow 0$$

Once again, $\mathcal{A}_G^3 \simeq \mathcal{R} := \mathbb{R}[a_{12}, a_{13}, a_{23}]$. Also $\mathcal{F}_G^3 = \mathbb{R}(a_{12}, a_{13}, a_{23})$.

3.1.3. $V^{\times 4}$

Here $\dim V^{\times 4} = 8$, $\dim G = 3$ and we have 6 invariants $a = \{a_{ij} : 1 \leq i < j \leq 4\}$. To obtain a relation, we try eliminating the variables $x_1, x_2, x_3, x_4, y_1, y_2, y_3, y_4$, but this fails with the standard MAPLE command. Yet, using the transitivity of the G-action we fix A_1 at $(1,0)$ and A_2 at $(0, a_{12})$, and then obtain the only relation

$$b_{1234} := a_{12}a_{34} - a_{13}a_{24} + a_{14}a_{23} = 0$$

that we identify as the *Plücker relation*. Thus, the first syzygy is a module over $\mathcal{R} := \mathbb{R}[a]$ with one generator, hence the minimal free resolution is:

$$\mathbb{R}[x, y] \supset \mathcal{A}_G^4 \leftarrow \mathbb{R}[a_{12}, a_{13}, a_{14}, a_{23}, a_{24}, a_{34}] \leftarrow \mathbb{R}[b_{1234}] \leftarrow 0.$$

For the field of rational invariants one of the generators is superfluous, for instance we can resolve the relation $b_{1234} = 0$ for $a_{34} = (a_{13}a_{24} - a_{14}a_{23})/a_{12}$, and get

$$\mathbb{R}(x_1, x_2, x_3, x_4, y_1, y_2, y_3, y_4) \supset \mathcal{F}_G^4 \simeq \mathbb{R}(a_{12}, a_{13}, a_{14}, a_{23}, a_{24}) \stackrel{5}{\supset} \mathbb{R}$$

3.1.4. $V^{\times 5}$

The algebra of invariants $\mathcal{A}_G^5$ is generated by $a = \{a_{ij} : 1 \leq i < j \leq 5\}$. This time the number of generators is 10, while codimension of the orbit is $10 - 3 = 7$. Using the same method we obtain that the first syzygy module is generated by the Plücker relations

$$b_{ijkl} := a_{ij}a_{kl} - a_{ik}a_{jl} + a_{il}a_{jk} = 0.$$

We have 5 of those: $b = \{b_{ijkl} : 1 \leq i < j < k < l \leq 5\}$. Thus, there should be relations among relations, or equivalently second syzygies. If $F_0 = \mathbb{R}[a] =: \mathcal{R}$ and $F_1 = \mathcal{R}[b]$ then this module is

$S_2 = \text{Ker}(\phi_1 : F_1 \to S_1 \subset F_0)$. Using elimination of parameters, we find that S_2 is generated by $c = \{c_i : 1 \leq i \leq 5\}$ with

$$c_i := \sum_{j=1}^{5} (-1)^j a_{ij} b_{1...\hat{j}...5}.$$

For instance, $c_1 = a_{12}b_{1345} - a_{13}b_{1245} + a_{14}b_{1235} - a_{15}b_{1234}$. Then we look for relations between the generators c of S_2, defining the third syzygy module S_3. It is generated by one element

$$d := (a_{23}a_{45} - a_{24}a_{35} + a_{25}a_{34})c_1 + (-a_{13}a_{45} + a_{14}a_{35} - a_{15}a_{34})c_2$$
$$+ (a_{12}a_{45} - a_{14}a_{25} + a_{15}a_{24})c_3 + (-a_{12}a_{35} + a_{13}a_{25} - a_{15}a_{23})c_4$$
$$+ (a_{12}a_{34} - a_{13}a_{24} + a_{14}a_{23})c_5 = 0.$$

Thus, the minimal free resolution of $\mathcal{A}_G^5$ is (note that here, as well as in our other examples, the length of the resolution is smaller than what the Hilbert theorem predicts):

$$\mathbb{R}[x,y] \supset \mathcal{A}_G^5 \leftarrow \mathcal{R}[a] \leftarrow \mathcal{R}[b] \leftarrow \mathcal{R}[c] \leftarrow \mathcal{R}[d] \leftarrow 0.$$

As before, to generate the field of rational invariants, we express superfluous generators in terms of the others using the first syzygies. Specifically, we express a_{34}, a_{35}, a_{45} from the relations $b_{1234}, b_{1235}, b_{1245}$; the other 2 syzygies follow from the higher syzygies. Removing these generators, we obtain a set of 7 independent generators $\bar{a} = a \setminus \{a_{34}, a_{35}, a_{45}\}$ whence

$$\mathbb{R}(x,y) \supset \mathcal{F}_G^5 \simeq \mathbb{R}(\bar{a}) \overset{7}{\supset} \mathbb{R}.$$

3.1.5. General $V^{\times m}$

The previous arguments generalize straightforwardly to conclude that $\mathcal{A}_G^m$ is generated by $a = \{a_{ij} : 1 \leq i < j \leq m\}$. The first syzygy module is generated by the Plücker relations $b = \{b_{ijkl} : 1 \leq i < j < k < l \leq m\}$. In other words we have:

$$\mathcal{A}_G^m = \langle a \mid b \rangle.$$

Similarly, the field of rational invariants is generated by a, yet all of them except for a_{1j}, a_{2j} can be expressed (rationally) through the rest via the Plücker relations b_{12kl}. Denote $\bar{a} := \{a_{12}, a_{13}, \ldots, a_{1m}, a_{23}, \ldots, a_{2m}\}$, $\#\bar{a} = 2m - 3$. Then we get for $m \geq 2$:

$$\mathbb{R}(x,p) \supset \mathcal{F}_G^m \simeq \mathbb{R}(\bar{a}) \overset{2m-3}{\supset} \mathbb{R}.$$

3.2. The General Case: Algebra of Polynomial Invariants

Minimal free resolutions can be computed in many examples for $n \geq 1$. However, in what follows we restrict our attention to describing generators/relations of $\mathcal{A}_G^m$.

Let us count the number of local smooth invariants. The action of G on V is almost transitive, so the stabilizer of a nonzero point A_1 has $\dim G_{A_1} = \binom{2n+1}{2} - 2n = \binom{2n}{2}$. For a generic A_2 there is only one invariant a_{12} (the orbit has codimension 1) and the stabilizer of A_2 in G_{A_1} has $\dim G_{A_1,A_2} = \binom{2n}{2} - (2n-1) = \binom{2n-1}{2}$. For a generic A_3 there are two more new invariants a_{13}, a_{23} (the orbit has codimension $2 + 1 = 3$) and the stabilizer of A_3 in G_{A_1,A_2} has $\dim G_{A_1,A_2,A_3} = \binom{2n-1}{2} - (2n-2) = \binom{2n-2}{2}$. By the same reason for $k \leq 2n$ the stabilizer of a generic k-tuple of points $A_1, \ldots, A_k$ has $\dim G_{A_1,\ldots,A_k} = \binom{2n-k+1}{2}$. Finally, for $k = 2n$ the stabilizer of generic $A_1, \ldots, A_{2n}$ is trivial.

Thus, we get the expected number of invariants a_{ij}. For $m \leq 2n+1$ there are no relations between them, and the first comes at $m = 2n+2$. These can be obtained by successively studying cases of increasing n resulting in the *Pfaffian relation*:

$$b_{i_1 i_2 \ldots i_{2n+1} i_{2n+2}} := \text{Pf}(a_{i_p i_q})_{1 \leq p,q \leq 2n+2} = 0.$$

Recall that the Pfaffian of a skew-symmetric operator S on V with respect to ω is $\text{Pf}(S) = \text{vol}_\omega(Se_1, \ldots, Se_{2n})$ for any symplectic basis e_i of V. The properties of the Pfaffian are: $\text{Pf}(S)^2 = \det(S)$, $\text{Pf}(TST^t) = \det(T) \text{Pf}(S)$. For $n = 1$ we get

$$b_{1234} = \text{Pf} \begin{pmatrix} 0 & a_{12} & a_{13} & a_{14} \\ -a_{12} & 0 & a_{23} & a_{24} \\ -a_{13} & -a_{23} & 0 & a_{34} \\ -a_{14} & -a_{24} & -a_{34} & 0 \end{pmatrix} = a_{12}a_{34} - a_{13}a_{24} + a_{14}a_{23}.$$

Similarly, for $n = 2$ we get

$$\begin{aligned} b_{123456} = & a_{12}a_{34}a_{56} - a_{12}a_{35}a_{46} + a_{12}a_{36}a_{45} - a_{13}a_{24}a_{56} + a_{13}a_{25}a_{46} - a_{13}a_{26}a_{45} + \\ & a_{14}a_{23}a_{56} - a_{14}a_{25}a_{36} + a_{14}a_{26}a_{35} - a_{15}a_{23}a_{46} + a_{15}a_{24}a_{36} - a_{15}a_{26}a_{34} + \\ & a_{16}a_{23}a_{45} - a_{16}a_{24}a_{35} + a_{16}a_{25}a_{34} = 0. \end{aligned}$$

Denote $\mathbf{b} = \{b_{i_1 i_2 \ldots i_{2n+1} i_{2n+2}} : 1 \leq i_1 < i_2 < \cdots < i_{2n+1} < i_{2n+2} \leq m\}$.

Theorem 1. *The algebra of G-invariants is generated by $\mathbf{a}$ with syzygies $\mathbf{b}$:*

$$\mathcal{A}_G^m = \langle \mathbf{a} \mid \mathbf{b} \rangle.$$

Proof. Let us first prove that the invariants a_{ij} generate the field $\mathcal{F}_G^m$ of rational invariants for $m = 2n$. We use the symplectic analog of Gram-Schmidt normalization: given points $A_1, \ldots, A_{2n}$ in general position, we normalize them using $G = \text{Sp}(2n, \mathbb{R})$ as follows.

Let $e_1, \ldots, e_{2n}$ be a symplectic basis of V, i.e., $\omega(e_{2k-1}, e_{2k}) = 1$ and $\omega(e_i, e_j) = 0$ else. At first A_1 can be mapped to the vector e_1. The point A_2 can be mapped to the line $\mathbb{R}e_2$, and because of $\omega(OA_1, OA_2) = a_{12}$ it is mapped to the vector $a_{12}e_2$. Next in mapping A_3 we have two constraints $\omega(OA_1, OA_3) = a_{13}$, $\omega(OA_2, OA_3) = a_{23}$, and the point can be mapped to the space spanned by e_1, e_2, e_3 satisfying those constraints. Continuing like this, we arrive to the following matrix with columns OA_i:

$$\begin{pmatrix} 1 & 0 & -\frac{a_{23}}{a_{12}} & -\frac{a_{24}}{a_{12}} & \cdots & -\frac{a_{2,2n-1}}{a_{12}} & -\frac{a_{2,2n}}{a_{12}} \\ 0 & a_{12} & a_{13} & a_{14} & \cdots & a_{1,2n-1} & a_{1,2n} \\ 0 & 0 & 1 & 0 & \cdots & * & * \\ 0 & 0 & 0 & \frac{b_{1234}}{a_{12}} & \vdots & * & * \\ \vdots & \vdots & \vdots & \vdots & \ddots & \vdots & \vdots \\ 0 & 0 & 0 & 0 & \cdots & 1 & 0 \\ 0 & 0 & 0 & 0 & \cdots & 0 & a_{2n-1,2n} \end{pmatrix}$$

where $b_{1234} = a_{12}a_{34} - a_{13}a_{24} + a_{14}a_{23}$ (this does not vanish in general if $n > 1$) and by $*$ we denote some rational expressions in a_{ij} that do not fit the table.

If $m < 2n$ then only the first m columns of this matrix have to be kept. If $m > 2n$ then the remaining points $A_{2n+1}, \ldots, A_m$ have all their coordinates invariant as the stabilizer of the first $2n$ points is trivial. Thus, the invariants are expressed rationally in a_{ij}.

To obtain polynomial invariants one clears the denominators in these rational expressions, and so $\mathcal{A}_G^m$ is generated by $\mathbf{a}$ as well.

Now the Pfaffian of the skew-symmetric matrix $(a_{ij})_{2k \times 2k}$ is the square root of the determinant of the Gram matrix of the vectors OA_i, $1 \leq i \leq k$, with respect to ω. If we take $k = n + 1$ then the vectors are linearly dependent and therefore the Pfaffian vanishes. Thus, b are syzygies among the generators a. That they form a complete set follows from the same normalization procedure as above. □

Remark 1. *Theorem 1 is basically known: H. Weyl described the generators a as the first fundamental theorem; his second fundamental theorem gives not only the syzygy denoted above by b, but also several different Pfaffians of larger sizes. Namely he lists in ([5], VI.1) the syzygies $b_{i_1...i_{2n+2k}} := \text{Pf}(a_{i_p i_q})_{1 \leq p,q \leq 2n+2k} = 0$, $1 \leq k \leq n$. Those however are abundant. For instance, in the simplest case $n = 2$*

$$b_{12345678} = a_{12}b_{345678} - a_{13}b_{245678} + a_{14}b_{235678} - a_{15}b_{234678} + a_{16}b_{234578} - a_{17}b_{234568} + a_{18}b_{234567}.$$

In general, the larger Pfaffians can be expressed via the smallest through the expansion by minors [15] (this fact was also noticed in [16]). Here is the corresponding Pfaffian identity (below we denote $S_{2n+1} = \{\sigma \in S_{2n+2} : \sigma(1) = 1\}$)

$$b_{i_1 i_2 ... i_{2n+1} i_{2n+2}} = \frac{1}{n!} \sum_{\sigma \in S_{2n+1}} (-1)^{\text{sgn}(\sigma)} a_{i_1 i_{\sigma(2)}} b_{i_{\sigma(3)}...i_{\sigma(2n+2)}}.$$

In ([3], §9.5) another set of syzygies was added: $q_{i_1...i_{4n+2}} = \det(a_{i_s, i_{t+2n+1}})_{s,t=1}^{2n+1} = 0$. These are also abundant, and should be excluded. For instance, for $n = 1$ we get

$$q_{123456} = a_{12}b_{3456} - a_{34}b_{1256} + a_{35}b_{1246} - a_{36}b_{1245}.$$

3.3. The General Case: Field of Rational Invariants

Since G is simple, the field of rational invariants is the field of fractions of the algebra of polynomial invariants: $\mathcal{F}_G^m = F(\mathcal{A}_G^m)$. To obtain its basis one can use the syzygies $b_{i_1...i_{2n+2}} = 0$ to express all invariants through $\bar{a} = \{a_{ij} : 1 \leq i \leq 2n; i < j \leq m\}$.

This can be done rationally (with $b_{1...2n} \neq 0$ in the denominator), for instance for $n = 2$ we can express a_{56} from the syzygy $b_{123456} = 0$ as follows:

$$a_{56} = (a_{12}a_{35}a_{46} - a_{12}a_{36}a_{45} - a_{13}a_{25}a_{46} + a_{13}a_{26}a_{45} + a_{14}a_{25}a_{36} - a_{14}a_{26}a_{35} + a_{15}a_{23}a_{46}$$
$$- a_{15}a_{24}a_{36} + a_{15}a_{26}a_{34} - a_{16}a_{23}a_{45} + a_{16}a_{24}a_{35} - a_{16}a_{25}a_{34})/(a_{12}a_{34} - a_{13}a_{24} + a_{14}a_{23}).$$

In general, we have $\#\bar{a} = 2nm - n(2n+1)$ for $m \geq 2n$, in summary:

$$\boxed{\mathbb{R}(x,y) \supset \mathcal{F}_G^m \simeq \mathbb{R}(\bar{a}) \overset{d(m,n)}{\supset} \mathbb{R},}$$

where

$$d(m,n) = \begin{cases} 2nm - n(2n+1) & \text{for } m \geq 2n \\ \binom{m}{2} & \text{for } m \leq 2n. \end{cases}$$

4. Variation on the Group and Space

Let us consider inclusion of symmetrization, scaling and translations to the transformation group G. We also discuss contactization of the action.

4.1. Symmetric Joint Invariants

Invariants of the extended group $\hat{G} = \mathrm{Sp}(2n,\mathbb{R}) \times S_m$ on $V^{\times m}$ are equivalent to G-invariants on configurations of unordered sets of points $V^{\times m}/S_m$ (which is an orbifold). Denote the algebra of polynomial $\hat{G}$-invariants on $V^{\times m}$ by $\mathcal{S}_G^m \subset \mathcal{A}_G^m$. The projection $\pi : \mathcal{A}_G^m \to \mathcal{S}_G^m$ is given by

$$\pi(f) = \frac{1}{m!} \sum_{\sigma \in S_m} \sigma \cdot f.$$

As a Noetherian algebra $\mathcal{S}_G^m$ is finitely generated, yet it is not easy to establish its generating set explicitly. All linear terms average to zero, $\pi(a_{ij}) = 0$, but there are several invariant quadratic terms in terms of the homogeneous decomposition $\mathcal{A}_G^m = \oplus_{k=0}^{\infty} \mathcal{A}_k^m$.

For example, for $n = 1$, $m = 4$ we have $\mathcal{A}_0^4 = \mathbb{R}$, $\mathcal{A}_1^4 = \mathbb{R}^6 = \langle a_{12}, a_{13}, a_{14}, a_{23}, a_{24}, a_{34}\rangle$, $\mathcal{A}_2^4 = \mathbb{R}^{20}$ (21 monomials $a_{ij}a_{kl}$ modulo 1 Plücker relation), etc. Then $\pi(\mathcal{A}_0^4) = \mathbb{R}$, $\pi(\mathcal{A}_1^4) = 0$, and $\pi(\mathcal{A}_2^4) = \mathbb{R}^2$ has generators

$$6\pi(a_{12}^2) = a_{12}^2 + a_{13}^2 + a_{14}^2 + a_{23}^2 + a_{24}^2 + a_{34}^2,$$
$$12\pi(a_{12}a_{13}) = a_{12}a_{13} + a_{12}a_{14} + a_{13}a_{14} - a_{12}a_{23} - a_{12}a_{24} + a_{23}a_{24}$$
$$+ a_{13}a_{23} - a_{13}a_{34} - a_{23}a_{34} + a_{14}a_{24} + a_{14}a_{34} + a_{24}a_{34}.$$

Theorem 2. *The field of symmetric rational invariants $\mathfrak{F}_G^m = \pi(\mathcal{F}_G^m)$ is the field of fractions $\mathfrak{F}_G^m = F(\mathcal{S}_G^m)$ and its transcendence degree is $d(m,n)$.*

Proof. This follows from general theorems ([17], §2.5) and discussion in Section 2. □

The last statement can be made more constructive: Let ℓ numerate indices (ij) of the basis $\tilde{a}$ of $\mathcal{F}_G^m$ as in Section 3.3, $1 \le \ell \le d = d(m,n)$. One can check that $q_k = \pi(\prod_{\ell \le k} a_\ell^2)$ are algebraically independent. Thus, denoting $q = (q_1, \ldots, q_d)$ we obtain the presentation

$$\mathbb{R}(x,y) \supset \mathfrak{F}_G^m \overset{\mathrm{alg}}{\supset} \mathbb{R}(q) \overset{d(m,n)}{\supset} \mathbb{R}.$$

Here is an algorithm to obtain generators of $\mathcal{S}_G^m$.

Proposition 1. *Fix an order on generators a_{ij} of $\mathcal{A}_G^m$, and induce the total lexicographic order on monomials $a^\sigma \in \mathcal{R} = \mathbb{R}[a]$. Let Σ be the Gröbner basis of the $\mathcal{R}$-ideal generated by $\pi(a^\sigma)$. Then elements $\pi(a^\sigma)$, contributing to Σ, generate $\mathcal{S}_G^m = \pi(\mathcal{A}_G^m)$.*

Proof. Please note that the algorithm proceeds in total degree of a^σ until the Gröbner basis stabilizes. That the involved $\pi(a^\sigma)$ generate $\mathcal{S}_G^m$ as an algebra (initially they generate the ideal $\mathcal{R} \cdot \pi(\mathcal{A}_G^m) \subset \mathcal{A}_G^m$) follows from the same argument as in the proof of Hilbert's theorem on invariants [1]. (The above π is the Reynolds operator used there.) □

Let us illustrate how this works in the first nontrivial case $m = 3$, for any n.

In this case, the graded components of $\mathcal{S}_G^3 = \pi(\mathcal{A}_G^3)$ have the following dimensions: $\dim \mathcal{S}_0^3 = 1$, $\dim \mathcal{S}_1^3 = 0$, $\dim \mathcal{S}_2^3 = 2$, $\dim \mathcal{S}_3^3 = 1$, $\dim \mathcal{S}_4^3 = 4$, $\dim \mathcal{S}_5^3 = 2$, $\dim \mathcal{S}_6^3 = 7$, etc., encoded into the Poincaré series

$$P_\mathcal{S}^3(z) = 1 + 2z^2 + z^3 + 4z^4 + 2z^5 + 7z^6 + 4z^7 + 10z^8 + 7z^9 + \ldots = \frac{1+z^4}{(1-z^2)^2(1-z^3)}.$$

For the monomial order $a_{12} > a_{13} > a_{23}$ the invariants

$$I_{2a} = 3\pi(a_{12}^2) = a_{12}^2 + a_{13}^2 + a_{23}^2, \quad I_{2b} = 3\pi(a_{12}a_{13}) = a_{12}a_{13} - a_{12}a_{23} + a_{13}a_{23},$$
$$I_3 = 6\pi(a_{12}^2 a_{13}) = a_{12}^2(a_{13} + a_{23}) - a_{23}^2(a_{12} + a_{13}) + a_{13}^2(a_{12} - a_{23}),$$
$$I_4 = 3\pi(a_{12}^2 a_{13}^2) = a_{12}^2 a_{13}^2 + a_{12}^2 a_{23}^2 + a_{13}^2 a_{23}^2$$

generate a Gröbner basis of the ideal $\mathcal{R} \cdot \pi(A_G^m)$ with the leading monomials of the corresponding Gröbner basis equal: $a_{12}^2, a_{12}a_{13}, a_{13}^3, a_{12}a_{23}^3, a_{13}^2 a_{23}^2, a_{13}a_{23}^3, a_{23}^4$.

The Gröbner basis also gives the following syzygy R_8:

$$(4I_{2a}^2 + 4I_{2a}I_{2b} + 3I_{2b}^2)I_{2b}^2 - (8I_{2a}^2 + 4I_{2a}I_{2b} + 14I_{2b}^2)I_4 + 4(I_{2a} - 2I_{2b})I_3^2 + 27I_4^2 = 0.$$

In other words, $\mathcal{S}_G^3 = \langle I_{2a}, I_{2b}, I_3, I_4 \mid R_8 \rangle$. We also derive a presentation of the field of rational invariants (2 : 1 means quadratic extension)

$$\mathbb{R}(x, y) \supset \mathfrak{F}_G^3 \overset{2:1}{\supset} \mathbb{R}(I_{2a}, I_{2b}, I_3) \overset{3}{\supset} \mathbb{R}.$$

4.2. Conformal and Affine Symplectic Groups

For the group $G_1 = \mathrm{CSp}(2n, \mathbb{R}) = \mathrm{Sp}(2n, \mathbb{R}) \times \mathbb{R}_+$ the scaling makes the invariants a_{ij} relative, yet of the same weight, so their ratios $[a_{12} : a_{13} : \cdots : a_{m-1,m}]$ or simply the invariants $I_{ij} = \frac{a_{ij}}{a_{12}}$ are absolute invariants. These generate the field of invariants of transcendence degree $d(m, n) - 1$.

For the group $G_2 = \mathrm{ASp}(2n, \mathbb{R}) = \mathrm{Sp}(2n, \mathbb{R}) \ltimes \mathbb{R}^{2n}$ the translations do not preserve the origin O and this makes a_{ij} non-invariant. However due to the formula $2\omega(A_1 A_2 A_3) = a_{12} + a_{23} - a_{13}$ (or more symmetrically: $a_{12} + a_{23} + a_{31}$), with the proper orientation of the triangle $A_1 A_2 A_3$, we easily recover the absolute invariants $a_{ij} + a_{jk} + a_{ki}$.

Alternatively, using the translational freedom, we can move the point A_1 to the origin O. Then its stabilizer in G_2 is $G = \mathrm{Sp}(2n, \mathbb{R})$ and we compute the invariants of $(m-1)$ tuples of points $A_2, \ldots, A_m$ as before. In particular they generate the field of invariants of transcendence degree $d(m-1, n)$.

4.3. Invariants in the Contact Space

Infinitesimal symmetries of the contact structure $\Pi = \mathrm{Ker}(\alpha)$, $\alpha = du - y\,dx$ in the contact space $M = \mathbb{R}^{2n+1}(x, y, u)$, where $x = (x_1, \ldots, x_n)$, $y = (y_1, \ldots, y_n)$, are given by the contact vector field X_H with the generating function $H = H(x, y, u)$. Taking quadratic functions H with weights $w(x) = 1$, $w(y) = 1$, $w(u) = 2$ results in the conformally symplectic Lie algebra, which integrates to the conformally symplectic group $G_1 = \mathrm{CSp}(2n, \mathbb{R})$ (taking H of degree ≤ 2 results in the affine extension of it by the Heisenberg group).

Alternatively, one considers the natural lift of the linear action of $G = \mathrm{Sp}(2n, \mathbb{R})$ on $V = \mathbb{R}^{2n}$ to the contactization M and makes a central extension of it. We will discuss the invariants of this action. Please note that this action is no longer linear, so the invariants cannot be taken to be polynomial, but can be assumed rational.

4.3.1. The Case $n = 1$

In the 3-dimensional case the group $G_1 = \mathrm{GL}(2, \mathbb{R})$ acts on $M = \mathbb{R}^3(x, y, u)$ as follows:

$$G_1 \ni g = \begin{pmatrix} \alpha & \beta \\ \gamma & \delta \end{pmatrix} : (x, y, u) \mapsto (\alpha x + \beta y, \gamma x + \delta y, f(x, y, u)),$$

$$\text{where } f(x, y, u) = (\alpha\delta - \beta\gamma)\left(u - \frac{xy}{2}\right) + \frac{(\alpha x + \beta y)(\gamma x + \delta y)}{2}.$$

This action is almost transitive (no invariants); however there are singular orbits and a relative invariant $R = xy - 2u$. Extending the action to multiple copies of M, i.e., considering the diagonal action of G_1 on $M^{\times m}$, results in m copies of this relative invariant, but also in the lifted invariants from various $V^{\times 2}$:

$$R_k = x_k y_k - 2u_k \ (1 \leq k \leq m), \quad R_{ij} = x_i y_j - x_j y_i \ (1 \leq i < j \leq m).$$

These are all relative invariants of the same weight, therefore their ratios are absolute invariants:

$$T_k = \frac{R_k}{R_m} \ (1 \leq k < m), \quad T_{ij} = \frac{R_{ij}}{R_m} \ (1 \leq i < j \leq m).$$

Since u_k enter only R_k there are no relations involving those, and the relations on T_{ij} are the same as for a_{ij}, namely they are Plücker relations (since those are homogeneous, they are satisfied by both R_{ij} and T_{ij}). As previously, we can use them to eliminate all invariants except for $\bar{T} = \{T_k, T_{1i}, T_{2i}\}$:

$$T_{kl} = \frac{T_{1k} T_{2l} - T_{1l} T_{2k}}{T_{12}}, \quad 3 \leq k < l \leq m.$$

The field of rational invariants for $m > 1$ is then described as follows:

$$\mathbb{R}(x, y, u) \supset \mathcal{F}_{G_1}^m \simeq \mathbb{R}(\bar{T}) \overset{3m-4}{\supset} \mathbb{R}.$$

4.3.2. The General Case

In general, we also have no invariants on M and the following relative invariants on $M^{\times m}$

$$R_k = x_k y_k - 2u_k \ (1 \leq k \leq m), \quad R_{ij} = x_i y_j - x_j y_i \ (1 \leq i < j \leq m)$$

resulting in absolute invariants T_k, T_{ij} given by the same formulae. Again, using the Pfaffian relations we can rationally eliminate superfluous generators, and denote the resulting set by $\bar{T} = \{T_k, T_{ij} : 1 \leq k < m, i < j \leq m, 1 \leq i \leq 2n\}$. This set is independent and contains $\bar{d}(m, n)$ elements, where

$$\bar{d}(m, n) = \begin{cases} (2n+1)m - n(2n+1) - 1 & \text{for } m \geq 2n \\ \binom{m}{2} + m - 1 = \binom{m+1}{2} - 1 & \text{for } m \leq 2n. \end{cases}$$

This $\bar{d}(m, n)$ is thus the transcendence degree of the field of rational invariants:

$$\mathbb{R}(x, y, u) \supset \mathcal{F}_{G_1}^m \simeq \mathbb{R}(\bar{T}) \overset{\bar{d}(m,n)}{\supset} \mathbb{R}.$$

5. From Joint to Differential Invariants

When we pass from finite to continuous objects the equivalence problem is solved through differential invariants. In [7] this was done for submanifolds and functions with respect to our groups G. After briefly recalling the results, we will demonstrate how to perform the discretization in several different cases.

5.1. Jets of Curves in Symplectic Vector Spaces

Locally a curve in $\mathbb{R}^{2n}$ is given as $u = u(t)$ for $t = x_1$ and $u = (x_2, \ldots, x_n, y_1, \ldots, y_n)$ in the canonical coordinates $(x_1, x_2, \ldots, x_n, y_1, \ldots, y_n)$, $\omega = dx_1 \wedge dy_1 + \cdots + dx_n \wedge dy_n$. The corresponding jet-space $J^\infty(V, 1)$ has coordinates $t, u, u_t, u_{tt}, \ldots$, and J^k is the truncation of it. For instance, $J^1(V, 1) = \mathbb{R}^{4n-1}(t, u, u_t)$. Please note that $\dim J^k(V, 1) = 2n + k(2n - 1)$.

In the case of dimension $2n = 2$, the jet-space is $J^k(V, 1) = \mathbb{R}^{k+2}(x, y, y_x, \ldots, y_{x..x})$. Here $G = \text{Sp}(2, \mathbb{R})$ has an open orbit in $J^1(V, 1)$, and the first differential invariant is of order 2:

$$I_2 = \frac{y_{xx}}{(xy_x - y)^3}.$$

There is also an invariant derivation ($\mathcal{D}_x$ is the total derivative with respect to x)

$$\nabla = \frac{1}{xy_x - y} \mathcal{D}_x.$$

By differentiation we get new differential invariants $I_3 = \nabla I_2$, $I_4 = \nabla^2 I_2$, etc. The entire algebra of differential invariants is free:

$$\mathcal{A}_G = \langle I_2; \nabla \rangle.$$

In the general case we denote the canonical coordinates on $V = \mathbb{R}^{2n}$ by (t, x, y, z), where x and z and $(n-1)$-dimensional vectors. $G = \text{Sp}(2n, \mathbb{R})$ acts on $J^\infty(V, 1)$. The invariant derivation is equal to

$$\nabla = \frac{1}{(ty_t - y + xz_t - x_t z)} \mathcal{D}_t.$$

and the first differential invariant of order 2 is

$$I_2 = \frac{x_t z_{tt} - x_{tt} z_t + y_{tt}}{(ty_t - y + xz_t - x_t z)^3}.$$

There is one invariant I_3 of order 3 independent of $I_2, \nabla(I_2)$, one invariant I_4 of order 4 independent of $I_2, \nabla(I_2), I_3, \nabla^2(I_2), \nabla(I_3)$, and so on up to order $2n$. Then the algebra of differential invariants of G is freely generated ([7], §4) so:

$$\mathcal{A}_G = \langle I_2, I_3, \ldots, I_{2n}; \nabla \rangle.$$

5.2. Symplectic Discretization

Consider first the case $n = 1$ with coordinates (x, y) on $V = \mathbb{R}^2$. Let $A_i = (x_i, y_i)$, $i = 0, 1, 2$, be three close points lying on the curve $y = y(x)$. We assume A_1 is in between A_0, A_2 and omit indices for its coordinates, i.e., $A_1 = (x, y)$.

Let $x_0 = x - \delta$ and $x_2 = x + \epsilon$. Denote also $y' = y'(x)$, $y'' = y''(x)$, etc. Then from the Taylor formula we have:

$$y_0 = y - \delta y' + \tfrac{1}{2}\delta^2 y'' - \tfrac{1}{6}\delta^3 y''' + o(\delta^3),$$
$$y_2 = y + \epsilon y' + \tfrac{1}{2}\epsilon^2 y'' + \tfrac{1}{6}\epsilon^3 y''' + o(\epsilon^3).$$

Therefore, the symplectic invariants $a_{ij} = x_i y_j - x_j y_i$ are:

$$a_{12} = \epsilon(xy' - y) + \tfrac{1}{2}\epsilon^2 xy'' + \tfrac{1}{6}\epsilon^3 xy''' + o(\epsilon^3),$$
$$a_{01} = \delta(xy' - y) - \tfrac{1}{2}\delta^2 xy'' + \tfrac{1}{6}\delta^3 xy''' + o(\delta^3),$$
$$a_{02} = (\epsilon + \delta)(xy' - y) + \tfrac{1}{2}(\epsilon^2 - \delta^2) xy''$$
$$+ \tfrac{1}{6}(\epsilon^3 + \delta^3) xy''' - \tfrac{1}{2}(\epsilon + \delta)\epsilon\delta y'' + o((|\delta| + |\epsilon|)^3).$$

This implies:

$$\frac{a_{01} - a_{02} + a_{12}}{a_{01} a_{02} a_{12}} = \frac{1}{2} \frac{y''}{(xy' - y)^3} + o(|\delta| + |\epsilon|).$$

Thus, we can extract the invariant exploiting no distance (like $\epsilon = \delta$) but only the topology ($\epsilon, \delta \to 0$) and the symplectic area. This works in any dimension n, and using the coordinates from the previous subsection we get

$$\lim_{A_0, A_2 \to A_1} \frac{\text{Area}_\omega(A_0 A_1 A_2)}{\text{Area}_\omega(OA_0 A_1)\, \text{Area}_\omega(OA_0 A_2)\, \text{Area}_\omega(OA_1 A_2)} = \frac{2(x_t z_{tt} - x_{tt} z_t + y_{tt})}{(ty_t - y + xz_t - x_t z)^3} = 2I_2.$$

Similarly, we obtain the invariant derivation (it uses only two points and hence is of the first order)

$$\lim_{A_0 \to A_1} \frac{\overrightarrow{A_0 A_1}}{\text{Area}_\omega(OA_0 A_1)} = \frac{2D_t}{(ty_t - y + xz_t - x_t z)} = 2\nabla.$$

The other generators $I_3, I_4, \ldots$ (important for $n > 1$) can be obtained by a higher order discretization, but the formulae become more involved.

5.3. Contact Discretization

Now we use joint invariants to obtain differential invariants of curves in contact 3-space $W = \mathbb{R}^3(x, y, u)$ with respect to the group $G = \text{GL}(2, \mathbb{R})$, acting as in §4.3. The curves will be given as $y = y(x), u = u(x)$ and their jet-space is $J^k(W, 1) = \mathbb{R}^{2k+3}(x, y, u, y_x, u_x, \ldots, y_{x..x}, u_{x..x})$. The differential invariants are generated in the Lie–Tresse sense ([7], §8.1) as

$$\mathcal{A}_G = \langle I_1, I_2; \nabla \rangle.$$

where

$$I_1 = \frac{u_x - y}{xy_x - y}, \quad I_2 = \frac{(xy - 2u)^2}{(xy_x - y)^3} y_{xx}, \quad \nabla = \frac{xy - 2u}{xy_x - y} D_x.$$

Instead of exploiting the absolute rational invariants T_i, T_{ij} we will work with the relative polynomial invariants R_i, R_{ij} from Section 4.3. To get absolute invariants we will then have to pass to weight zero combinations.

Consider three close points $\hat{A}_i = (x_i, y_i, u_i)$, $i = 0, 1, 2$, lying on the curve. We again omit indices for the middle point, so $x_0 = x - \delta$, $x_1 = x$ and $x_2 = x + \epsilon$. Using the Taylor decomposition as in the preceding subsection, we obtain

$$R_1 = xy - 2u, \quad R_0 - R_1 = \delta(2u' - y - xy') + o(\delta),$$
$$R_{01} = \delta(xy' - y) + o(\delta), \quad R_{02} = (\epsilon + \delta)(xy' - y) + o(|\epsilon| + |\delta|),$$
$$R_{12} = \epsilon(xy' - y) + o(\epsilon), \quad R_{01} + R_{12} - R_{02} = \tfrac{1}{2}\epsilon\delta(\epsilon + \delta)y'' + o((|\epsilon| + |\delta|)^3)$$

as well as

$$\overrightarrow{A_0 A_1} = \delta(\partial_x + y'\partial_y + u'\partial_y) + o(\delta).$$

Passing to jet-notations, we obtain the limit formulae for basic differential invariants:

$$I_1 = \lim_{A_0 \to A_1} \frac{R_0 - R_1}{2R_{01}} + \frac{1}{2} = \lim_{A_0 \to A_1} \frac{T_0 - 1 + T_{01}}{2T_{01}},$$

$$\frac{1}{2} I_2 = \lim_{A_0, A_2 \to A_1} \frac{R_1^2 (R_{01} + R_{12} - R_{12})}{R_{01} R_{02} R_{12}} = \lim_{A_0, A_2 \to A_1} \frac{T_{01} + T_{12} - T_{12}}{T_{01} T_{02} T_{12}},$$

$$\nabla = \lim_{A_0 \to A_1} \frac{R_1}{R_{01}} \overrightarrow{A_0 A_1} = \lim_{A_0 \to A_1} \frac{\overrightarrow{A_0 A_1}}{T_{01}}.$$

These formulae straightforwardly generalize to invariants of jets of curves in contact manifolds of dimension $2n + 1$, $n > 1$, in which case there are also other generators obtained by higher order discretizations.

5.4. Functions and Other Examples

Let us discuss invariants of jets of functions on the symplectic plane. The action of $G = \mathrm{Sp}(2,\mathbb{R})$ on $J^0 V = V \times \mathbb{R}(u) \simeq \mathbb{R}^3(x,y,u)$, with $I_0 = u$ invariant, prolongs to $J^\infty(V) = \mathbb{R}^\infty(x,y,u,u_x,u_y,u_{xx},u_{xy},u_{yy},\dots)$. Please note that functions can be identified as surfaces in $J^0 V$ through their graphs.

For any finite set of points $\hat{A}_k = (x_k, y_k, u_k)$ the values u_k are invariant, and the other invariants a_{ij} are obtained from the projections $A_k = (x_k, y_k)$. In this way we get the basic first order invariant (as before we omit indices $x_1 = x$, $y_1 = y$, $u_1 = y$ for the reference point A_1 in the right-hand side)

$$I_1 = \lim_{A_0, A_2 \to A_1} \frac{a_{01}(u_1 - u_2) + a_{12}(u_1 - u_0)}{a_{01} - a_{02} + a_{12}} = x u_x + y u_y$$

as well as two invariant derivations

$$\nabla_1 = \overrightarrow{OA_1} = x \mathcal{D}_x + y \mathcal{D}_y, \quad \nabla_2 = \lim_{A_0 \to A_1} \frac{I_1}{a_{01}} \overrightarrow{A_0 A_1} - \frac{u_1 - u_0}{a_{01}} \overrightarrow{OA_1} = u_x \mathcal{D}_y - u_y \mathcal{D}_x.$$

To obtain the second order invariant $I_{2c} = u_x^2 u_{yy} - 2 u_x u_y u_{xy} + u_y^2 u_{xx}$ let A_0 belong to the line through A_1 in the direction ∇_2 (this constraint reduces the second order formula to depend on only two points), i.e., $A_0 = (x + \epsilon u_y, y - \epsilon u_x)$, $A_1 = (x,y)$. Then $u_0 - u_1 = \frac{\epsilon^2}{2} I_{2c} + o(\epsilon^2)$, $a_{01} = \epsilon I_1$ and letting $\epsilon \to 0$ we obtain

$$\lim_{\substack{A_0 \to A_1 \\ A_0 A_1 \| \nabla_2}} \frac{u_0 - u_1}{a_{01}^2} = \frac{I_{2c}}{2 I_1^2}.$$

In the same way we get $I_{2a} = x^2 u_{xx} + 2 x y u_{xy} + y^2 u_{yy}$ and $I_{2b} = x u_y u_{xx} - y u_x u_{yy} + (y u_y - x u_x) u_{xy}$. These however are not required as the algebra of differential invariants is generated as follows ([7], §3.1) for some differential syzygies $\mathcal{R}_i$:

$$\mathcal{A}_G = \langle I_0, I_{2c}; \nabla_1, \nabla_2 \mid \mathcal{R}_1, \mathcal{R}_2, \mathcal{R}_3 \rangle.$$

Similarly, one can consider surfaces in the contact 3-space (with the same coordinates x, y, u but different lift of $\mathrm{Sp}(2,\mathbb{R})$ extended to $\mathrm{GL}(2,\mathbb{R})$) and higher-dimensional cases. The idea of discretization of differential invariants applies to other problems treated in [7].

6. Relation to Binary and Higher Order Forms

According to the Sylvester theorem [9] a general binary form $p \in \mathbb{C}[x,y]$ of odd degree $2m-1$ with complex coefficients can be written as

$$p(x,y) = \sum_{i=1}^m (\alpha_i x + \beta_i y)^{2m-1}.$$

This decomposition is determined up to permutation of linear factors and independent multiplication of each of them by a $(2m-1)$-th root of unity.

In other words, we have the branched cover of order $k_m = (2m-1)^m m!$

$$\times^m(\mathbb{C}^2) \to S^{2m-1} \mathbb{C}^2$$

and the deck group of this cover is $S_m \ltimes \mathbb{Z}_{2m-1}^{\times m}$.

Please note that in the real case, due to uniqueness of the odd root of unity, the corresponding cover over an open subset of the base

$$\times^m(\mathbb{R}^2) \to S^{2m-1} \mathbb{R}^2$$

has the deck group S_m.

With this approach the invariants of real binary forms are precisely the joint symmetric invariants studied in this paper, and for complex forms one must additionally quotient by $\mathbb{Z}_{2m-1}^{\times m}$, which is equivalent to passing from a_{ij} to a_{ij}^{2m-1} and other invariant combinations (example for $m = 4$: $a_{12}^3 a_{13}^2 a_{14}^2 a_{23}^2 a_{24}^2 a_{34}^3$) and subsequently averaging by the map π.

Other approaches to classification of binary forms, most importantly through differential invariants [2,8], can be related to this via symplectic discretization.

Remark 2. *Please note that the standard "root cover"* $\mathbb{C}^{2m} \to S^{2m-1}\mathbb{C}^2$:

$$(a_0, a_1, \ldots, a_{2m-1}) \mapsto (p_0, p_1, \ldots, p_{2m-1}), \quad \sum_{i=0}^{2m-1} p_i x^i y^{2m-i-1} = a_0 \prod_{i=1}^{2m-1}(x - a_i y)$$

has order $(2m-1)! < k_m$. *Polynomial* $\mathrm{SL}(2, \mathbb{C})$-*invariants of binary forms with this approach correspond to functions on the orbifold* $\mathbb{C}^{2m}/S_{2m}$.

The above idea extends further to ternary and higher valence forms (see [18] for the differential invariants approach and [19] for an approach using joint differential invariants) with the Waring decompositions [10] as the cover, but here the group G is no longer symplectic. We expect all the ideas of the present paper to generalize to the linear and affine actions of other reductive groups G.

Author Contributions: Both authors are equally responsible for all results in this paper. All authors have read and agreed to the published version of the manuscript.

Funding: The publication charges for this article have been funded by a grant from the publication fund of UiT the Arctic University of Norway.

Acknowledgments: FA thanks Jørn Olav Jensen for stimulating conversations and feedback. BK thanks Pavel Bibikov, Eivind Schneider and Boris Shapiro for useful discussions.

Conflicts of Interest: The authors declare no conflict of interest.

References

1. Hilbert, D. *Theory of Algebraic Invariants*; Cambridge University Press: Cambridge, UK, 1993.
2. Olver, P.J. Classical invariant theory. In *London Mathematical Society Student Texts*; Cambridge University Press: Cambridge, UK, 1999; Volume 44.
3. Popov, V.L.; Vinberg, E.B. Invariant theory. In *Algebraic Geometry IV, Encyclopaedia of Mathematical Sciences*; Springer: Berlin, Germany, 1994; Volume 55.
4. Rosenlicht, M. Some basic theorems on algebraic groups. *Am. J. Math.* **1956**, *78*, 401–443. [CrossRef]
5. Weyl, H. *Classical Groups*; Princeton University Press: Princeton, NJ, USA, 1946.
6. Kruglikov, B.; Lychagin, V. Global Lie-Tresse theorem. *Selecta Math.* **2016**, *22*, 1357–1411. [CrossRef]
7. Jensen, J.O.; Kruglikov, B. Differential Invariants of Linear Symplectic Actions. *Symmetry* **2020**, *12*, 2023. [CrossRef]
8. Bibikov, P.; Lychagin, V. $\mathrm{GL}_2(\mathbb{C})$-orbits of Binary Rational Forms. *Lobachevskii J. Math.* **2011**, *32*, 95–102. [CrossRef]
9. Sylvester, J.J. *An Essay on Canonical Forms*; George Bell: 1851; *On a Remarkable Discovery in the Theory of Canonical Forms and of Hyperdeterminants*; Philosophical Magazine: 1851; *Mathematical Papers*; Chelsea: New York, NY, USA, 1973; pp. 34, 203–216, 41, 265–283.
10. Alexander, J.; Hirschowitz, A. Polynomial interpolation in several variables. *J. Algebr. Geom.* **1995**, *4*, 201–222
11. Andreassen, F. Joint Invariants of of Symplectic and Contact Lie Algebra Actions. Master's Thesis in Mathematics, UiT the Arctic University of Norway, Tromsø, Norway, June 2020. Available online: https://hdl.handle.net/10037/19003 (accessed on 20 August 2020).
12. Mumford, D.; Fogarty, J.; Kirwan, F. *Geometric Invariant Theory*; Springer: Berlin/Heidelberg, Germany, 1994.
13. Eisenbud, D. The geometry of syzygies: A second course in algebraic geometry and commutative algebra. In *Graduate Texts in Mathematics*; Springer: Berlin/Heidelberg, Germany, 2006; Volume 229.

14. Olver, P.J. Joint invariant signatures. *Found. Comput. Math.* **2001**, *1*, 3–68. [CrossRef]
15. Ishikawa, M.; Wakayama, M. Minor summation formula of Pfaffians, Survey and a new identity. *Adv. Stud. Pure Math.* **2000**, *28*, 133–142.
16. Vust, T. Sur la théorie des invariants des groupes classiques. *Ann. Inst. Fourier* **1976**, *26*, 1–31. [CrossRef]
17. Springer, T.A. Invariant Theory. In *Lecture Notes in Math*; Springer: Berlin/Heidelberg, Germany, 1977; Volume 585.
18. Bibikov, P.; Lychagin, V. Classification of linear actions of algebraic groups on spaces of homogeneous forms. *Dokl. Math.* **2012**, *85*, 109–112. [CrossRef]
19. Gün Polat, G.; Olver, P.J. Joint differential invariants of binary and ternary forms. *Port. Math.* **2019**, *76*, 169–204. [CrossRef]

Publisher's Note: MDPI stays neutral with regard to jurisdictional claims in published maps and institutional affiliations.

© 2020 by the authors. Licensee MDPI, Basel, Switzerland. This article is an open access article distributed under the terms and conditions of the Creative Commons Attribution (CC BY) license (http://creativecommons.org/licenses/by/4.0/).

Article

Nonlocal Conservation Laws of PDEs Possessing Differential Coverings †

Iosif Krasil′shchik

V.A. Trapeznikov Institute of Control Sciences RAS, Profsoyuznaya 65, 117342 Moscow, Russia; josephkra@gmail.com
† To the memory of Alexandre Vinogradov, my teacher.

Received: 22 September 2020; Accepted: 20 October 2020; Published: 23 October 2020

Abstract: In his 1892 paper, L. Bianchi noticed, among other things, that quite simple transformations of the formulas that describe the Bäcklund transformation of the sine-Gordon equation lead to what is called a nonlocal conservation law in modern language. Using the techniques of differential coverings, we show that this observation is of a quite general nature. We describe the procedures to construct such conservation laws and present a number of illustrative examples.

Keywords: nonlocal conservation laws; differential coverings

MSC: 37K10

1. Introduction

In [1], L. Bianchi, dealing with the celebrated Bäcklund auto-transformation (I changed the original notation slightly)

$$\frac{\partial(u-w)}{\partial x} = \sin(u+w), \quad \frac{\partial(u+w)}{\partial y} = \sin(u-w) \tag{1}$$

for the sine-Gordon equation

$$\frac{\partial^2(2u)}{\partial x \partial y} = \sin(2u) \tag{2}$$

in the course of intermediate computations (see ([1], p. 10)) notices that the function

$$\psi = \ln \frac{\partial u}{\partial C},$$

where C is an arbitrary constant on which the solution u may depend, enjoys the relations

$$\frac{\partial \psi}{\partial x} = \cos(u+w), \quad \frac{\partial \psi}{\partial y} = \cos(u-w).$$

Reformulated in modern language, this means that the 1-form

$$\omega = \cos(u+w)\,dx + \cos(u-w)\,dy$$

is a nonlocal conservation law for Equation (1).

It became clear much later, some 100 years after the publication of [1], that nonlocal conservation laws are important invariants of PDEs and are used in numerous applications, e.g.,: numerical methods [2,3], sociological models [4,5], integrable systems [6], electrodynamics [7,8], mechanics [9–11], etc.

Actually, Bianchi's observation is of a very general nature and this is shown below.

In Section 2, I shortly introduce the basic constructions in nonlocal geometry of PDEs, i.e., the theory of differential coverings, [12]. Section 3 contains an interpretation of the result by L. Bianchi in the most general setting. In Section 4, a number of examples is discussed.

Everywhere below we use the notation $\mathscr{F}(\cdot)$ for the $\mathbb{R}$-algebra of smooth functions, $D(\cdot)$ for the Lie algebra of vector fields, and $\Lambda^*(\cdot) = \oplus_{k \geq 0} \Lambda^k(\cdot)$ for the exterior algebra of differential forms.

2. Preliminaries

Following [13], we deal with infinite prolongations $\mathscr{E} \subset J^\infty(\pi)$ of smooth submanifolds in $J^k(\pi)$, where $\pi \colon E \to M$ is a smooth locally trivial vector bundle over a smooth manifold M, $\dim M = n$, $\operatorname{rank} \pi = m$. These $\mathscr{E}$ are differential equations for us. Solutions of $\mathscr{E}$ are graphs of infinite jets that lie in $\mathscr{E}$. In particular, $\mathscr{E} = J^\infty(\pi)$ is the tautological equation $0 = 0$.

The bundle $\pi_\infty \colon \mathscr{E} \to M$ is endowed with a natural flat connection $\mathscr{C} \colon D(M) \to D(\mathscr{E})$ called the *Cartan connection*. Flatness of $\mathscr{C}$ means that $\mathscr{C}_{[X,Y]} = [\mathscr{C}_X, \mathscr{C}_Y]$ for all $X, Y \in D(M)$. The distribution on $\mathscr{E}$ spanned by the fields of the form $\mathscr{C}_X$ (the *Cartan distribution*) is Frobenius integrable. We denote it by $\mathscr{C} \subset D(\mathscr{E})$ as well.

A (higher infinitesimal) *symmetry* of $\mathscr{E}$ is a π_∞-vertical vector field $S \in D(\mathscr{E})$ such that $[X, \mathscr{C}] \subset \mathscr{C}$.

Consider the submodule $\Lambda_h^k(\mathscr{E})$ generated by the forms $\pi_\infty^*(\theta)$, $\theta \in \Lambda^k(M)$. Elements $\omega \in \Lambda_h^k(\mathscr{E})$ are called *horizontal k-forms*. Generalizing slightly the action of the Cartan connection, one can apply it to the de Rham differential $d \colon \Lambda^k(M) \to \Lambda^{k+1}(M)$ and obtain the *horizontal de Rham* complex

$$0 \longrightarrow \mathscr{F}(\mathscr{E}) \longrightarrow \ldots \longrightarrow \Lambda_h^k(\mathscr{E}) \xrightarrow{d_h} \Lambda_h^{k+1}(\mathscr{E}) \longrightarrow \ldots \longrightarrow \Lambda_h^n(\mathscr{E}) \longrightarrow 0$$

on $\mathscr{E}$. Elements of its $(n-1)$st cohomology group $H_h^{n-1}(\mathscr{E})$ are called *conservation laws* of $\mathscr{E}$. We always assume $\mathscr{E}$ to be *differentially connected* which means that $H_h^0(\mathscr{E}) = \mathbb{R}$.

Remark 1. *The concept of a differentially connected equation reflects Vinogradov's correspondence principle [14], (p. 195): when 'secondary dimension' (dimension of the Cartan distribution) $\operatorname{Dim} \to 0$, the objects of PDE geometry degenerate to their counterparts in geometry of finite-dimensional manifolds. Following this principle, we informally have*

$$\lim_{\operatorname{Dim} \to 0} H_h^i(\mathscr{E}) = H_{\mathrm{dR}}^i(M).$$

Since $H_{\mathrm{dR}}^0(M)$ is responsible for topological connectedness of M, the group $H_h^0(\mathscr{E})$ stands for differential one.

Coordinates. Consider a trivialization of π with local coordinates $x^1, \ldots, x^n$ in $\mathscr{U} \subset M$ and $u^1, \ldots, u^m$ in the fibers of $\pi|_{\mathscr{U}}$. Then in $\pi_\infty^{-1}(\mathscr{U}) \subset J^\infty(\pi)$ the adapted coordinates u_σ^i arise and the Cartan connection is determined by the total derivatives

$$\mathscr{C} \colon \frac{\partial}{\partial x^i} \mapsto D_i = \frac{\partial}{\partial x^i} + \sum_{j,\sigma} u_{\sigma i}^j \frac{\partial}{\partial u_\sigma^j}.$$

Let $F = (F^1, \ldots, F^r)$, where F^j are smooth functions on $J^k(\pi)$. The the infinite prolongation of the locus

$$\{ z \in J^k(\pi) \mid F^1(z) = \cdots = F^r(z) = 0 \} \subset J^k(\pi)$$

is defined by the system

$$\mathscr{E} = \mathscr{E}_F = \{ z \in J^\infty(\pi) \mid D_\sigma(F^j)(z) = 0,\ j = 1, \ldots, r,\ |\sigma| \geq 0 \},$$

where D_σ denotes the composition of the total derivatives corresponding to the multi-index σ. The total derivatives, as well as all differential operators in total derivatives, can be restricted to infinite prolongations and we preserve the same notation for these restrictions. Given an $\mathscr{E}$, we always choose internal local coordinates in it for subsequent computations. To restrict an operator to $\mathscr{E}$ is to express this operator in terms of internal coordinates.

Any symmetry of $\mathscr{E}$ is an evolutionary vector field

$$\mathbf{E}_\varphi = \sum D_\sigma(\varphi^j) \frac{\partial}{\partial u^j_\sigma}$$

(summation on internal coordinates), where the functions $\varphi^1, \ldots, \varphi^m \in \mathscr{F}(\mathscr{E})$ satisfy the system

$$\sum_{\sigma,\alpha} \frac{\partial F^j}{\partial u^\alpha_\sigma} D_\sigma(\varphi^\alpha) = 0, \quad j = 1, \ldots, r.$$

A horizontal $(n-1)$-form

$$\omega = \sum_i a_i \, dx^1 \wedge \cdots \wedge dx^{i-1} \wedge dx^{i+1} \wedge \cdots \wedge dx^n$$

defines a conservation law of $\mathscr{E}$ if

$$\sum_i (-1)^{i+1} D_i(a_i) = 0.$$

We are interested in nontrivial conservation laws, i.e., such that ω is not exact.

Finally, $\mathscr{E}$ is differentially connected if the only solutions of the system

$$D_1(f) = \cdots = D_n(f) = 0, \quad f \in \mathscr{F}(\mathscr{E}),$$

are constants.

Consider now a locally trivial bundle $\tau: \tilde{\mathscr{E}} \to \mathscr{E}$ such that there exists a flat connection $\tilde{\mathscr{C}}$ in $\pi_\infty \circ \tau: \tilde{\mathscr{E}} \to M$. Following [12], we say that τ is a (*differential*) *covering* over $\mathscr{E}$ if one has

$$\tau_*(\tilde{\mathscr{C}}_X) = \mathscr{C}_X$$

for any vector field $X \in D(M)$. Objects existing on $\tilde{\mathscr{E}}$ are nonlocal for $\mathscr{E}$: e.g., symmetries of $\tilde{\mathscr{E}}$ are *nonlocal symmetries* of $\mathscr{E}$, conservation laws of $\tilde{\mathscr{E}}$ are *nonlocal conservation laws* of $\mathscr{E}$, etc. A derivation $S: \mathscr{F}(\mathscr{E}) \to \mathscr{F}(\tilde{\mathscr{E}})$ is called a *nonlocal shadow* if the diagram

$$\begin{array}{ccc} \mathscr{F}(\mathscr{E}) & \xrightarrow{\mathscr{C}_X} & \mathscr{F}(\mathscr{E}) \\ S \downarrow & & \downarrow S \\ \mathscr{F}(\tilde{\mathscr{E}}) & \xrightarrow{\tilde{\mathscr{C}}_X} & \mathscr{F}(\tilde{\mathscr{E}}) \end{array}$$

is commutative for any $X \in D(M)$. In particular, any symmetry of the equation $\mathscr{E}$, as well as restrictions $\tilde{S}|_{\mathscr{F}(\mathscr{E})}$ of nonlocal symmetries may be considered as shadows. A nonlocal symmetry is said to be *invisible* if its shadow $\tilde{S}|_{\mathscr{F}(\mathscr{E})}$ vanishes.

A covering τ is said to be *irreducible* if $\tilde{\mathcal{E}}$ is differentially connected. Two coverings are equivalent if there exists a diffeomorphism $g \colon \tilde{\mathcal{E}}_1 \to \tilde{\mathcal{E}}_2$ such that the diagrams

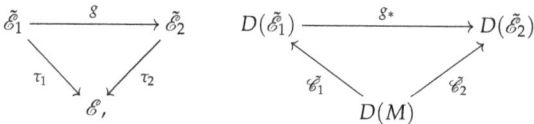

are commutative. Note also that for any two coverings their *Whitney product* is naturally defined. A covering is called *linear* if τ is a vector bundle and the action of vector fields $\mathscr{C}_X$ preserves the subspace of fiber-wise linear functions in $\mathscr{F}(\tilde{\mathcal{E}})$.

In the case of 2D equations, there exists a fundamental relation between special type of coverings over $\mathcal{E}$ and conservation laws of the latter. Let τ be a covering of rank $l < \infty$. We say that τ is an *Abelian covering* if there exist l independent conservation laws $[\omega_i] \in H^1_h(\mathcal{E})$, $i = 1, \ldots, l$, such that the forms $\tau^*(\omega_i)$ are exact. Then equivalence classes of such coverings are in one-to-one correspondence with l-dimensional $\mathbb{R}$-subspaces in $H^1_h(\mathcal{E})$.

Coordinates. Choose a trivialization of the covering τ and let $w^1, \ldots, w^l, \ldots$ be coordinates in fibers (the are called nonlocal variables). Then the covering structure is given by the extended total derivatives

$$\tilde{D}_i = D_i + X_i, \quad i = 1, \ldots, n,$$

where

$$X_i = \sum_\alpha X_i^\alpha \frac{\partial}{\partial w^\alpha}$$

are τ-vertical vector fields (nonlocal tails) enjoying the condition

$$D_i(X_j) - D_j(X_i) + [X_i, X_j] = 0, \quad i < j. \tag{3}$$

Here $D_i(X_j)$ denotes the action of D_i on coefficients of X_j. Relations (3) (flatness of $\tilde{\mathscr{C}}$) amount to the fact that the manifold $\tilde{\mathcal{E}}$ endowed with the distribution $\tilde{\mathscr{C}}$ coincides with the infinite prolongation of the overdetermined system

$$\frac{\partial w^\alpha}{\partial x^i} = X_i^\alpha,$$

which is compatible modulo $\mathcal{E}$.

Irreducible coverings are those for which the system of vector fields $\tilde{D}_1, \ldots, \tilde{D}_n$ has no nontrivial integrals. If $\bar{\tau}$ is another covering with the nonlocal tails $\bar{X}_i = \sum \bar{X}_i^\beta \partial / \partial \bar{w}^\beta$, then the Whitney product $\tau \oplus \bar{\tau}$ of τ and $\bar{\tau}$ is given by

$$\tilde{D}_i = D_i + \sum_\alpha X_i^\alpha \frac{\partial}{\partial w^\alpha} + \sum_\beta \bar{X}_i^\beta \frac{\partial}{\partial \bar{w}^\beta}.$$

A covering is Abelian if the coefficients X_i^α are independent of nonlocal variables w^j. If $n = 2$ and $\omega_\alpha = X_1^\alpha dx^1 + X_2^\alpha dx^2$, $\alpha = 1, \ldots, l$, are conservation laws of $\mathcal{E}$ then the corresponding Abelian covering is given by the system

$$\frac{\partial w^\alpha}{\partial x^i} = X_i^\alpha, \quad i = 1, 2, \quad \alpha = 1, \ldots, l,$$

or

$$\tilde{D}_i = D_i + \sum_\alpha X_i^\alpha \frac{\partial}{\partial w^\alpha}.$$

Vice versa, if such a covering is given, then one can construct the corresponding conservation law.

The horizontal de Rham differential on $\tilde{\mathcal{E}}$ is $\tilde{d}_h = \sum_i dx^i \wedge \tilde{D}_i$. A covering is linear if

$$X_i^\alpha = \sum_\beta X_{i,\beta}^\alpha w^\beta, \tag{4}$$

where $X_{i,\beta}^\alpha \in \mathcal{F}(\mathcal{E})$.

Remark 2. *Denote by $\mathbf{X}_i$ the $\mathcal{F}(\mathcal{E})$-valued matrix $(X_{i,\beta}^\alpha)$ that appears in (4). Then Equation (3) may be rewritten as*

$$D_i(\mathbf{X_j}) - D_j(\mathbf{X_i}) + [\mathbf{X_i}, \mathbf{X_j}] = 0.$$

for linear coverings. Thus, a linear covering defines a zero-curvature representation for $\mathcal{E}$ and vice versa.

A nonlocal symmetry in τ is a vector field

$$S_{\varphi,\psi} = \sum \tilde{D}_\sigma(\varphi^j) \frac{\partial}{\partial u_\sigma^j} + \sum \psi^\alpha \frac{\partial}{\partial w^\alpha},$$

where the vector functions $\varphi = (\varphi^1, \ldots, \varphi^m)$ and $\psi = (\psi^1, \ldots, \psi^\alpha, \ldots)$ on $\tilde{\mathcal{E}}$ satisfy the system of equations

$$\sum \frac{\partial F^j}{\partial u_\sigma^j} \tilde{D}_\sigma(\varphi^j) = 0, \tag{5}$$

$$\tilde{D}_i(\psi^\alpha) = \sum \frac{\partial X_i^\alpha}{\partial u_\sigma^j} \tilde{D}_\sigma(\varphi^j) + \sum \frac{\partial X_i^\alpha}{\partial w^\beta} \psi^\beta. \tag{6}$$

Nonlocal shadows are the derivations

$$\tilde{\mathbf{E}}_\varphi = \sum \tilde{D}_\sigma(\varphi^j) \frac{\partial}{\partial u_\sigma^j},$$

where φ satisfies Equation (5), invisible symmetries are

$$S_{0,\psi} = \sum \psi^\alpha \frac{\partial}{\partial w^\alpha},$$

where ψ satisfies

$$\tilde{D}_i(\psi^\alpha) = \sum \frac{\partial X_i^\alpha}{\partial w^\beta} \psi^\beta. \tag{7}$$

In what follows, we use the notation $\tau^{\mathbf{I}} \colon \tilde{\mathcal{E}}^{\mathbf{I}} \to \tilde{\mathcal{E}}$ for the covering defined by Equation (7).

Remark 3. *Equation (7) defines a linear covering over $\tilde{\mathcal{E}}$. Due to Remark 2, we see that for any non-Abelian covering we obtain in such a way a nonlocal zero-curvature representation with the matrices $\mathbf{X}_i = (\partial X_i^\alpha / \partial w^\beta)$.*

Remark 4. *The covering $\tau^{\mathbf{I}} \colon \tilde{\mathcal{E}}^{\mathbf{I}} \to \tilde{\mathcal{E}}$ is the vertical part of the tangent covering $t \colon \mathcal{T}\tilde{\mathcal{E}} \to \tilde{\mathcal{E}}$, see the definition in [15].*

3. The Main Result

From now on we consider two-dimensional scalar equations with the independent variables x and y. We shall show that any such an equation that admits an irreducible covering possesses a (nonlocal) conservation law.

Example 1. Let us revisit the Bianchi example discussed in the beginning of the paper. Equation (1) define a one-dimensional non-Abelian covering $\tau\colon \tilde{\mathcal{E}} = \mathcal{E} \times \mathbb{R} \to \mathcal{E}$ over the sine-Gordon Equation (2) with the nonlocal variable w. Then the defining Equation (7) for invisible symmetries in this covering are

$$\frac{\partial \psi}{\partial x} = -\cos(u+w)\psi, \quad \frac{\partial \psi}{\partial y} = -\cos(u-w)\psi.$$

This is a one-dimensional linear covering over $\tilde{\mathcal{E}}$ which is equivalent to the Abelian covering

$$\frac{\partial \tilde{\psi}}{\partial x} = -\cos(u+w), \quad \frac{\partial \tilde{\psi}}{\partial y} = -\cos(u-w),$$

where $\tilde{\psi} = \ln \psi$. Thus, we obtain the nonlocal conservation law

$$\omega = -\cos(u+w)\, dx - \cos(u-w)\, dy$$

of the sine-Gordon equation.

The next result shows that Bianchi's observation is of a quite general nature.

Proposition 1. Let $\tau\colon \tilde{\mathcal{E}} \to \mathcal{E}$ be a one-dimensional non-Abelian covering over $\mathcal{E}$. Then, if τ is irreducible, $\tau^{\mathrm{I}}\colon \tilde{\mathcal{E}}^{\mathrm{I}} \to \tilde{\mathcal{E}}$ defines a nontrivial conservation law of the equation $\tilde{\mathcal{E}}$ (and, consequently, of $\mathcal{E}$ too).

Proof. Consider the total derivatives

$$D_x^{\mathrm{I}} = \tilde{D}_x + \frac{\partial X}{\partial w}\psi\frac{\partial}{\partial \psi} = D_x + X\frac{\partial}{\partial w} + \frac{\partial X}{\partial w}\psi\frac{\partial}{\partial \psi}$$

$$D_y^{\mathrm{I}} = \tilde{D}_y + \frac{\partial Y}{\partial w}\psi\frac{\partial}{\partial \psi} = D_y + Y\frac{\partial}{\partial w} + \frac{\partial Y}{\partial w}\psi\frac{\partial}{\partial \psi}$$

on $\tilde{\mathcal{E}}^{\mathrm{I}}$ and assume that $a \in \mathcal{F}(\tilde{\mathcal{E}})$ is a common nontrivial integral of these fields:

$$D_x^{\mathrm{I}}(a) = D_y^{\mathrm{I}}(a) = 0, \quad a \neq \mathrm{const}. \tag{8}$$

Choose a point in $\tilde{\mathcal{E}}^{\mathrm{I}}$ and assume that the formal series

$$a_0 + a_1\psi + \cdots + a_j\psi^j + \ldots, \quad a_j \in \mathcal{F}(\tilde{\mathcal{E}}), \tag{9}$$

converges to a in a neighborhood of this point. Substituting relations (9) to (8) and equating coefficients at the same powers of ψ, we get

$$\tilde{D}_x(a_j) + j\frac{\partial X}{\partial w}a_j = 0, \quad \tilde{D}_y(a_j) + j\frac{\partial Y}{\partial w}a_j = 0, \quad j = 0,1,\ldots,$$

and, since τ is irreducible, this implies that $a_0 = k_0 = \mathrm{const}$ and

$$\frac{\tilde{D}_x(a_j)}{a_j} = j\frac{\tilde{D}_x(a_1)}{a_1}, \quad \frac{\tilde{D}_y(a_j)}{a_j} = j\frac{\tilde{D}_y(a_1)}{a_1}.$$

Hence, $a_j = k_j(a_1)^j$, $j > 0$. Substituting these relations to (9), we see that $a = a(\theta)$, where $\theta = a_1\psi$, $a_1 \in \mathcal{F}(\tilde{\mathcal{E}})$. Then Equation (8) take the form

$$\dot{a}\psi\left(\tilde{D}_x(a_1) + \frac{\partial X}{\partial w}\right) = 0, \quad \dot{a}\psi\left(\tilde{D}_y(a_1) + \frac{\partial Y}{\partial w}\right) = 0, \quad \dot{a} = \frac{da}{d\theta}.$$

Thus
$$\frac{\partial X}{\partial w} = -\tilde{D}_x(a_1), \quad \frac{\partial Y}{\partial w} = -\tilde{D}_y(a_1),$$
and the function $w + a_1$ is a nontrivial integral of $\tilde{D}_x$ and $\tilde{D}_y$. Contradiction.

Finally, repeating the scheme of Example 1, we pass to the equivalent covering by setting $\bar{\psi} = \ln \psi$ and obtain the nontrivial conservation law
$$\omega = \frac{\partial X}{\partial w} dx + \frac{\partial Y}{\partial w} dy$$
on $\tilde{\mathcal{E}}^{\mathrm{I}}$. □

Indeed, Bianchi's result has a further generalization. To formulate the latter, let us say that a covering $\tau\colon \tilde{\mathcal{E}} \to \mathcal{E}$ is *strongly non-Abelian* if for any nontrivial conservation law ω of the equation $\mathcal{E}$ its lift $\tau^*(\omega)$ to the manifold $\tilde{\mathcal{E}}$ is nontrivial as well. Now, a straightforward generalization of Proposition 1 is

Proposition 2. *Let $\tau\colon \tilde{\mathcal{E}} \to \mathcal{E}$ be an irreducible covering over a differentially connected equation. Then τ is a strongly non-Abelian covering if and only if the covering τ^{I} is irreducible.*

We shall now need the following construction. Let $\tau\colon \tilde{\mathcal{E}} \to \mathcal{E}$ be a linear covering. Consider the fiber-wise *projectivization* $\tau^{\mathbf{P}}\colon \tilde{\mathcal{E}}^{\mathbf{P}} \to \mathcal{E}$ of the vector bundle τ. Denote by $\mathbf{p}\colon \tilde{\mathcal{E}} \to \tilde{\mathcal{E}}^{\mathbf{P}}$ the natural projection. Then, obviously, the projection $\mathbf{p}_*(\tilde{\mathcal{C}})$ is well defined and is an n-dimensional integrable distribution on $\tilde{\mathcal{E}}^{\mathbf{P}}$. Thus, we obtain the following commutative diagram of coverings

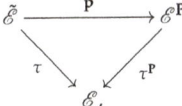

where $\mathrm{rank}(\mathbf{p}) = 1$ and $\mathrm{rank}(\tau^{\mathbf{P}}) = \mathrm{rank}(\tau) - 1$.

Proposition 3. *Let $\tau\colon \tilde{\mathcal{E}} \to \mathcal{E}$ be an irredicible covering. Then the covering $\tau^{\mathbf{P}}$ is irreducible as well.*

Coordinates. Let $\mathrm{rank}(\tau) = l > 1$ and
$$w^\alpha_{x^i} = \sum_{\beta=1}^{l} X^\alpha_{i,\beta} w^\beta, \quad i = 1, \ldots, n, \quad \alpha = 1, \ldots, l, \qquad (10)$$
be the defining equations of the covering τ, see Equation (4). Choose an affine chart in the fibers of $\tau^{\mathbf{P}}$. To this end, assume for example that $w^l \neq 0$ and set
$$\bar{w}^\alpha = \frac{w^\alpha}{w^l}, \quad l = 1, \ldots, l-1,$$
in the domain under consideration. Then from Equation (10) it follows that the system
$$\bar{w}^\alpha_{x^i} = X^\alpha_{i,l} - X^l_{i,l}\bar{w}^\alpha + \sum_{\beta=1}^{l-1} X^\alpha_{i,\beta}\bar{w}^\beta - \bar{w}^\alpha \sum_{\beta=1}^{l-1} X^l_{i,\beta}\bar{w}^\beta, \quad i = 1, \ldots, n, \quad \alpha = 1, \ldots, l-1.$$

locally provides the defining equation for the covering $\tau^{\mathbf{P}}$.

We are now ready to state and prove the main result.

Theorem 1. *Assume that a differentially connected two-dimensional equation $\mathcal{E}$ admits a nontrivial covering $\tau \colon \tilde{\mathcal{E}} \to \mathcal{E}$ of finite rank. Then it possesses at least one nontrivial (nonlocal) conservation law.*

Proof. Actually, the proof is a description of a procedure that allows one to construct the desired conservation law.

Note first that we may assume the covering τ to be irreducible. Indeed, otherwise the space $\tilde{\mathcal{E}}$ is foliated by maximal integral manifolds of the distribution $\mathcal{C}$. Let l_0 denote the codimension of the generic leaf and $l = \mathrm{rank}(\tau)$. Then

- $l > l_0$, because τ is a nontrivial covering;
- the integral leaves project to $\mathcal{E}$ surjectively, because $\mathcal{E}$ is a differentially connected equation.

This means that in vicinity of a generic point we can consider τ as an l_0-parametric family of irreducible coverings whose rank is $r = l - l_0 > 0$. Let us choose one of them and denote it by $\tau_0 \colon \mathcal{E}_0 \to \mathcal{E}$.

If τ_0 is not strongly non-Abelian, then this would mean that $\mathcal{E}$ possesses at least one nontrivial conservation law and we have nothing to prove further. Assume now that the covering τ_0 is strongly non-Abelian. Then due to Proposition 2 the linear covering τ_0^I is irreducible and by Proposition 3 its projectivization $\tau_1 = (\tau_0^I)^P$ possesses the same property and $\mathrm{rank}(\tau_1) = r - 1$. Repeating the construction, we arrive to the diagram

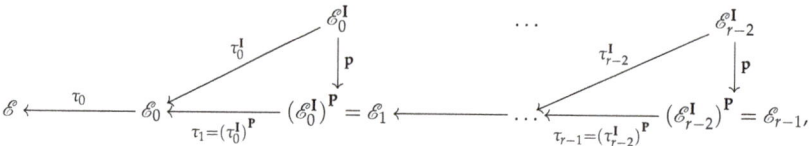

where $\mathrm{rank}(\tau_i) = l - i$. Thus, in $r - 1$ steps at most we shall arrive to a one-dimensional irreducible covering and find ourselves in the situation of Proposition 1 and this finishes the proof. □

4. Examples

Let us discuss several illustrative examples.

Example 2. *Consider the Korteweg-de Vries equation in the form*

$$u_t = u u_x + u_{xxx} \tag{11}$$

and the well known Miura transformation [16]

$$u = w_x - \frac{1}{6}w^2.$$

The last formula is a part of the defining equations for the non-Abelian covering

$$\begin{aligned} w_x &= u + \frac{1}{6}w^2, \\ w_t &= u_{xx} + \frac{1}{3}w u_x + \frac{1}{3}u^2 + \frac{1}{18}w^2 u, \end{aligned}$$

the covering equation being

$$w_t = w_{xxx} - \frac{1}{6}w^2 w_x,$$

i.e., the modified KdV equation. Then the corresponding covering τ^I is defined by the system

$$\psi_x = \frac{1}{3}w\psi,$$
$$\psi_t = \frac{1}{3}\left(u_x + \frac{1}{3}wu\right)\psi$$

that, after relabeling $\psi \mapsto 3\ln\psi$ gives us the nonlocal conservation law

$$\omega = w\,dx + \left(u_x + \frac{1}{3}wu\right)dt$$

of the KdV equation.

Example 3. *The well known Lax pair, see [17], for the KdV equation may be rewritten in terms of zero-curvature representation*

$$D_x(\mathbf{T}) - D_t(\mathbf{X}) + [\mathbf{X}, \mathbf{T}] = 0.$$

The (2×2) matrices $\mathbf{X}$ and $\mathbf{T}$ become much simpler if we present the equation in the form

$$u_t = 6uu_x - u_{xxx}.$$

In this case, they are

$$\mathbf{X} = \begin{pmatrix} 0 & 1 \\ u - \lambda & 0 \end{pmatrix}, \quad \mathbf{T} = \begin{pmatrix} -u_x & 2(u + 2\lambda) \\ 2u^2 - u_{xx} + 2\lambda u - 4\lambda^2 & u_x \end{pmatrix},$$

$\lambda \in \mathbb{R}$ being a real parameter. As it follows from Remark 2, this amounts to existence of the two-dimensional linear covering τ given by the system

$$w_{1,x} = w_2,$$
$$w_{1,t} = -u_x w_1 + 2(u + 2\lambda)w_2,$$
$$w_{2,x} = (u - \lambda)w_1,$$
$$w_{2,t} = (2u^2 - u_{xx} + 2\lambda u - 4\lambda^2)w_1 + u_x w_2.$$

Let us choose for the affine chart the domain $w_2 \neq 0$ and set $\psi = w_1/w_2$. Then the covering τ^P is described by the system

$$\psi_x = 1 - (u - \lambda)\psi,$$
$$\psi_t = 2(u + 2\lambda) - 2u_x\psi - (2u^2 - u_{xx} + 2\lambda u - 4\lambda^2)\psi^2,$$

while $\tau_1 = (\tau^P)^I$ is given by

$$\tilde{\psi}_x = (\lambda - u)\tilde{\psi},$$
$$\tilde{\psi}_t = -2(u_x + (2u^2 - u_{xx} + 2\lambda u - 4\lambda^2)\psi)\tilde{\psi}.$$

Thus, we obtain the conservation law

$$\omega = (\lambda - u)\,dx - 2(u_x + (2u^2 - u_{xx} + 2\lambda u - 4\lambda^2)\psi)\,dt$$

that depends on the nonlocal variable ψ.

Example 4. Consider the potential KdV equation in the form

$$u_t = 3u_x^2 + u_{xxx}$$

Its Bäcklund auto-transformation is associated to the covering τ

$$w_x = \lambda - u_x - \frac{1}{2}(w-u)^2,$$
$$w_t = 2\lambda^2 - 2\lambda u_x - u_x^2 - u_{xxx} + 2u_{xx}(w-u) - (\lambda + u_x)(w-u)^2,$$

where $\lambda \in \mathbb{R}$, see [18]. Then the covering $\tau^{\mathbf{I}}$ is

$$\psi_x = -(w-u)\psi,$$
$$\psi_t = 2(u_{xx}\psi - (\lambda + u_x)(w-u))\psi,$$

which leads to the nonlocal conservation law

$$\omega = -(w-u)\,dx + 2(u_{xx}\psi - (\lambda + u_x)(w-u))\,dt$$

of the potential KdV equation.

Example 5. The Gauss-Mainardi-Codazzi equations read

$$u_{xy} = \frac{g - fh}{\sin u}, \quad f_y = g_x + \frac{h - g\cos u}{\sin u}u_x, \quad g_y = h_x - \frac{f - g\cos u}{\sin u}u_y, \quad (12)$$

see [19]. This is an under-determined system, and imposing additional conditions on the unknown functions u, f, g, and h one obtains equations that describe various types of surfaces in $\mathbb{R}^2$, cf. [20]. System (12) always admits the following $\mathbb{C}$-valued zero-curvature representation

$$D_x(\mathbf{Y}) - D_y(\mathbf{X}) + [\mathbf{X}, \mathbf{Y}] = 0$$

with the matrices

$$\mathbf{X} = \frac{i}{2}\begin{pmatrix} u_x & \dfrac{e^{iu}f - g}{\sin u} \\ \dfrac{e^{-iu}f - g}{\sin u} & -u_x \end{pmatrix}, \quad \mathbf{Y} = \frac{i}{2}\begin{pmatrix} 0 & \dfrac{e^{iu}g - h}{\sin u} \\ \dfrac{e^{-iu}g - h}{\sin u} & 0 \end{pmatrix}$$

The corresponding two-dimensional linear covering τ is defined by the system

$$w_x^1 = u_x w^1 + \frac{e^{iu}f - g}{\sin u}w^2, \qquad w_x^2 = \frac{e^{-iu}f - g}{\sin u}w^1 - u_x w^2,$$
$$w_y^1 = \frac{e^{iu}g - h}{\sin u}w^2, \qquad w_y^2 = \frac{e^{-iu}g - h}{\sin u}w^1.$$

Hence, the covering $\tau^{\mathbf{P}}$ in the domain $w^2 \neq 0$ is

$$\psi_x = \frac{e^{iu}f - g}{\sin u} + 2u_x\psi - \frac{e^{-iu}f - g}{\sin u}\psi^2, \quad \psi_y = \frac{e^{iu}g - h}{\sin u} - \frac{e^{-iu}g - h}{\sin u}\psi^2.$$

Thus, the covering $(\tau^{\mathbf{P}})^{\mathbf{I}}$, given by

$$\tilde{\psi}_x = 2\left(u_x - \frac{e^{-iu}f - g}{\sin u}\psi\right)\tilde{\psi}, \quad \tilde{\psi}_y = -2\frac{e^{-iu}g - h}{\sin u}\psi\tilde{\psi},$$

defines the nonlocal conservation law

$$\omega = \left(u_x - \frac{e^{-iu}f - g}{\sin u}\psi \right) dx - \frac{e^{-iu}g - h}{\sin u}\psi \, dy$$

of the Gauss-Mainardi-Codazzi equations.

Example 6. *The last example shows that the above described techniques fail for infinite-dimensional coverings (such coverings are typical for equations of dimension greater than two).*
Consider the equation

$$u_{yy} = u_{tx} + u_y u_{xx} - u_x u_{xy}$$

that arises in the theory of integrable hydrodynamical chains, see [21]. This equation admits the covering τ with the nonlocal variables w^i, $i = 0, 1, \ldots$, that enjoy the defining relations

$$w_t^0 + u_y w_x^1 = 0, \quad w_y^0 + u_x w_x^1 = 0,$$
$$w_x^i = w^{i+1}, \quad i \geq 0,$$
$$w_t^i + D_x^i(u_y w_x^1) = 0, \quad w_y^i + D_x^i(u_x w_x^1) = 0, \quad i \geq 1.$$

see [22]. This is a linear covering, but its projectivization does not lead to construction of conservation laws.

5. Discussion

We described a procedure that allows one to associate, in an algorithmic way, with any nontrivial finite-dimensional covering over a differentially connected equation a nonlocal conservation law. Nevertheless, this method fails in the case of infinite-dimensional coverings. It is unclear, at the moment at least, whether this is an immanent property of such coverings or a disadvantage of the method. I hope to clarify this in future research.

Funding: The work was partially supported by the RFBR Grant 18-29-10013 and IUM-Simons Foundation.

Acknowledgments: I am grateful to Michal Marvan, who attracted my attention to the paper by Luigi Bianchi [1], and to Raffaele Vitolo, who helped me with Italian. I am also grateful to Valentin Lychagin for a fruitful discussion.

Conflicts of Interest: The authors declare no conflict of interest.

References

1. Bianchi, L. Sulla trasformazione di Bäcklund per le superfici pseudosferiche. *Rend. Mat. Acc. Lincei,* **1892**, *1*, 3–12.
2. Chatterjee, N.; Fjordholm, U.S. A convergent finite volume method for the Kuramoto equation and related nonlocal conservation laws. *IMA J. Numer. Anal.* **2020**, *40*, 405–421. [CrossRef]
3. Naz, R. Potential systems and nonlocal conservation laws of Prandtl boundary layer equations on the surface of a sphere. *Z. Naturforschung A* **2017**, *72*, 351–357. [CrossRef]
4. Aggarwal, A.; Goatin, P. Crowd dynamics through non-local conservation laws. *Bull. Braz. Math. Soc. New Ser.* **2016**, *47*, 37–50. [CrossRef]
5. Keimer, A.; Pflug, L. Nonlocal conservation laws with time delay. *Nonlinear Differ. Equ. Appl.* **2019**, *26*, 54. [CrossRef]
6. Sil, S.; Sekhar, T.R.; Zeidan, D. Nonlocal conservation laws, nonlocal symmetries and exact solutions of an integrable soliton equation. *Chaos Solitons Fractals* **2020**, *139*, 110010 [CrossRef]
7. Anco, S.C.; Bluman, G. Nonlocal symmetries and nonlocal conservation laws of Maxwell's equations. *J. Math. Phys.* **1997**, *38*, 350. [CrossRef]
8. Anco, S.C.; Webb, G.M. Conservation laws in magnetohydrodynamics and fluid dynamics: Lagrangian approach. *AIP Conf. Proc.* **2019**, *2153*, 020024.
9. Betancourt, F.; Bürger, R.; Karlsen, K.; Tory, E.M. On nonlocal conservation laws modelling sedimentation. *Nonlinearity* **2011**, *24*, 855–885. [CrossRef]

10. Christoforou, C. Nonlocal conservation laws with memory. In *Hyperbolic Problems: Theory, Numerics, Applications*; Benzoni-Gavage S., Serre D., Eds.; Springer: Berlin/Heidelberg, Germany, 2008.
11. Ibragimov, N.; Karimova, E.N.; Galiakberova, L.R. Chaplygin gas motions associated with nonlocal conservation laws. *J. Coupled Syst. Multiscale Dyn.* **2017**, *5*, 63–68. [CrossRef]
12. Krasil'shchik, I.S.; Vinogradov, A.M. Nonlocal trends in the geometry of differential equations: symmetries, conservation laws, and Bäcklund transformations. *Acta Appl. Math.* **1989**, *15*, 161–209.
13. Bocharov, A.V.; Chetverikov, V.N.; Duzhin, S.V.; Khor'kova, N.G.; Krasil'shchik, I.S.; Samokhin, A.V.; Torkhov, Yu.N.; Verbovetsky, A.M.; Vinogradov, A.M. *Symmetries of Differential Equations in Mathematical Physics and Natural Sciences*; English translation: Amer. Math. Soc., 1999; Vinogradov, A.M., Krasil'shchik, I.S., Eds.; Factorial Publ. House: Moscow, 1997. (In Russian)
14. Vinogradov, A.M. *Cohomological Analysis of Partial Differential Equations and Secondary Calculus*; Translations of Mathematical Monographs; American Mathematical Society: Providence, RI, USA, 2001; Volume 204.
15. Krasil'shchik, I.S.; Verbovetskiy, A.M.; Vitolo, R. *The Symbolic Computation of Integrability Structures for Partial Differential Equations*; Texts & Monographs in Symbolic Computation; Springer: Berlin, Germnay, 2017.
16. Gardner, C.S.; Green, J.M.; Kruskal, M.D.; Miura, R.M. Method for solving the Korteweg-de Vries equation. *Phys. Rev. Lett.* **1967**, *19*, 1095–1097. [CrossRef]
17. Lax, P.D. Integrals of nonlinear equations of evolution and solitary waves. *Commun. Pure Appl. Math.* **1968**, *21*, 467. [CrossRef]
18. Wahlquist, H.B.; Estabrook, F.B. Bäcklund transformation for solutions to the Korteweg-de Vries equation. *Phys. Rev. Lett.* **1973**, *31*, 1386–1390. [CrossRef]
19. Sym, A. Soliton surfaces and their applications (soliton geometry from spectral problems). In *Geometric Aspects of the Einstein Equations and Integrable Systems, Proceedings of the Conference Scheveningen, The Netherlands, 26–31 August 1984*; Lecture Notes in Physics; Martini, R., Ed., Springer: Berlin, Germany, 1985; Volume 239, pp. 154–231.
20. Krasil'shchik, I.S.; Marvan, M. Coverings and integrability of the Gauss-Mainardi-Codazzi equations. *Acta Appl. Math.* **1999**, *56*, 217–230.
21. Pavlov, M.V. Integrable hydrodynamic chains. *J. Math. Phys.* **2003**, *44*, 4134. [CrossRef]
22. Baran, H.; Krasil'shchik, I.S.; Morozov, O.I.; Vojčák, P. Nonlocal symmetries of integrable linearly degenerate equations: A comparative study. *Theoret. Math. Phys.* **2018**, *196*, 169–192. [CrossRef]

Publisher's Note: MDPI stays neutral with regard to jurisdictional claims in published maps and institutional affiliations.

 © 2020 by the authors. Licensee MDPI, Basel, Switzerland. This article is an open access article distributed under the terms and conditions of the Creative Commons Attribution (CC BY) license (http://creativecommons.org/licenses/by/4.0/).

Article

Geometrical Formulation for Adjoint-Symmetries of Partial Differential Equations

Stephen C. Anco *,† and Bao Wang †,‡

Department of Mathematics and Statistics, Brock University, St. Catharines, ON L2S3A1, Canada; wangbao@lsec.cc.ac.cn
* Correspondence: sanco@brocku.ca
† These authors contributed equally to this work.
‡ Current address: Department of Mathematics, Ningbo University, Ningbo 315211, China.

Received: 3 September 2020; Accepted: 14 September 2020; Published: 19 September 2020

Abstract: A geometrical formulation for adjoint-symmetries as one-forms is studied for general partial differential equations (PDEs), which provides a dual counterpart of the geometrical meaning of symmetries as tangent vector fields on the solution space of a PDE. Two applications of this formulation are presented. Additionally, for systems of evolution equations, adjoint-symmetries are shown to have another geometrical formulation given by one-forms that are invariant under the flow generated by the system on the solution space. This result is generalized to systems of evolution equations with spatial constraints, where adjoint-symmetry one-forms are shown to be invariant up to a functional multiplier of a normal one-form associated with the constraint equations. All of the results are applicable to the PDE systems of interest in applied mathematics and mathematical physics.

Keywords: adjoint-symmetry; one-form; symmetry; vector field; geometrical formulation

1. Introduction

Symmetries are a fundamental coordinate-free structure of a partial differential equation (PDE). In geometrical terms, an infinitesimal symmetry is an evolutionary (vertical) vector field that is tangent to the solution space of a PDE, where the components of the vector field are the solutions of the linearization of the PDE on its solution space (see, e.g., [1–4]).

Knowledge of the symmetries of a PDE can be used to map given solutions into other solutions, find invariant solutions, detect and find mappings in a target class of PDEs, detect integrability, and find conservation laws through Noether's theorem when a PDE has a variational (Lagrangian) structure.

Solutions of the adjoint linearization of a PDE on its solution space are known as adjoint-symmetries. This terminology was first introduced and explored for ordinary differential equations (ODEs) in [5–8] and then generalized to PDEs in [9,10] (see [11] for a recent overview for PDEs). When a PDE lacks a variation structure, then its adjoint-symmetries will differ from its symmetries.

Knowledge of the adjoint-symmetries of a PDE can be used for several purposes just as symmetries can. Specifically, solutions of the PDE can be found analogously to the invariant surface condition associated with a symmetry; mappings into a target class of PDEs can be detected and found analogously to characterizing the symmetry structure of the target class; integrability can be detected analogously to the existence of higher order symmetries; and conservation laws can be determined analogously to symmetries that satisfy a variational condition. In particular, the counterpart of variational symmetries for a general PDE is provided by multipliers, which are well known to be adjoint-symmetries that satisfy a Euler–Lagrange condition.

However, a simple geometrical meaning (apart from abstract formulations) for adjoint-symmetries has yet to be developed in general for PDEs. Several significant new steps toward this goal will be taken in the present paper.

Firstly, for general PDE systems, adjoint-symmetries will be shown to correspond to evolutionary (vertical) one-forms that functionally vanish on the solution space of the system. This formulation has two interesting applications. It will provide a geometrical derivation of a well-known formula that generates a conservation law from a pair consisting of a symmetry and an adjoint-symmetry [9,12]. It also will yield three different actions of symmetries on adjoint-symmetries from Cartan's formula for the Lie derivative, providing a geometrical formulation of some recent work that used an algebraic viewpoint [13].

Secondly, for evolution systems, these adjoint-symmetry one-forms will be shown to have the structure of a Lie derivative of a simpler underlying one-form, utilizing the flow generated by the system. As a result, adjoint-symmetries of evolution systems will geometrically correspond to one-forms that are invariant under the flow on the solution space of the system. This directly generalizes the geometrical meaning of adjoint-symmetries known for ODEs [8].

Thirdly, a bridge between the preceding results for general PDE systems and evolution systems will be developed by considering evolution systems with spatial constraints. These systems are ubiquitous in applied mathematics and mathematical physics, for example: Maxwell's equations, incompressible fluid equations, magnetohydrodynamical equations, and Einstein's equations. For such systems, invariance of the adjoint-symmetry one-form under the constrained flow will be shown to hold up to a functional multiple of the normal one-form associated with the constraint equations.

Throughout, the approach will be concrete, rather than abstract, so that the results can be readily understood and applied to specific PDE systems of interest in applied mathematics and mathematical physics.

The rest of the paper is organized as follows. Section 2 discusses the evolutionary form of vector fields and its counterpart for one-forms in the mathematical framework of calculus in jet space, which will underlie all of the main results. Section 3 reviews the geometrical formulation of symmetries and presents the counterpart geometrical formulation of adjoint-symmetries. In addition, some examples of adjoint-symmetries of physically interesting PDE systems are discussed. Section 4 gives the two applications of adjoint-symmetry one-forms. Section 5 develops the main results for adjoint-symmetries of evolution systems and extends these results to constrained evolution systems. Some concluding remarks are made in Section 6.

2. Vector Fields, One-Form Fields, and Their Evolutionary Form

To begin, some essential tools [3,11,14] from calculus in jet space will be reviewed. This will set the stage for a discussion of the evolutionary form of vector fields and its counterpart for one-forms, as needed for the main results in the subsequent sections.

Independent variables are denoted x^i, $i = 1, \ldots, n$, and dependent variables are denoted u^α, $\alpha = 1, \ldots, m$. Derivative variables are indicated by subscripts employing a multi-index notation: $I = \{i_1, \ldots, i_N\}$, $u_I^\alpha = u_{i_1 \cdots i_N}^\alpha := \partial_{x^{i_1}} \cdots \partial_{x^{i_N}} u^\alpha$, $|I| = N$; $I = \emptyset$, $u_I^\alpha := u^\alpha$, $|I| = 0$. Some useful notation is as follows: $\partial^k u$ will denote the set $\{u_I^\alpha\}_{|I|=k}$ of all derivative variables of order $k \geq 0$; $u^{(k)}$ will denote the set $\{u_I^\alpha\}_{0 \leq |I| \leq k}$ of all derivative variables of all orders up to $k \geq 0$. The summation convention of summing over any repeated (multi-)index in an expression is used throughout.

Jet space is the coordinate space $J = (x^i, u^\alpha, u_j^\alpha, \ldots)$. A smooth function $u^\alpha = \phi^\alpha(x) : \mathbb{R}^n \to \mathbb{R}^m$ determines a point in J: at any $x^i = (x_0)^i$; the values $(u_0)^\alpha := \phi^\alpha(x_0)$ and the derivative values $(u_0)_J^\alpha := \partial_{j_1} \cdots \partial_{j_N} \phi^\alpha(x_0)$ for all orders $N \geq 1$ give a map,

$$u^\alpha = \phi^\alpha(x) \xrightarrow{x_0} ((x_0)^i, (u_0)^\alpha, (u_0)_j^\alpha, \ldots) \in J. \tag{1}$$

In jet space, the primitive geometric objects consist of partial derivatives ∂_{x^i}, $\partial_{u_J^\alpha}$, and differentials dx^i, du_J^α. They are related by duality (hooking) relations:

$$\partial_{x^i} \rfloor dx^j = \delta_i^j, \qquad (2)$$

$$\partial_{u_I^\alpha} \rfloor du_J^\beta = \delta_\alpha^\beta \delta_J^I. \qquad (3)$$

It will be useful to also introduce the geometric contact one-forms:

$$\Theta_I^\alpha = du_I^\alpha - u_{Ii}^\alpha dx^i. \qquad (4)$$

Under the evaluation map (1), the pull back of a contact one-form vanishes.

Total derivatives are given by $D_i = \partial_{x^i} + u_{ij}^\alpha \partial_{u_J^\alpha}$, which corresponds to the chain rule under the evaluation map (1). Higher total derivatives are defined by $D_J = D_{j_1} \cdots D_{j_N}$, $J = \{j_1, \ldots, j_N\}$, $|J| = N$. For $J = \emptyset$, $D_\emptyset = \mathrm{id}$ is the identity operator, where $|\emptyset| = 0$. In particular, $D_J u^\alpha = u_J^\alpha$, and $D_J du^\alpha = du_J^\alpha$.

A differential function is a function $f(x, u^{(k)})$ defined on a finite jet space $J^{(k)} = (x^i, u^\alpha, u_j^\alpha, \ldots, u_{j_1 \cdots j_k}^\alpha)$ of order $k \geq 0$. The Frechet derivative of a differential function f is given by

$$f' = f_{u_I^\alpha} D_I \qquad (5)$$

which acts on (differential) functions F^α. The adjoint-Frechet derivative of a differential function f is given by

$$(f'^*)_\alpha = (-1)^{|I|} D_I f_{u_I^\alpha} \qquad (6)$$

which acts on (differential) functions F, where the right-hand side is viewed as a composition of operators.

The Frechet second-derivative is given by

$$f''(F_1, F_2) = f_{u_I^\alpha u_J^\beta}(D_I F_1^\alpha)(D_J F_2^\beta). \qquad (7)$$

This expression is symmetric in the pair of functions (F_1^α, F_2^α).

The commutator of two differential functions f_1 and f_2 is given by $[f_1, f_2] = f_2'(f_1) - f_1'(f_2)$.

The Euler operator (variational derivative) is given by

$$E_{u^\alpha} = (-1)^{|I|} D_I \partial_{u_I^\alpha}. \qquad (8)$$

It characterizes total divergence expressions: $E_{u^\alpha}(f) = 0$ holds identically iff $f = D_i F^i$ for some differential vector function $F^i(x, u^{(k)})$. The product rule takes the form:

$$E_{u^\alpha}(f_1 f_2) = f_1'^*(f_2)_\alpha + f_2'^*(f_1)_\alpha. \qquad (9)$$

The higher Euler operators

$$E_{u^\alpha}^I = \binom{I}{J}(-1)^{|J|} D_J \partial_{u_{IJ}^\alpha} \qquad (10)$$

characterize higher order total derivative expressions: $E_{u^\alpha}^I(f) = 0$ holds identically iff $f = D_{i_1} \cdots D_{i_{|I|}} F^{i_1 \cdots i_{|I|}}$ for some differential tensor function $F^{i_1 \cdots i_{|I|}}(x, u^{(k)})$.

The Frechet derivative is related to the Euler operator by:

$$f'(F) = F^\alpha E_{u^\alpha}(f) + D_i \Gamma^i(F; f), \quad \Gamma^i(F; f) = (D_J F^\alpha) E_{u_{iJ}^\alpha}(f). \qquad (11)$$

The Frechet derivative and its adjoint are related by

$$F_2 f'(F_1) - F_1^\alpha f'^*(F_2)_\alpha = D_i \Psi^i(F_1, F_2; f), \quad \Psi^i(F_1, F_2; f) = (D_K F_2)(D_J F_1^\alpha) E_{u_{ij}^\alpha}^K(f). \tag{12}$$

Evolutionary Vector Fields and One-Form Fields

A vector field in jet space is defined as the geometric object,

$$P^i \partial_{x^i} + P_I^\alpha \partial_{u_I^\alpha} \tag{13}$$

whose components are differential functions. Similarly, a one-form field in jet space is defined as the geometric object,

$$Q_i dx^i + Q_\alpha^I du_I^\alpha \tag{14}$$

whose components are differential functions. Total derivatives $D_i = \partial_{x^i} + u_{iI}^\alpha \partial_{u_I^\alpha}$ represent trivial vector fields that annihilate contact one-forms: $D_i \lrcorner \Theta_J^\alpha = 0$.

Geometric counterparts of partial derivatives $\partial_{u_J^\alpha}$ are evolutionary (vertical) differentials du_J^α, where d is the evolutionary version of d: $d^2 = 0$, $dx^i = 0$. They satisfy the duality (hooking) relation:

$$\partial_{u_I^\alpha} \lrcorner du_J^\beta = \delta_\alpha^\beta \delta_J^I. \tag{15}$$

An evolutionary (vertical) vector field is the geometric object

$$P_I^\alpha \partial_{u_I^\alpha} \tag{16}$$

whose components are differential functions. Every vector field $\mathbf{X} = P^i \partial_{x^i} + P_I^\alpha \partial_{u_I^\alpha}$ has a unique evolutionary form $\hat{\mathbf{X}} = \mathbf{X} - P^i D_i = \hat{P}_I^\alpha \partial_{u_I^\alpha}$ given by the components $\hat{P}_I^\alpha = P_I^\alpha - P^i u_{iI}^\alpha$. Its dual counterpart is an evolutionary (vertical) one-form field,

$$Q_\alpha^I du_I^\alpha \tag{17}$$

whose components are differential functions.

For later developments, it will be useful to define the functional pairing relation,

$$\langle P_I^\alpha \partial_{u_I^\alpha}, Q_\alpha^I du_I^\alpha \rangle = \int P_I^\alpha Q_\alpha^I \, dx \tag{18}$$

between evolutionary vector fields and evolutionary one-form fields. In the local form, this pairing is given by the expression:

$$P_I^\alpha Q_\alpha^I \text{ mod total } D. \tag{19}$$

Two evolutionary one-forms will be considered functionally equivalent iff their pairings with an arbitrary evolutionary vector field agree,

$$\langle P_I^\alpha \partial_{u_I^\alpha}, Q_{1\beta}^J du_J^\beta \rangle = \langle P_I^\alpha \partial_{u_I^\alpha}, Q_{2\beta}^J du_J^\beta \rangle, \tag{20}$$

or in the local form,

$$P_I^\alpha (Q_{1\alpha}^I - Q_{2\alpha}^I) = 0 \text{ mod total } D. \tag{21}$$

The functional equivalence of one-forms is closely related to the notion of functional one-forms in the variational bi-complex. See [3] for details.

3. Geometric Formulation of Symmetries and Adjoint-Symmetries

Consider a general PDE system of order N consisting of M equations,

$$G^A(x, u^{(N)}) = 0, \quad A = 1, \ldots, M \tag{22}$$

where x^i, $i = 1, \ldots, n$, are the independent variables and u^α, $\alpha = 1, \ldots, m$, are the dependent variables. The space of formal solutions $u^\alpha(x)$ of the PDE system will be denoted $\mathcal{E}$.

There are many equivalent starting points for the formulation of infinitesimal symmetries. For the present purpose, the most useful one is given by evolutionary vector fields and utilizes only the Frechet derivative. A symmetry is a vector field,

$$\mathbf{X}_P = P^\alpha(x, u^{(k)}) \partial_{u^\alpha} \tag{23}$$

whose component functions $P^\alpha(x, u^{(k)})$ are non-singular on $\mathcal{E}$ and satisfy the linearization of the PDE system on $\mathcal{E}$,

$$(\mathrm{pr}\mathbf{X}_P G^A)|_\mathcal{E} = G'(P)^A|_\mathcal{E} = 0. \tag{24}$$

This is the *symmetry determining equation*, and the functions P^α are called the characteristic of the symmetry.

In this setting, an *adjoint-symmetry* consists of functions $Q_A(x, u^{(l)})$ that are non-singular on $\mathcal{E}$ and that satisfy the adjoint linearization of the PDE system on $\mathcal{E}$,

$$G'^*(Q)_\alpha|_\mathcal{E} = 0. \tag{25}$$

This is the *adjoint-symmetry determining equation*.

In particular, the two determining equations (24) and (25) are formal adjoints of each other. They coincide only in two cases: either $G' = G'^*$, which is the necessary and sufficient condition for a PDE system to be a Euler–Lagrange equation (namely, possess a variational structure) [1,3,11]; or $G' = -G'^*$, which is the necessary and sufficient condition for a PDE system to be a linear, constant-coefficient system of odd order [10].

Since P^α has the geometrical status as the components of the vector field (23), a natural question is whether Q_A has any status given by the components of some other geometrical object [11,12].

It will be useful to work with a coordinate-free description of the PDE system (22) in jet space. Such a system of equations $(G^1(x, u^{(N)}), \ldots, G^M(x, u^{(N)})) = 0$ describes a set of M surfaces in the finite space $J^{(N)}(x, u, \partial u, \ldots, \partial^N u)$. Total derivatives of these equations, $(D_I G^1(x, u^{(N)}), \ldots, D_I G^M(x, u^{(N)})) = 0$, correspondingly describe sets of surfaces in the higher derivative finite spaces $J^{(N+|I|)}(x, u, \partial u, \ldots, \partial^{N+|I|} u)$. Altogether, the set comprised by the equations and the derivative equations for all orders $|I| \geq 0$ corresponds to an infinite set of surfaces in jet space, which can be identified with the solution space $\mathcal{E}$.

As is well known, symmetry vector fields geometrically describe tangent vector fields with respect to $\mathcal{E}$. To see this explicitly, first consider the identities:

$$dG^A = (G^A)_{u^\alpha_I} du^\alpha_I, \tag{26}$$

$$G'(P)^A = \mathrm{pr}\mathbf{X}_P G^A = \mathrm{pr}\mathbf{X}_P \,\lrcorner\, dG^A. \tag{27}$$

Now, observe that dG^A is the normal one-form to the surfaces $G^A = 0$. The symmetry determining equation (24) then shows that the prolonged vector field $\mathrm{pr}\mathbf{X}_P$ is annihilated by the normal one-form and hence is tangent to these surfaces iff $\mathbf{X}_P$ is a symmetry of the PDE system.

This normal one-form (26) provides a natural way to associate a one-form to an adjoint-symmetry via:

$$\omega_Q = Q_A(x, u^{(l)}) dG^A. \tag{28}$$

A functionally equivalent one-form is obtained through integration by parts:

$$Q_A dG^A = Q_A(G^A)'(du) = G'^*(Q)_\alpha du^\alpha \text{ mod total } D. \tag{29}$$

Evaluating this one-form on the solution space $\mathcal{E}$ then gives

$$\omega_Q|_\mathcal{E} = 0 \text{ mod total } D. \tag{30}$$

Thus, a one-form ω_Q functionally vanishes on the surfaces $\mathcal{E}$ iff its components Q_A are an adjoint-symmetry.

This establishes a main geometrical result.

Theorem 1. *Adjoint-symmetries describe evolutionary one-forms $Q_A dG^A$ that functionally vanish on the solution space $\mathcal{E}$ of a PDE system (22).*

These developments have used evolutionary (vertical) vector fields and evolutionary one-forms. It is straightforward to reformulate everything in terms of full vector fields and full one-forms.

First, consider the normal one-form

$$\begin{aligned} dG^A &= (G^A)_{x^i} dx^i + (G^A)'(du) \\ &= (G^A)'(\Theta) + ((G^A)_{x^i} + (G^A)'(u_i)) dx^i \\ &= (G^A)'(\Theta) + D_i G^A dx^i \end{aligned} \tag{31}$$

which yields the relation

$$dG^A|_\mathcal{E} = (G^A)'(\Theta)|_\mathcal{E}. \tag{32}$$

Then, observe:

$$\begin{aligned} Q_A dG^A|_\mathcal{E} &= Q_A(G^A)'(\Theta)|_\mathcal{E} \\ &= (G^A)'^*(Q_A)_\alpha|_\mathcal{E} \Theta^\alpha \text{ mod total } D. \end{aligned} \tag{33}$$

As a consequence, $Q_A dG^A|_\mathcal{E}$ vanishes mod total D iff Q_A satisfies the adjoint-symmetry determining Equation (25). Moreover, the determining equation itself can be expressed directly in terms of the one-form $Q_A dG^A|_\mathcal{E}$ by $E_{\Theta^\alpha}(Q_A dG^A)|_\mathcal{E} = (G^A)'^*(Q_A)|_\mathcal{E} = 0$.

Proposition 1. *The adjoint-symmetry determining Equation (25) can be expressed geometrically as:*

$$E_{\Theta^\alpha}(Q_A dG^A)|_\mathcal{E} = 0. \tag{34}$$

Examples of Adjoint-Symmetries

To illustrate the results, some examples of PDEs that possess non-trivial adjoint-symmetries will be given.

The Korteweg–de Vries (KdV) equation

$$u_t + u u_x + u_{xxx} = 0 \tag{35}$$

for shallow water waves is an example of an evolutionary wave equation. Its symmetries $\mathbf{X} = P \partial_u$ are the solutions of the determining equation

$$G'(P)|_\mathcal{E} = (D_t P + D_x(uP) + D_x^3 P)|_\mathcal{E} = 0, \tag{36}$$

with $G' = D_t + D_x u + D_x^3$ being the Frechet derivative of the KdV equation, where P is a non-singular function of t, x, u, and derivatives of u on the space of KdV solutions $\mathcal{E}$. The determining equation for adjoint-symmetries $\omega = QG'(du)$ is the adjoint equation

$$G'^{*}(Q)|_{\mathcal{E}} = (-D_t Q - u D_x Q - D_x^3 Q)|_{\mathcal{E}} = 0, \quad (37)$$

where Q is a non-singular function of t, x, u, and derivatives of u on $\mathcal{E}$.

KdV adjoint-symmetries up to first-order $Q(t, x, u, u_t, u_x)$ are given by [9] the span of

$$Q^{(1)} = 1, \quad Q^{(2)} = u, \quad Q^{(3)} = tu - x. \quad (38)$$

The first two are part of a hierarchy of higher order adjoint-symmetries generated by a recursion operator $\mathcal{R} = D_x^2 + \frac{1}{3}u + \frac{1}{3}D_x^{-1}uD_x$ applied to $Q = u$. The third one along with all of the ones in the hierarchy are related to symmetries of the KdV equation through the Hamiltonian operator $\mathcal{H} = D_x$. If a linear combination of the lowest order adjoint-symmetries is used like an invariant surface condition, $c_1 + c_2(tu - x) + c_3 u = 0$, then this yields $u = (c_2 x - c_1)/(c_2 t + c_3)$, which is a similarity solution of the KdV equation.

An example of a non-evolutionary equation is,

$$\Delta \phi_t + \phi_x \Delta \phi_y - \phi_y \Delta \phi_x = 0 \quad (39)$$

which governs the vorticity $\Omega = \Delta \phi$ for incompressible inviscid fluid flow in two spatial dimensions, where the fluid velocity has the components $\vec{v} = (-\phi_y, \phi_x)$. The symmetries $\mathbf{X} = P\partial_\phi$ of this equation are the solutions of the determining equation,

$$G'(P)|_{\mathcal{E}} = (D_t \Delta P + \phi_x D_y \Delta P + \Delta \phi_y D_x P - \phi_y D_x \Delta P - \Delta \phi_x D_y P)|_{\mathcal{E}} = 0, \quad (40)$$

where P is a non-singular function of t, x, y, ϕ, and derivatives of ϕ on the space of vorticity solutions $\mathcal{E}$, with $G' = D_t \Delta + \phi_x D_y \Delta + \Delta \phi_y D_x - \phi_y D_x \Delta - \Delta \phi_x D_y$ being the Frechet derivative of the vorticity equation given in terms of the total Laplacian operator $\Delta = D_x^2 + D_y^2$. The determining equation for adjoint-symmetries $\omega = QG'(d\phi)$ is the adjoint equation,

$$G'^{*}(Q)|_{\mathcal{E}} = -(D_t \Delta Q + D_y \Delta(\phi_x Q) + D_x(\Delta \phi_y Q) - D_x \Delta(\phi_y Q) - D_y(\Delta \phi_x Q))|_{\mathcal{E}} = 0, \quad (41)$$

where Q is a non-singular function of t, x, y, ϕ, and derivatives of ϕ on $\mathcal{E}$.

The first-order adjoint-symmetries $Q(t, x, y, \phi, \phi_t, \phi_x, \phi_y)$ are given by [13] the span of,

$$Q^{(1)} = x^2 + y^2, \quad Q^{(2)} = \phi, \quad Q^{(3)} = f(t), \quad Q^{(4)} = xf(t), \quad Q^{(5)} = yf(t), \quad (42)$$

where $f(t)$ is an arbitrary smooth function. If a linear combination of these adjoint-symmetries is used like an invariant surface condition, $c_1(x^2 + y^2) + c_2 \phi + c_3 f(t) + c_4 x f(t) + c_5 y f(t) = 0$, then taking $c_2 = -1$ gives $\phi = c_1(x^2 + y^2) + (c_3 + c_4 x + c_5 y)f(t)$, which is a constant vorticity solution, with $\Omega = 2c_1$ and $\vec{v} = (-2c_1 y + c_5 f(t), 2c_1 x + c_4 f(t))$.

Maxwell's equations in free space are an example of an evolution system with spatial constraints:

$$\vec{E}_t - \nabla \times \vec{B} = 0, \quad \vec{B}_t + \nabla \times \vec{E} = 0, \quad \nabla \cdot \vec{E} = \nabla \cdot \vec{B} = 0 \quad (43)$$

(in relativistic units with the speed of light set to one). The symmetries $\mathbf{X} = \vec{P}^E \cdot \partial_{\vec{E}} + \vec{P}^B \cdot \partial_{\vec{B}}$ of this system are the solutions of the determining equations

$$G'\begin{pmatrix} \vec{P}^E \\ \vec{P}^B \end{pmatrix}\bigg|_{\mathcal{E}} = \begin{pmatrix} (D_t \vec{P}^E - \nabla \times \vec{P}^B)|_{\mathcal{E}} \\ (D_t \vec{P}^B + \nabla \times \vec{P}^E)|_{\mathcal{E}} \\ (\nabla \cdot \vec{P}^E)|_{\mathcal{E}} \\ (\nabla \cdot \vec{P}^B)|_{\mathcal{E}} \end{pmatrix} = 0, \quad (44)$$

where $\vec{P}^E$ and $\vec{P}^B$ are non-singular vector functions of $t, x, y, z, \vec{E}, \vec{B}$, and derivatives of $\vec{E}, \vec{B}$ on the space of Maxwell solutions $\mathcal{E}$, with $G' = \begin{pmatrix} D_t & -\nabla\times \\ \nabla\times & D_t \\ \nabla\cdot & 0 \\ 0 & \nabla\cdot \end{pmatrix}$ being the Frechet derivative of the system in terms of the total derivative operator $\nabla = (D_x, D_y, D_z)$. The determining equation for adjoint-symmetries $\omega = \begin{pmatrix} \vec{Q}^E & \vec{Q}^B & Q^E & Q^B \end{pmatrix} G' \begin{pmatrix} d\vec{E} \\ d\vec{B} \end{pmatrix}$ is the adjoint equation

$$G'^* \begin{pmatrix} \vec{Q}^E & \vec{Q}^B & Q^E & Q^B \end{pmatrix}\Big|_{\mathcal{E}} = \begin{pmatrix} (-D_t\vec{Q}^E + \nabla\times\vec{Q}^B - \nabla Q^E)|_{\mathcal{E}} \\ (-D_t\vec{Q}^B - \nabla\times\vec{Q}^E - \nabla Q^B)|_{\mathcal{E}} \end{pmatrix} = 0, \tag{45}$$

where the vectors $\vec{Q}^E, \vec{Q}^B$, and the scalars Q^E, Q^B, are non-singular functions of $t, x, y, z, \vec{E}, \vec{B}$, and derivatives of $\vec{E}, \vec{B}$ on $\mathcal{E}$. Note that the adjoint $*$ here includes a matrix transpose applied to the row matrix comprising the adjoint-symmetry vector and scalar functions.

Because Maxwell's equations are a linear system and contain constraints, it possesses three types of adjoint-symmetries [15,16]: elementary adjoint-symmetries such that $\vec{Q}^E, \vec{Q}^B, Q^E, Q^B$ are functions only of t, x, y, z; gauge adjoint-symmetries given by $\vec{Q}^E = \nabla\chi^E, \vec{Q}^B = \nabla\chi^B, Q^E = -D_t\chi^E, Q^B = -D_t\chi^B$ in terms of scalars χ^E and χ^B that are arbitrary non-singular functions of $t, x, y, z, \vec{E}, \vec{B}$, and derivatives of $\vec{E}, \vec{B}$ on $\mathcal{E}$; and a hierarchy of linear adjoint-symmetries. The linear adjoint-symmetries of zeroth order are given by the span of

$$\vec{Q}^E = \vec{\xi}\times\vec{B} + \zeta\vec{E}, \quad \vec{Q}^B = -\vec{\xi}\times\vec{E} + \zeta\vec{B}, \quad Q^E = \vec{\xi}\cdot\vec{E}, \quad Q^B = \vec{\xi}\cdot\vec{B} \tag{46}$$

and

$$\vec{Q}^E = \vec{\xi}\times\vec{E} - \zeta\vec{B}, \quad \vec{Q}^B = \vec{\xi}\times\vec{B} + \zeta\vec{E}, \quad Q^E = -\vec{\xi}\cdot\vec{B}, \quad Q^B = \vec{\xi}\cdot\vec{E} \tag{47}$$

where

$$\begin{aligned} \vec{\xi} &= \vec{a}_0 + \vec{a}_1\times\vec{x} + \vec{a}_2 t + a_3\vec{x} + a_4 t\vec{x} + (\vec{a}_5\cdot\vec{x})\vec{x} - \tfrac{1}{2}\vec{a}_5(\vec{x}\cdot\vec{x} + t^2), \\ \zeta &= a_0 + \vec{a}_2\cdot\vec{x} + a_3 t + \tfrac{1}{2}a_4(\vec{x}\cdot\vec{x} + t^2) + (\vec{a}_5\cdot\vec{x})t, \end{aligned} \tag{48}$$

in terms of arbitrary constant scalars a_0, a_3, a_4 and arbitrary constant vectors $\vec{a}_0, \vec{a}_1, \vec{a}_2, \vec{a}_5$, with $\vec{x} = (x, y, z)$. The pair $(\vec{\xi}, \zeta)$ represents a conformal Killing vector in Minkowski space $\mathbb{R}^{3,1}$.

These two zeroth-order adjoint-symmetries are related by the duality symmetry $(\vec{E}, \vec{B}) \to (\vec{B}, -\vec{E})$. The linear first-order adjoint-symmetries are more complicated and involve conformal Killing–Yano tensors. All higher order adjoint-symmetries can be obtained from the zeroth and first-order adjoint-symmetries by taking Lie derivatives with respect to conformal Killing vectors. Their explicit description can be found in [15,16]. An unexplored question is whether the lowest order adjoint-symmetries can be used like an invariant surface condition to produce solutions of Maxwell's equations.

4. Some Applications

Two geometrical applications of Theorem 1 will be presented. The first application is a geometrical derivation of a well-known formula that generates a conservation law from a pair consisting of a symmetry and an adjoint-symmetry. This derivation will use the functional pairing (18). The second application is a geometrical derivation of three actions of symmetries on adjoint-symmetries. These symmetry actions have been obtained in recent work using an algebraic point of view [13]. They will be shown here to arise from Cartan's formula for the Lie derivative of an adjoint-symmetry one-form (28).

It will be useful to work with the determining equations for symmetries and adjoint-symmetries off of the solution space $\mathcal{E}$ of a given PDE system (22). More precisely, the determining equations will be expressed in the full jet space containing $\mathcal{E}$.

Remark 1. *A PDE system (22) will be assumed to be regular [11], so that Hadamard's lemma holds: a differential function f satisfies $f|_\mathcal{E} = 0$ iff $f = R_f(G)$, where R_f is a linear differential operator whose coefficients are non-singular on $\mathcal{E}$.*

Consequently, for symmetries, $G'(P)^A|_\mathcal{E} = 0$ holds iff

$$G'(P)^A = R_P(G)^A, \qquad (49)$$

and likewise for adjoint-symmetries, $G'^*(Q)_\alpha|_\mathcal{E} = 0$ holds iff

$$G'^*(Q)_\alpha = R_Q(G)_\alpha, \qquad (50)$$

where R_P and R_Q are linear differential operators whose coefficients are non-singular on $\mathcal{E}$.

4.1. Conservation Laws from Symmetries and Adjoint-Symmetries

The functional pairing (18) between a symmetry vector field (23) and an adjoint-symmetry one-form (28) is given by,

$$\langle \mathrm{pr}\mathbf{X}_P, \omega_Q \rangle = \langle \mathrm{pr} P^\alpha \partial_{u^\alpha}, Q_A \mathrm{d} G^A \rangle = \int Q_A G'(P)^A \, dx \qquad (51)$$

from identity (27). This pairing in local form (19) is the expression,

$$Q_A G'(P)^A \text{ mod total } D. \qquad (52)$$

There are two different ways to evaluate it.

First, since $\mathbf{X}_P$ is a symmetry, $Q_A G'(P)^A = Q_A R_P(G)^A$. Second, since ω_Q is an adjoint-symmetry, $Q_A G'(P)^A = G'^*(Q)_\alpha P^\alpha + D_i \Psi^i(P,Q)_G = P^\alpha R_Q(G)_\alpha + D_i \Psi^i(P,Q;G)$, where

$$\Psi^i(P,Q;G) = (D_K Q_A)(D_J P^\alpha) E^K_{u^\alpha_{ij}}(G^A). \qquad (53)$$

Hence, on $\mathcal{E}$, $Q_A G'(P)^A|_\mathcal{E} = D_i \Psi^i(P,Q)_G|_\mathcal{E} = 0$, which is equivalent to $\langle \mathrm{pr}\mathbf{X}_P, \omega_Q \rangle|_\mathcal{E} = 0$. This establishes the following conservation law.

Theorem 2. *Vanishing of the functional pairing (51) for any symmetry (23) and any adjoint-symmetry (28) corresponds to a conservation law*

$$D_i \Psi^i(P,Q;G)|_\mathcal{E} = 0 \qquad (54)$$

holding for the PDE system $G^A = 0$, where the conserved current $\Psi^i(P,Q;G)$ is given by expression (53).

4.2. Action of symmetries on adjoint-symmetries

For any PDE system (22), its set of adjoint-symmetries is a linear space, and as shown in [13], symmetries of the PDE system have three different actions on this space.

The primary symmetry action can be derived from the Lie derivative of an adjoint-symmetry one-form with respect to a symmetry vector field.

Proposition 2. *If ω_Q is an adjoint-symmetry one-form (28), namely $\omega_Q|_\mathcal{E} = 0$ (mod total D), then its Lie derivative with respect to any symmetry vector $\mathbf{X}_P = P^\alpha \partial_{u^\alpha}$ yields an adjoint-symmetry one-form,*

$$\mathcal{L}_{\mathbf{X}_P}\omega_Q|_{\mathcal{E}} = \omega_{S_P(Q)}|_{\mathcal{E}} = 0 \;(\text{mod total } D) \tag{55}$$

where

$$S_P(Q)_A = Q'(P)_A + R_P^*(Q)_A \tag{56}$$

are its components.

Here and throughout, R_P and R_Q are the linear differential operators determined by Equations (49) and (50). The adjoints of these operators are denoted R_P^* and R_Q^*.

Proof. Recall that the Lie derivative has the following properties: it acts as a derivation; it commutes with the differential d; it reduces to the Frechet derivative when acting on a differential function.

By the use of these properties,

$$\begin{aligned}\mathcal{L}_{\mathbf{X}_P}\omega_Q &= \mathcal{L}_{\mathbf{X}_P}(Q_A dG^A) \\ &= (\mathcal{L}_{\mathbf{X}_P} Q_A) dG^A + Q_A \mathcal{L}_{\mathbf{X}_P}(dG^A) \\ &= Q'(P)_A dG^A + Q_A d(G'(P)^A) \\ &= Q'(P)_A dG^A + Q_A d(R_P(G)^A).\end{aligned} \tag{57}$$

The last term can be simplified on $\mathcal{E}$: $Q_A d(R_P(G)^A)|_{\mathcal{E}} = Q_A R_P(dG)^A|_{\mathcal{E}} = R_P^*(Q)_A dG^A \;(\text{mod total } D)$. This yields

$$\mathcal{L}_{\mathbf{X}_P}\omega_Q|_{\mathcal{E}} = ((Q'(P)_A + R_P^*(Q)_A) dG^A)|_{\mathcal{E}} \;(\text{mod total } D), \tag{58}$$

completing the derivation. □

There is an elegant formula, due to Cartan, for the Lie derivative in terms of the operations d and $\lrcorner$. This formula gives rise to two additional symmetry actions.

Theorem 3. *The terms in Cartan's formula*

$$\mathcal{L}_{\mathbf{X}_P}\omega_Q = d(\text{pr}\mathbf{X}_P \lrcorner\, \omega_Q) + \text{pr}\mathbf{X}_P \lrcorner\, (d\omega_Q) \tag{59}$$

evaluated on $\mathcal{E}$ each yield an action of symmetries on adjoint symmetries. The action produced by the Lie derivative term has the components (56), and the actions produced by the differential term and the hook term respectively have the components

$$S_{1P}(Q) = R_P^*(Q)_A - R_Q^*(P)_A, \tag{60}$$
$$S_{2P}(Q) = Q'(P)_A + R_Q^*(P)_A. \tag{61}$$

Proof. Consider the first term on right-hand side in the formula (59). It can be evaluated in two different ways. Firstly, $\text{pr}\mathbf{X}_P \lrcorner (Q_A dG^A) = Q_A G'(P)^A = Q_A R_P(G)^A$ yields

$$d(\text{pr}\mathbf{X}_P \lrcorner (Q_A dG^A))|_{\mathcal{E}} = d(Q_A R_P(G)^A)|_{\mathcal{E}} = (Q_A R_P(dG^A))|_{\mathcal{E}} = (R_P^*(Q)_A dG^A)|_{\mathcal{E}}. \tag{62}$$

Secondly, $Q_A dG^A = R_Q(G)_\alpha \Theta^\alpha + Q_A(D_i G^A) dx^i \;(\text{mod total } D)$ gives $\text{pr}\mathbf{X}_P \lrcorner (Q_A dG^A) = \text{pr}\mathbf{X}_P \lrcorner (R_Q(G)_\alpha \Theta^\alpha + Q_A(D_i G^A) dx^i \;(\text{mod total } D)) = R_Q(G)_\alpha P^\alpha \;(\text{mod total } D)$. This yields

$$\begin{aligned}d(\text{pr}\mathbf{X}_P \lrcorner (Q_A dG^A))|_{\mathcal{E}} &= d(R_Q(G)_\alpha P^\alpha \;(\text{mod total } D))|_{\mathcal{E}} \\ &= (R_Q(dG)_\alpha P^\alpha \;(\text{mod total } D))|_{\mathcal{E}} \\ &= (R_Q^*(P)_A dG^A \;(\text{mod total } D))|_{\mathcal{E}}.\end{aligned} \tag{63}$$

Then, equating expressions (62) and (63) leads to the result:

$$((R_P^*(Q)_A - R_Q^*(P)_A)dG^A)|_{\mathcal{E}} = 0 \;(\text{mod total } D)|_{\mathcal{E}}. \tag{64}$$

This equation shows that the symmetry action (60) produces an adjoint-symmetry.

Now, consider the second term on the right-hand side in formula (59). Similarly to the first term, it can be evaluated in two different ways. Firstly, $d\omega_Q = dQ_A \wedge dG^A$ yields

$$\text{pr}\mathbf{X}_P \rfloor (dQ_A \wedge dG^A) = Q'(P)_A dG^A - G'(P)^A dQ_A = Q'(P)_A dG^A - R_P(G)^A dQ_A. \tag{65}$$

Hence, on $\mathcal{E}$,

$$(\text{pr}\mathbf{X}_P \rfloor (dQ_A \wedge dG^A))|_{\mathcal{E}} = (Q'(P)_A dG^A)|_{\mathcal{E}}. \tag{66}$$

Secondly, $d\omega_Q = d(R_Q(G)_\alpha \Theta^\alpha + Q_A(D_i G^A) dx^i)$ (mod total D) gives

$$d\omega_Q|_{\mathcal{E}} = (R_Q(dG)_\alpha \wedge \Theta^\alpha + Q_A(D_i dG^A) \wedge dx^i)|_{\mathcal{E}} \;(\text{mod total } D). \tag{67}$$

This yields

$$\begin{aligned}
(\text{pr}\mathbf{X}_P \rfloor & (R_Q(dG)_\alpha \wedge \Theta^\alpha + Q_A(D_i dG^A) \wedge dx^i))|_{\mathcal{E}} \\
&= (R_Q(G'(P))_\alpha \Theta^\alpha - P^\alpha R_Q(dG)_\alpha + Q_A(D_i G'(P)^A) dx^i)|_{\mathcal{E}} \\
&= -(R_Q^*(P)_A dG^A)|_{\mathcal{E}} \;(\text{mod total } D).
\end{aligned} \tag{68}$$

Equating expressions (66) and (68) then gives the equation

$$((Q'(P)_A + R_Q^*(P)_A)dG^A)|_{\mathcal{E}} = 0 \;(\text{mod total } D)|_{\mathcal{E}}, \tag{69}$$

showing that the symmetry action (61) produces an adjoint-symmetry. □

Observe that the three actions (56), (60) and (61) are related by:

$$S_{1P}(Q) + S_{2P}(Q) = S_P(Q). \tag{70}$$

Each action is a mapping on the linear space of adjoint-symmetries Q_A. The algebraic properties of these actions can be found in [13].

5. Geometrical Adjoint-Symmetries of Evolution Equations

A general system of evolution equations of order N has the form

$$u_t^\alpha = g^\alpha(x, u, \partial_x u, \ldots, \partial_x^N u) \tag{71}$$

where t is the time variable, x^i, $i = 1, \ldots, n$, are now the space variables, and u^α, $\alpha = 1, \ldots, m$, are the dependent variables. The space of formal solutions $u^\alpha(t, x)$ of the system will be denoted $\mathcal{E}$.

The developments for general PDE systems can be specialized to evolution systems, with $G^\alpha = u_t^\alpha - g^\alpha$ via identifying the indices $A = \alpha$ ($M = m$). On $\mathcal{E}$, since u_t^α can be eliminated through the evolution equations, the components of symmetries and adjoint-symmetries can be assumed to contain only u^α and its spatial derivatives in addition to t and x^i. Hereafter, multi-indices will refer to spatial derivatives.

A symmetry is thereby an evolutionary vector field,

$$\mathbf{X}_P = P^\alpha(t, x, \partial_x u, \ldots, \partial_x^k u) \partial_{u^\alpha} \tag{72}$$

satisfying the linearization of the evolution system on $\mathcal{E}$:

$$(\text{pr}\mathbf{X}_P(u_t^\alpha - g^\alpha))|_{\mathcal{E}} = (D_t P^\alpha - g'(P)^\alpha)|_{\mathcal{E}} = 0. \tag{73}$$

Off of $\mathcal{E}$, $D_t P^\alpha = (P_t + P'(g))^\alpha + P'(G)^\alpha$, whereby $R_P = P'$. Consequently, the symmetry determining equation (73) can be expressed simply as:

$$(P_t + [g, P])^\alpha = 0. \tag{74}$$

The determining equation for adjoint-symmetries $Q_\alpha(t, x, \partial_x u, \ldots, \partial_x^l u)$ is given by the adjoint linearization of the evolution system on $\mathcal{E}$:

$$(-D_t Q - g'^*(Q))_\alpha|_\mathcal{E} = 0. \tag{75}$$

Similar to the symmetry case, here, $R_Q = -Q'$ off of $\mathcal{E}$, and the adjoint-symmetry determining equation simply becomes

$$(Q_t + Q'(g) + g'^*(Q))_\alpha = 0. \tag{76}$$

These two determining equations have a geometrical formulation given by a Lie derivative defined in terms of a flow arising from the evolution system, similar to the situation for ODEs [8]. Specifically, observe that $D_t u^\alpha|_\mathcal{E} = g^\alpha$, and hence, $D_t f|_\mathcal{E} = f_t + f'(g)$ for any differential function f. This motivates introducing the flow vector field,

$$\mathbf{Y} = \partial_t + g^\alpha \partial_{u^\alpha} \tag{77}$$

which is related to the total time derivative by prolongation,

$$\mathrm{pr}\mathbf{Y} = D_t|_\mathcal{E} = \partial_t + (D_I g^\alpha) \partial_{u_I^\alpha}. \tag{78}$$

Associated with this flow vector field is the Lie derivative

$$\mathcal{L}_t := \mathcal{L}_{\mathrm{pr}\mathbf{Y}} \tag{79}$$

which acts on differential functions by $\mathcal{L}_t f = \mathrm{pr}\mathbf{Y}(f) = D_t f|_\mathcal{E}$. On evolutionary vector fields (72), this Lie derivative acts in the standard way as a commutator:

$$\begin{aligned}
\mathcal{L}_t \mathrm{pr}\mathbf{X}_P &= \mathrm{pr}((\mathrm{pr}\mathbf{Y}(P) - \mathrm{pr}\mathbf{X}_P(g))^\alpha \partial_{u^\alpha}) \\
&= \mathrm{pr}((P_t + P'(g) - g'(P))^\alpha \partial_{u^\alpha}) \\
&= \mathrm{pr}((P_t + [g, P])^\alpha \partial_{u^\alpha}).
\end{aligned} \tag{80}$$

Thus, the symmetry determining equation (74) can be formulated as the vanishing of the Lie derivative expression (80). This establishes the following well-known geometrical result.

Proposition 3. *A symmetry of an evolution system (71) is an evolutionary vector field (72) that is invariant under the associated flow (79).*

In particular, the resulting Lie-derivative vector field

$$\mathcal{L}_t \mathbf{X}_P = (P_t + [g, P])^\alpha \partial_{u^\alpha} \tag{81}$$

vanishes iff the functions P_α are the components of a symmetry.

A similar characterization will now be given for adjoint-symmetries, based on viewing the adjoint relation between the determining equations (74) and (76) as a duality relation between vectors and one-forms.

Introduce the evolutionary one-form:

$$\omega_Q = Q_\alpha(t, x, \partial_x u, \ldots, \partial_x^l u) du^\alpha. \tag{82}$$

Its Lie derivative is given by

$$\begin{aligned}
\mathcal{L}_t \omega_Q &= (\mathcal{L}_t Q_\alpha) du^\alpha + Q_\alpha \mathcal{L}_t(du^\alpha) \\
&= (Q_t + Q'(g))_\alpha du^\alpha + Q_\alpha d(\mathcal{L}_t u^\alpha) \\
&= (Q_t + Q'(g))_\alpha du^\alpha + Q_\alpha dg^\alpha \\
&= (Q_t + Q'(g) + g'^*(Q))_\alpha du^\alpha \pmod{\text{total } D}.
\end{aligned} \quad (83)$$

This shows that the adjoint-symmetry determining equation (76) can be formulated as the functional vanishing of the Lie derivative expression (83).

Theorem 4. *An adjoint-symmetry of an evolution system (71) is an evolutionary one-form (82) that is functionally invariant under the associated flow (79).*

In particular, the resulting Lie-derivative one-form

$$\mathcal{L}_t \omega_Q = (Q_t + Q'(g) + g'^*(Q))_\alpha du^\alpha \pmod{\text{total } D} \quad (84)$$

functionally vanishes iff the functions Q_α are the components of an adjoint-symmetry. This one-form (84) is functionally equivalent to the adjoint-symmetry one-form (28) introduced for a general PDE system. To see the relationship in detail, observe that:

$$\begin{aligned}
\omega_Q &= Q_\alpha dG^\alpha = Q_\alpha d(u_t^\alpha - g^\alpha) \\
&= Q_\alpha (D_t(du^\alpha) - g'(du)^\alpha) \\
&= -(D_t Q_\alpha + g'^*(Q)_\alpha) du^\alpha \pmod{\text{total } D} \\
&= -\mathcal{L}_t \omega_Q \pmod{\text{total } D}.
\end{aligned} \quad (85)$$

An interesting question is how to extend this relationship to more general PDE systems.

Evolution Equations with Spatial Constraints

A wide generalization of evolution systems occurring in applied mathematics and mathematical physics is given by systems comprised of evolution equations with spatial constraints. Some notable examples are Maxwell's equations, incompressible fluid equations, magnetohydrodynamical equations, and Einstein's equations.

The constraints in such systems in general consist of spatial equations

$$C^Y(x, u, \partial_x u, \dots, \partial_x^{N'} u) = 0, \quad Y = 1, \dots, M' \quad (86)$$

that are compatible with the evolution equation (71). Compatibility means that the time derivative of the constraints vanishes on the solution space $\mathcal{E}$ of the whole system, $(D_t C^Y)|_\mathcal{E} = 0$. For systems that are regular [11], Hadamard's lemma implies that the system obeys a differential identity,

$$D_t C^Y = C'(G)^Y + \mathcal{D}(C)^Y \quad (87)$$

where $G^\alpha = u_t^\alpha - g^\alpha$ denotes the evolution equation (71), and where $\mathcal{D}$ is a linear differential spatial operator whose coefficients are non-singular on $\mathcal{E}$. Equivalently, the constraints must obey the identity $C'(g)^Y = \mathcal{D}(C)^Y$. A comparison of the differential order of each side of this identity shows that $\mathcal{D}$ is of the same order N as the evolution equations, namely:

$$\mathcal{D} = \sum_{0 \leq |I| \leq N} R_\Lambda^{IY} D_I. \quad (88)$$

The full system consists of $n + M'$ equations $G^\alpha = 0$, $C^Y = 0$. Note that, in the previous notation (22), $(G^\alpha, C^Y) = (G^A)$ with $A = (\alpha, Y)$.

The symmetry determining equation is given by the linearization of the full system on $\mathcal{E}$, which is comprised by the evolution part (73) and the constraint part

$$(\text{pr}\mathbf{X}_P C^Y)|_\mathcal{E} = C'(P)^Y|_\mathcal{E} = 0. \tag{89}$$

Off of $\mathcal{E}$, $C'(P)^Y = R_C(C)^Y$, where R_C is a linear differential spatial operator whose coefficients are non-singular on $\mathcal{E}$. Hence, the determining equations (73) and (89) can be stated as:

$$(P_t + [g, P])^\alpha|_{\mathcal{E}_C} = 0, \quad C'(P)^Y|_{\mathcal{E}_C} = 0 \tag{90}$$

where $\mathcal{E}_C$ denotes the solution space of the spatial constraint equation (86).

The adjoint-symmetry determining equation is given by the adjoint linearization of the full system on $\mathcal{E}$, which comprises evolution terms and additional constraint terms:

$$(-D_t Q - g'^*(Q) + C'^*(q))_\alpha|_\mathcal{E} = 0. \tag{91}$$

Here, the components of an adjoint-symmetry consist of

$$(Q_\alpha(t, x, \partial_x u, \ldots, \partial_x^l u), q_Y(t, x, \partial_x u, \ldots, \partial_x^{l'} u)) \tag{92}$$

with Q_α being associated with the evolution equations as before, while q_Y is associated with the constraint equations. Similar to the symmetry case, the determining equation can be stated as:

$$(Q_t + Q'(g) + g'^*(Q) - C'^*(q))_\alpha|_{\mathcal{E}_C} = 0. \tag{93}$$

These determining equations for symmetries and adjoint-symmetries have a geometrical formulation in terms of a constrained flow (77), generalizing the previous formulation for evolution systems as follows.

Theorem 5. *A symmetry of a constrained evolution system (71) and (86) is an evolutionary vector field (72) that is invariant under the associated constrained flow (79) and that preserves the constraints.*

The proof of this result is simply the observation that, first, the determining Equation (89) corresponds to the constraints being preserved, and second, the Lie derivative of the symmetry vector field (81) along the flow vanishes on the constraint solution space.

Theorem 6. *An adjoint-symmetry of a constrained evolution system (71) and (86) is an evolutionary one-form (82) that is functionally invariant under the associated constrained flow (79), up to a functional multiple of the normal one-form dC^Y arising from the constraints.*

The proof is given by the earlier computation (84) for the Lie derivative of the adjoint-symmetry one-form. This computation shows that the adjoint-symmetry determining Equation (93) now can be expressed as:

$$\mathcal{L}_t \omega_Q|_{\mathcal{E}_C} = (C'^*(q)_\alpha du^\alpha)|_{\mathcal{E}_C} = (q_Y dC^Y)|_{\mathcal{E}_C} \ (\text{mod total } D) \tag{94}$$

where dC^Y is the normal one-form given by the constraints viewed as surfaces in jet space.

The Lie-derivative one-form (94) is functionally equivalent to the adjoint-symmetry one-form (28) introduced for a general PDE system. In the present notation, the full system of evolution and constraint equations (71) and (86) consists of $(G^\alpha, C^Y) = 0$, and the corresponding one-form associated with this system is given by $\omega_{Q,q} = Q_\alpha dG^\alpha + q_Y dC^Y$. Now, using the relation (85), observe that:

$$\omega_{Q,q} = q_Y dC^Y - \mathcal{L}_t \omega_Q \pmod{\text{total } D}. \tag{95}$$

There is a class of adjoint-symmetries arising from the summed product of arbitrary functions $\chi_Y(t, x)$ and the components of the the differential identity (87). This yields, after integration by parts,

$$\begin{aligned} 0 &= \chi_Y(D_t C^Y - C'(G)^Y - \mathcal{D}(C)^Y) \\ &= D_t(\chi_Y C^Y) + D_i \Psi^i(\chi, G; C) - D_i \Phi^i(\chi, C; R) - (D_t \chi + \mathcal{D}^*(\chi))_Y C^Y - C'^*(\chi)_\alpha G^\alpha \end{aligned} \tag{96}$$

where $\Phi^i(\chi, C; R) = \sum_{0 \le |I| \le N-1} (-1)^{|J|} D_J(\chi_Y R^{iIY}_\Lambda) D_{I/J} C^\Lambda$ from expression (88). Hence,

$$D_t(\chi_Y C^Y) + D_i(\Psi^i(\chi, G; C) - \Phi^i(\chi, C; R)) = C'^*(\chi)_\alpha G^\alpha + (D_t \chi + \mathcal{D}^*(\chi))_Y C^Y \tag{97}$$

has the form of a conservation law off $\mathcal{E}$, with $(C'^*(\chi)_\alpha, (D_t \chi + \mathcal{D}^*(\chi))_Y)$ being the multiplier. As is well known, every multiplier for a regular PDE system is an adjoint-symmetry [1,3,11,17,18]. This can be proven here by applying the Euler operator E_{u^α} and using its product rule. Consequently,

$$Q_\alpha = C'^*(\chi)_\alpha, \quad q_Y = (D_t \chi + \mathcal{D}^*(\chi))_Y \tag{98}$$

are components of an adjoint-symmetry, involving the arbitrary functions $\chi_Y(t, x)$. Such adjoint-symmetries are a counterpart of gauge symmetries, and accordingly are called gauge adjoint-symmetries [11].

The corresponding gauge adjoint-symmetry one-form is given by

$$\omega_\chi = C'^*(\chi)_\alpha du^\alpha = \chi_Y dC^Y \pmod{\text{total } D} \tag{99}$$

and satisfies the geometrical relation

$$\mathcal{L}_t \omega_\chi |_{\mathcal{E}_C} = ((D_t \chi + \mathcal{D}^*(\chi))_Y dC^Y)|_{\mathcal{E}_C} \pmod{\text{total } D}. \tag{100}$$

This establishes the following geometrical result.

Theorem 7. *A gauge adjoint-symmetry (98) is functionally equivalent to a normal one-form ω_χ associated with the constraint equation (86). Under the evolution flow, it is mapped into another normal one-form.*

The preceding developments for general systems of evolution equations with spatial constraints have used the classical notion of symmetries and adjoint-symmetries. It would be interesting to extend the formulation and the results by considering a notion of conditional symmetries and corresponding conditional adjoint-symmetries based on the spatial constraints.

Specifically, on the solution space of the full system, consider a symmetry given by an evolutionary vector field (72) that satisfies

$$(P_t + [g, P])^\alpha|_{\mathcal{E}_C} = 0 \tag{101}$$

where $\mathcal{E}_C$ denotes the solution space of the spatial constraint Equation (86). Such conditional symmetries (101) differ from classical symmetries (90) by relaxing the condition that the constraints are preserved. Their natural adjoint counterpart is given by an evolutionary one-form (82) satisfying

$$(Q_t + Q'(g) + g'^*(Q))_\alpha|_{\mathcal{E}_C} = 0. \tag{102}$$

which is the adjoint of the determining Equation (101). Such conditional adjoint-symmetries (102) differ from classical adjoint-symmetries (93) by excluding the terms arising from the spatial constraints.

This notion of conditional symmetries and adjoint-symmetries is more general than the classical notion because the conditional determining equations hold on $\mathcal{E}_C$ instead of the whole jet space.

6. Concluding Remarks

The main results showing how adjoint-symmetries correspond to evolutionary one-forms with certain geometrical properties provides a first step towards giving a fully geometrical interpretation for adjoint-symmetries. In particular, for systems of evolution equations, adjoint-symmetries can be geometrically described as one-forms that are invariant under the flow generated by the system on the solution space. This interesting result has a straightforward generalization to systems of evolution equations with spatial constraints. Consequently, the results presented here are applicable to all PDE systems of interest in applied mathematics and mathematical physics.

One direction for future work will be to translate and generalize these results into the abstract geometrical setting of secondary calculus [2,19] developed by Vinogradov and Krasil'shchik and their co-workers.

It will also be interesting to fully develop the use of adjoint-symmetries in the study of specific PDE systems, as outlined in the Introduction: finding exact solutions, detecting and finding mappings into a target class of PDEs, and detecting integrability, which are the counterparts of some important uses of symmetries. Another use of adjoint-symmetries, which has been introduced very recently [20], is for finding pre-symplectic operators.

Author Contributions: Conceptualization, S.C.A. and B.W.; methodology, S.C.A. and B.W.; writing, original draft preparation, S.C.A.; writing, review and editing, S.C.A. and B.W. All authors have read and agreed to the published version of the manuscript.

Funding: This research received no external funding.

Acknowledgments: S.C.A. is supported by an NSERC Discovery grant. W.B. thanks the Department of Mathematics & Statistics, Brock University, for support during a research visit when this work was completed.

Conflicts of Interest: The authors declare no conflict of interest.

References

1. Bluman, G.W.; Cheviakov, A.; Anco, S.C. *Applications of Symmetry Methods to Partial Differential Equations*; Springer: New York, NY, USA, 2009.
2. Krasil'shchik, I.S.; Vinogradov, A.M. (Eds.) *Symmetries and Conservation Laws for Differential Equations of Mathematical Physics*; Translations of Math. Monographs 182; American Mathematical Society: Providence, RI, USA, 1999.
3. Olver, P.J. *Applications of Lie Groups to Differential Equations*; Springer: New York, NY, USA, 1993.
4. Ovsiannikov, L.V. *Group Analysis of Differential Equations*; Academic Press: New York, NY, USA, 1982.
5. Bluman, G.W.; Anco, S.C. *Symmetry and Integration Methods for Differential Equations*; Springer: New York, NY, USA, 2002.
6. Sarlet, W.; Cantrijn, F.; Crampin, M. Pseudo-symmetries, Noether's theorem and the adjoint equation. *J. Phys. A Math. Gen.* **1987**, *20*, 1365–1376. [CrossRef]
7. Sarlet, W.; Bonne, J.V. REDUCE procedures for the study of adjoint symmetries of second-order differential equations. *J. Symb. Comput.* **1992**, *13*, 683–693. [CrossRef]
8. Sarlet, W. Construction of adjoint symmetries for systems of second-order and mixed first- and second-order ordinary differential equations. *Math. Comput. Model.* **1997**, *25*, 39–49. [CrossRef]
9. Anco, S.C.; Bluman, G. Direct construction of conservation laws from field equations. *Phys. Rev. Lett.* **1997**, *78*, 2869–2873. [CrossRef]
10. Anco, S.C.; Bluman, G. Direct construction method for conservation laws of partial differential equations Part II: General treatment. *Eur. J. Appl. Math.* **2002**, *41*, 567–585. [CrossRef]
11. Anco, S.C. Generalization of Noether's theorem in modern form to non-variational partial differential equations. In *Recent Progress and Modern Challenges in Applied Mathematics, Modeling and Computational Science*; Springer: New York, NY, USA, 2017; Volume 79, pp. 119–182,

12. Anco, S.C. On the incompleteness of Ibragimov's conservation law theorem and its equivalence to a standard formula using symmetries and adjoint-symmetries. *Symmetry* **2017**, *9*, 33. [CrossRef]
13. Anco, S.C.; Wang, B. Algebraic structures for adjoint-symmetries and symmetries of partial differential equations. *arXiv* **2020**, arXiv:2008.07476.
14. Nestruev, J. *Smooth Manifolds and Observables*; Graduate Texts in Mathematics 220; Springer: Berlin, Germany, 2002.
15. Anco, S.C.; Pohjanpelto, J. Classification of local conservation laws of Maxwell's equations. *Acta Appl. Math.* **2001**, *69*, 285–327. [CrossRef]
16. Anco, S.C.; Pohjanpelto, J. Symmetries and currents of massless neutrino fields, electromagnetic and graviton fields. In *CRM Proceedings and Lecture Notes (Workshop on Symmetry in Physics)*; American Mathematical Society: Providence, RI, USA, 2004; Volume 34, pp. 1–12.
17. Vinogradov, A.M. The C-spectral sequence, Lagrangian formalism, and conservation laws I. The linear theory. *J. Math. Anal. Appl.* **1984**, *100*, 1–40. [CrossRef]
18. Vinogradov, A.M. The C-spectral sequence, Lagrangian formalism, and conservation laws II. The nonlinear theory. *J. Math. Anal. Appl.* **1984**, *100*, 41–129. [CrossRef]
19. Vinogradov, A.M. Introduction to Secondary Calculus. In *Proceedings of the Conference Secondary Calculus and Cohomology Physics*; Henneaux, M., Krasil'shchik, I.S., Vinogradov, A.M., Eds.; Contemporary Mathematics; American Mathematical Society: Providence, RI, USA, 1998.
20. Anco, S.C.; Wang, B. A formula for symmetry recursion operators from non-variational symmetries of partial differential equations. *arXiv* **2020**, arXiv:2004.03743.

© 2020 by the authors. Licensee MDPI, Basel, Switzerland. This article is an open access article distributed under the terms and conditions of the Creative Commons Attribution (CC BY) license (http://creativecommons.org/licenses/by/4.0/).

MDPI
St. Alban-Anlage 66
4052 Basel
Switzerland
Tel. +41 61 683 77 34
Fax +41 61 302 89 18
www.mdpi.com

Symmetry Editorial Office
E-mail: symmetry@mdpi.com
www.mdpi.com/journal/symmetry

www.ingramcontent.com/pod-product-compliance
Lightning Source LLC
LaVergne TN
LVHW070726100526
838202LV00013B/1183